AF605851

THE ART OF OBJECTS

The Birth of Italian Industrial Culture, 1878–1928

LUCA COTTINI

The Art of Objects

The Birth of Italian Industrial Culture, 1878–1928

University of Toronto Press
Toronto Buffalo London

Toronto Buffalo London
utorontopress.com
Printed in the U.S.A.

ISBN 978-1-4875-0283-6

♾ Printed on acid-free paper with vegetable-based inks.

Toronto Italian Studies

Library and Archives Canada Cataloguing in Publication

Cottini, Luca, author
The art of objects : the birth of Italian industrial culture, 1878–1928 / Luca Cottini.

(Toronto Italian studies)
Includes bibliographical references and index.
ISBN 978-1-4875-0283-6 (cloth)

1. Manufacturing industries – Italy – History – 19th century. 2. Manufacturing industries – Italy – History – 20th century. 3. Industries – Italy – History – 19th century. 4. Industries – Italy – History – 20th century. 5. Industrial arts – Italy – History – 19th century. 6. Industrial arts – Italy – History – 20th century. I. Title. II. Series: Toronto Italian studies

HD9735.I8C68 2018 338.40945 C2018-900370-7

This book has been published with the assistance of ISSNAF (Italian Scientists and Scholars of North America Foundation), the Subvention of Publication Program of Villanova University, and the Richard and Mary Anne Francisco Endowed Fund for Italian Studies.

University of Toronto Press acknowledges the financial assistance to its publishing program of the Canada Council for the Arts and the Ontario Arts Council, an agency of the Government of Ontario.

Canada Council for the Arts
Conseil des Arts du Canada

Funded by the Government of Canada
Financé par le gouvernement du Canada

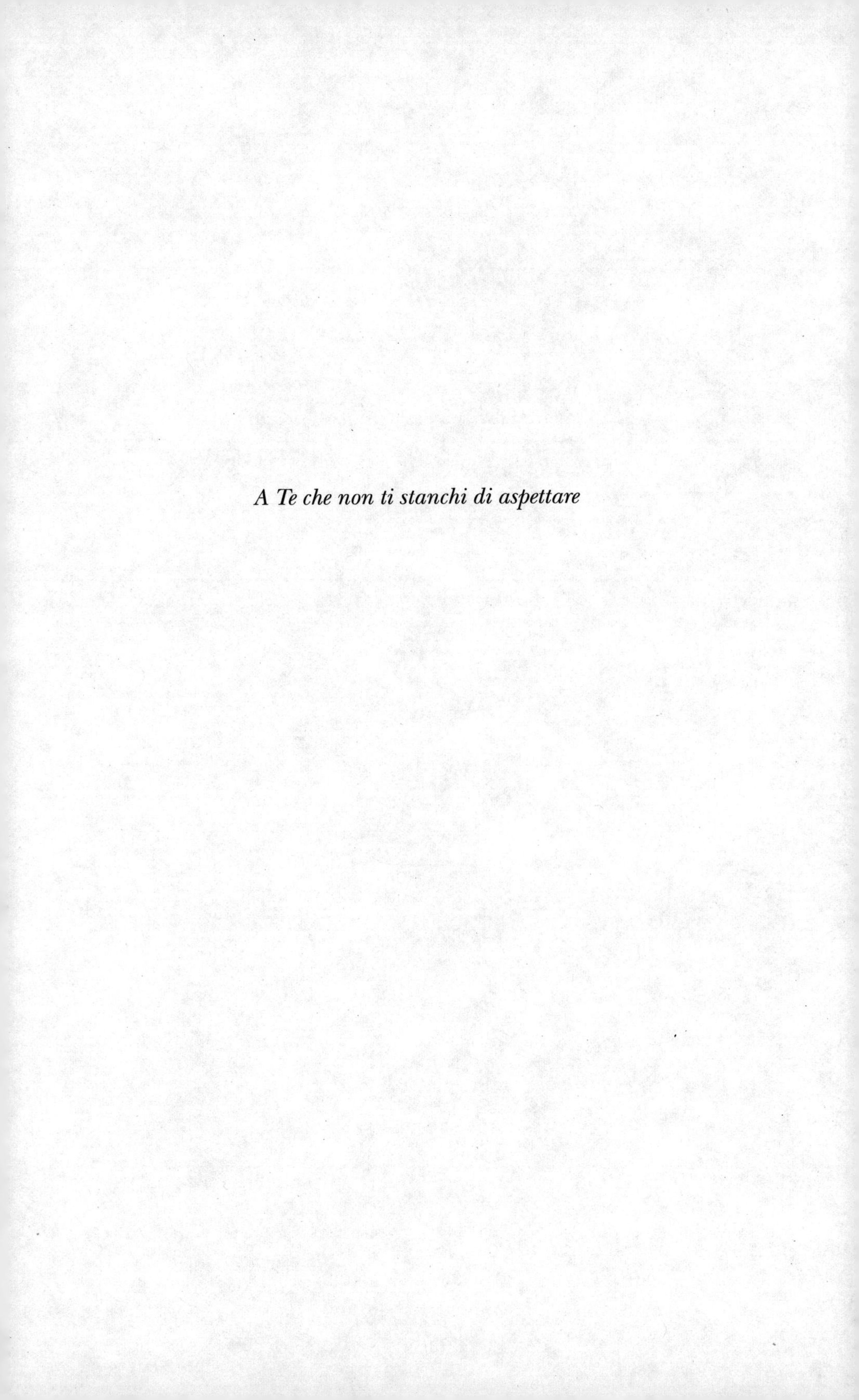

A Te che non ti stanchi di aspettare

Contents

Illustrations

Figures

Plates

Acknowledgments

This publication has been made possible thanks to the generous support of ISSNAF (Italian Scientists and Scholars of North America Foundation), the Subvention of Publication Program of Villanova University, and the Richard and Mary Anne Francisco Endowed Fund for Italian Studies.

At the professional level, this project has taken shape thanks to the invaluable contributions of the many people who have surrounded, inspired, and supported me over the years. In addition to all the scholars and authors with whom I dialogued throughout the writing of the book, I owe a great debt of gratitude to two specific *maestri* who encouraged me to risk this journey and taught me how to take my passion seriously: John Welle, who sowed the first seeds of this book; and Giuliana Minghelli, who accompanied it with care and rigour in its early stages. Throughout this project the conversations with colleagues like Jeffrey Schnapp, Francesco Erspamer, Zygmunt Baranski, Giuseppe Mazzotta, Millicent Marcus, Mary Watt, Simone Marchesi, Cristina Della Coletta, Maria Grazia Lolla, Ara Merjian, Teresa Fiore, Fabio Finotti, and Mauro Calcagno further deepened my understanding of my interests and increasingly persuaded me that this path was worth pursuing. Along with them, the daily dialogue with my students at McGill and Villanova universities kept alive my excitement for learning and discovery and taught me simplicity in complexity, as well as discipline in passion. Finally, a special recognition goes to Alexandra Ferretti and Angela Wingfield for the editing support, to Siobhan McMenemy and Mark Thompson for the continuous trust in my work, and to Jennifer Testa for the help with all translations from Italian to English, which, unless otherwise indicated, are my own.

At the personal level, immense thanks go to my parents and my friends, who kept me sane during this process; to my children, Stefano, Caterina, and Teresa, who always make me look at reality with curiosity and amazement; and to my beloved wife, Jen, who never ceases to support and accompany me, reminding me of the meaning of my scholarly vocation, and opening up my studium to a much broader horizon.

THE ART OF OBJECTS

The Birth of Italian Industrial Culture, 1878–1928

Introduction

The Art of Objects explores the initial stages of the Italian industrial project, against the political, social, and cultural background of post-unification Italy in the fifty years between the Universal Exposition of Paris in 1878 and Gió Ponti's first theorization of industrial design in 1928. By combining the perspective of the arts and industry, this volume offers a comprehensive account, not bound to any political periodization, of the creative workshop of Italy's industrial culture. By bridging the parallel (and often unrelated) fields of Italian studies and design studies, this work also proposes a novel and broader vision of Italy's transition to industrial modernity, from post-Risorgimento to early Fascism.

The study asserts both the existence and the relevance of early Italian industrialism (often discarded or overlooked because of its irregular evolution), through the observation of the age's extraordinary concatenation of scientific and technological advances vis-à-vis contemporary culture. By deliberately tracing a visible link between industrial innovations and their related effects (in the material conditions of living and in the inner psychology of modern man),[1] this book explores their "cultural record" (Kern 9) upon society and intellectual life, and the resulting growth upon them of "distinctive new modes of thinking about and experiencing time and space" (Kern 1).[2]

As an alternative to the compartmentalized histories of the age – individually focused on its new technologies, media (e.g., photography, cinema, radio), products (e.g., collectors' handbooks), art, or literature – this cultural study aims to open up a multidimensional vision of the new "world of objects" (Baudrillard 3) that was brought forth by industrialism, through the experimental synthesis of its many perspectives. This broad overview of the period offers a complex narrative of Italy's path to modernization, thus overcoming the exclusive reliance on one singular field of inquiry and, for the first time, including industrialism as a relevant factor in understanding the nation's modernity. Far from compiling an impossible compendium of the age or an analytical study of all its artefacts in different fields, this investigation of the early contacts between industrial production and intellectual life reveals the different and often hidden hermeneutics of Italy's transition to modernity. This is accomplished in two main ways. First, against the back-drop of Italy's political, social, and intellectual history, the book reconstructs the irregular nature of the Italian industrial venture through fragmentary symbolic episodes and exemplary cases revealing the interplay of high-, middle-, and lowbrow cultures. Second, it attempts to bridge the research in Italian literary or cultural studies (which have often overlooked the impact of material cultures on Italy's modern identity) and the investigation of design studies (which have often failed to see the influence of high culture on the evolution of new industrial forms).

In the field of Italian studies the advent of industrialization and the formation of new industrial languages (e.g., photography, cinema, advertising, and design) have been traditionally considered as secondary or even degenerating factors in Italy's cultural life. The influence of Benedetto Croce's aesthetic purism and his peremptory condemnation of the early industrial age as a period of political crisis and intellectual decadence that created the abnormal form of Fascism in Italy (as postulated a posteriori in his 1928 *History of Italy from 1870 to 1915*) led to a depreciation of material cultures and to the success for more than half a century of the critical notion of "decadentism" (after the publication of Walter Binni's book *La poetica del decadentismo* in 1936).[3] In spite of this critical background, scholarly studies over the last twenty-five years have started to include and re-evaluate the impact of the industrial arts in the definition of Italian modern culture. In *Italian Culture in the Industrial Era, 1880–1980* (1990), David Forgacs disputed the anti-industrial narrative and the tendency to "treat cultural history as history of 'intellectuals'" (5), by analysing the structures of Italy's culture industry in relation to broadcasts (radio and television) and the production and consumption of books, newspapers, magazines, and films. In the *Cambridge Companion to Modern Italian Culture* (2001), Zygmunt Baranski and Rebecca West pioneered a broader cultural history of Italy's transition to modernity, from unification to the twenty-first century, including essays on fashion, media, religion, folk music, and design. In *Italian Modernism* (2004), Mario Moroni and Luca Somigli opened a new field of investigation in the literary studies of the period by tracing the visible connections of Italian contemporary literature to the European modernist experience and reconfiguring as an applicable literary category the term *modernism* (previously discarded in Italy for its reference to the debate taking place within the Catholic Church in the early 1900s). Lastly, in *Modernitalia* (2012), Jeffrey Schnapp attempted to bridge the re-evaluated literary forms of "modernism" and the parallel experience of the Modern movement in architecture and the visual arts.

In the field of design studies the evolution of applied arts and an original industrial culture in Italy have been relegated to material cultures and mostly limited to accounts of industrial design. In *Italian Design: 1870 to the Present* (1988), Penny Sparke presented an evolution of design as an expression of Italy's peculiar "bid for modernity" (7), as a prolonged quest for a "unique approach" (7) in the manufacturing of objects, and as an appendage of the architectural culture of the Modern movement of the 1920s. In *Made in Italy: Rethinking a Century of Italian Design* (2014), Grace Lees-Maffei and Kjetil Fallan approached the phenomenon of Italian design without narrating its historical development but by presenting its core structures, protagonists, spaces, and industries through "episodic" essays. In contrast with traditional studies that trace the birth of mature Italian design in the late-1920s and of Italian industrialism during the economic miracle of the 1950s and 1960s, two more recent exhibits have proposed new reflections on their preliminary phases in the first decades of the twentieth century. In *Modo italiano* (Montreal, 2006), Giampiero Bosoni located an early source of Italian design in the so-called Second Futurism,[4] whereas in *Una dolce vita?* (Paris, 2015) Guy Cogeval and Beatrice Avanzi traced a fundamental background for its later development in the Italian applied arts and the Liberty style. Along with these studies, generally excluding a methodical reflection on the parallel evolution of high cultures (literature and painting), Italian scholars have focused on the philosophy of design by exploring its ideological and theoretical background (as in Renato De Fusco's *Filosofia del design*, 2012), by observing its unique obsessions (as in Silvana Annicchiarico and Andrea Branzi's *Che cosa è il design italiano? Le sette ossessioni del design italiano*, 2008), or by investigating its "genetic" imprint or experiential roots (as in Branzi's *Introduzione al design italiano: Una modernità incompleta*, 2008).[5]

By connecting literary, cultural, and design studies, I contend that Italian industry and the arts not only influence each other but also intertwine in discontinuous and experimental ways to gradually create a new hybrid and autonomous culture. By focusing on the apparently disconnected or "imperfect" cultural spaces elaborated in between industrial production and the humanistic tradition, I maintain that the age of Italy's first industrialization (1878–1928) constitutes not a battlefield of tensions (as implied in the mutually exclusive categories of *decadentismo* and *modernismo*) but rather a creative laboratory of new cultures and styles, as well as fertile soil for the birth of design in the late-1920s. In this light I subsequently claim that the "irregular" expansion of the Italian industry and the "unfinished" elaboration of a modern Italian culture – Italy's so-called incompleteness – do not equal a flawed modernization but rather constitute the core feature of its experimental modernity, characterized by a constant state of self-adjustment, eclectic synthesis, and aesthetic invention of new forms. In contrast with other contemporary models (such as those in England, Germany, France, and the United States) that focused on large-scale investments and centralized marketing strategies,[6] I see the industrial success of Italy after the Second World War, and its twentieth-century Renaissance in design and fashion, as the outcome of the country's polycentric system, which is able to mediate tradition and modernization (or artisanal small-scale and factory large-scale production)[7] and to transform "the absence of a unifying planning methodology into a great internal diversification of languages and trends" (Branzi 14).

In order to provide a larger cultural history of Italian modernization (involving the interaction of society, industry, and the arts) and a coherent account of the uneven birth of industrial culture in Italy, the book examines the five decades from 1878 to 1928 through the unusual perspective of objects. As products and culture-makers, objects reveal, on a micro-scale, a dual viewpoint on the age, encompassing global and Italian industrialization, a collective and an individual experience of modernity, and entrepreneurship and art. They appear in expositions, advertisements, and newspaper articles in connection with new mass practices (e.g., tourism, sports, smoking) and social debates (e.g., on physical education or timekeeping). At the same time they also break into literature and art, inspiring a critical reflection on and the experimental quest for an Italian way to modernity. In making visible the collaborative network uniting Italian entrepreneurs with writers, painters, artisans, designers, and architects, objects overcome a merely functionalist dynamic of production and consumption and acquire a new dimension as "mediating devices" (Lees-Maffei, "Production" 354)[8] and as "vehicles of meaning" (Attfield 75), absorbing, negotiating, and recreating the nation's tensions in an elastic style. In the absence, or rather in the impossibility, of an organic narration of Italy's industrial beginnings, the different cultures surrounding objects then asystematically illuminate the hidden stages of the nation's collective discernment of modernity, and assemble from the Italian conglomerate of "industries without industrialization" (Mori, "Industrie" 608) an organic picture of early Italian industrialism, which is endowed with identifiable features and diversified patterns. In their network of relations and expressive energy, objects also reveal an experimental site for the development of new art forms and entrepreneurial attitudes, which precedes the evolution of design and informs the maturation of Italy's autonomous model of industrial culture.

Against this general perspective, the study lingers on the significant cases of a few selected objects: wristwatches, cameras and photographs, bicycles, gramophones, radios, cigarettes, toys, clothes, and furniture. Amidst the age's plethora of life-changing inventions (also including the elevator, telephone, cinema, car, and refrigeration), these objects obey the following criteria: first, products growing from small-size retail to an industrial scale over this period;[9] and, second, items engendering or reinventing new social experiences, entrepreneurial models, aesthetic languages, and modern concepts (e.g., time, vision and memory, body and space, simultaneity, lightness, and play) during the same years.[10] In observing these objects as repositories of an "affective and intellectual investment" (Bodei 26),[11] I reconstruct not just their emotional impact on society and on the synchronistic formations of art and entrepreneurial models but also their "vis formativa" (Dorfles 158) of a new industrial culture.[12]

The Art of Objects is organized in eight chapters. In the first and last chapters I offer a history of the origins of Italian industry (from 1878 to 1918, and from 1918 to 1928), and I outline the trajectory from the early *art of objects* (in Italian crafts or in the representation of items in contemporary art) to design as an oxymoronic *industrial art* and as a self-conscious philosophy of creating "things with attitude" (Attfield 11). In the six central chapters I consider the *experimental art* or aesthetic discourse developing around specific objects, documenting the tangible foundation of Italy's industrial culture. In each of these chapters, organized in a similar pattern, I observe one or more items from an industrial, social, aesthetic, and conceptual point of view. Starting with an episode taken from current events, I outline the network of relations surrounding these objects (industries and markets), their impact on the nation-making process (habits and practices),[13] their transformation from commercial items into aesthetic symbols (in art and philosophical debates),[14] and their elaboration of new languages and concepts.

In the first chapter I present the evolution of the Italian industrial project from its early impact (in the perceived "invasion" of items produced by industrialism) to the development of an aesthetic gaze around objects (as they appear in literature and painting), and I outline the shift from Italy's tradition of crafts (dating back to the Renaissance) to the early theorization of modern Italian craft. Chapter 2 investigates the relationship between the establishment of world standard time (1884) and the launch of Italian industrialism after the National Exposition of Milan in 1881. Starting from the debates surrounding the Geodetic Conference of Rome (1883), I observe the impact of the transformation of time on literary and visual representations of timepieces, and the parallel development of two manufacturing models in the Italian timekeeping industry, relating clocks to craftsmanship and jewellery (Bulgari) or to factory production (Borletti). Chapter 3 considers the industrialization of photography following the 1888 release of the first portable camera (Kodak no. 1). I focus on the impact of photographs on public opinion, society, and art (in the works of contemporary photographers, writers, and painters) and document the growth of a photographic culture in Florence (around the Alinari firm) and in Turin (in connection with the birth of the cinematographic industry). Chapter 4 explores the impact of bicycles on social debates (cyclomania versus cyclophobia), and the development of an autonomous cycling literature at the turn of the twentieth century.. Starting from Luigi Masetti's cycling trip from Milan to Chicago in 1893, I observe bicycles as instigators of a new idea of culture in motion, of new modern experiences (tourism and sports), of Italy's mechanical industry (out of the expansion of Bianchi and Pirelli), and of corporeity. Chapter 5 analyses the parallel development of sound recording and transmitting technologies (phonographs, gramophones, and radios). Starting from Enrico Caruso's first recording in 1902, I reconstruct the impact of gramophones on creating a music industry, on ending the golden age of Italian opera, and on spurring the birth of poster design in Italy (on the initiative of Giulio Ricordi). Starting from Guglielmo Marconi's first transatlantic radio transmission (1903), I also analyse the impact of wireless telegraphy on Italian politics and the origins of broadcasting in Italy. Chapter 6 focuses on the cultural symbolization of smoke and cigarette smoking in the early twentieth century. Starting from the 1912 portrait of the Futurists smoking in Paris, the increased production and consumption of cigarettes in the Italian market (from tobacco import to state monopoly) is related to the contemporary symbolizations of a "smoky" or light modern being, in Filippo Tommaso Marinetti's *Zang Tumb Tumb* (1914), Aldo Palazzeschi's *Il codice di Perelà* (*Man of Smoke*, 1911), and Italo Svevo's *La coscienza di Zeno* (*Zeno's Conscience*, 1923). Chapter 7 investigates the impact of the Futurist idea of play on the development of the Italian toy, fashion, and furniture industries in the post-war period. Starting from the manifesto "The Futurist Reconstruction of the Universe" (1915), I present the Futurist work on puppetry, clothing, and furniture (conceived as a way to strip them of their association with decorative aestheticism) as a creative platform for reconfiguring Italian artisans as style makers and for developing Italian advertising (as seen in Fortunato Depero's pioneering collaboration with Campari). In the concluding chapter 8 is documented the Italian industrial expansion of the 1920s (with the launch of the automobile and aeronautic sectors), and I consider the relationship between Fascism, the development of the applied arts (e.g., the Novecento movement, rationalism), and the creation of industrial design as an independent culture after Gió Ponti's foundation of *Domus* magazine in 1928. In conclusion, after highlighting the elements of Ponti's theory (industry as a new culture maker, and the designer as a new modern artist), I define the political and behavioural origins of the "design vision" by relating the concepts and features that have emerged in the case studies on objects to the vision's successive developments.

Chapter One

At the Origin of Italian Industrialism

The Italian season of the Risorgimento came to a symbolic end in 1878, with the deaths of King Victor Emmanuel II (9 January) and Pope Pius IX (7 February), a month apart. In the same year, the excitement raised by the Universal Exposition of Paris (1 May–10 November)[1] helped to usher in the new age of Italian industrialization.

Within this context, the Italian writer Edmondo De Amicis (author of the celebratory book on Risorgimento, *Vita militare*; *Military Life*, 1868) travelled to the world's fair and provided early documentation of Italy's initial involvement with industrialism. In his report of the 1878 event, published the following year by Treves as *Ricordi di Parigi* (*Memories of Paris,* 1879), De Amicis reflected on the spreading phenomenon of universal expositions (along the lines of previous contributions by Karl Marx and Charles Baudelaire) and, at the same time, divulged to Italian readers a vision of their new cosmopolitan culture. As a fascinated and disillusioned visitor, De Amicis detected in the city of Paris the first impression of industrial modernity as a new chaotic civilization of objects, and he mirrored his readership's attraction towards and fear of the ongoing process of global industrialization. In his narration of the visit, De Amicis also gave voice to the opposing urges of an old and yet young nation like Italy, which both conformed to the global progress of civilization (in response to the country's economic backwardness) and separated itself from the progress (in light of Italy's long and rich past).

During De Amicis's walk from Boulevard Beaumarchais to the fair's pavilions the city appeared to be "an endless open theatre" (De Amicis, *Ricordi* 52) and an immense expositional space, teeming with an "infinite variety of treasures, objects of desire, playthings, works of art, ruinous trifles, and temptations of every sort" (39).[2] Likewise, De Amicis represented the pavilions as "an enormous museum," parading "all the gold, gems, lace, flowers, crystal, bronze, paintings, all the masterpieces of industries, all the seduction of the arts, all the galas of wealth, all the whims of fashion" (40).[3] The items accumulated in the theatre-museum of the exposition visualized the idea of modernity as a "huge international university" (78), providing new, universal knowledge for a few cents.[4] At the same time, the grandiose outdoor life of the city and the excited turmoil of products, emotions, and activities inside the pavilions reflected the idea of modernity as a feverish feast, moved by an inebriating energy of transformation.[5]

Although transported by the thrill of this city event, De Amicis also observed it with ironic critical distance. While chaotically accumulating objects on the virtual showcase of his page (as a way to charge his report with an excited sense of immediacy), he also interrogated them (in the dialogue with his readers) with an underlying humour, questioning their supposed equivalence to art (as exposed pieces in a new museum), challenging their idea of all-inclusiveness, and implying their future ephemerality.[6] His captivated and yet distant gaze towards objects extended to Paris itself, portrayed in the end of his book as a "beautiful and tremendous sinful woman" (178), a city with both luxurious allure and intrinsic fragility: "Is this the great Paris? What if an earthquake shatters all of the shop windows, and a burning rain wipes out all of the gold-plating? What would remain of it?" (167).[7] In his portrayal of objects (or Paris) as both trendy attractions and perspectively outmoded ruins De Amicis ultimately represented Italian culture's fascination with and resistance to modernity itself.

His sketch of the Italian pavilion at the exposition captured the bulk of this ambivalence. By picturing the nation's façade as "a crowd of pure white statues, a spreading blaze of crystals, a glittering of silk and mosaics, a joyful laughter of colours and shapes, in front of which all faces brighten, all hearts expand, and all mouths proclaim, 'Italy' – even before the eyes have read the pavilion's sign" (61),[8] De Amicis equated the nation's image to the glimmering theatre of Paris's lights and shapes. By praising the nation's peculiar display of statues, silk, and mosaics, however, he also affirmed the recognizable difference of the nation, identified by its continuity with the past and by its unique style (implicitly raising the issue of its true definition). In his outlook on the national pavilion De Amicis thus outlined the traits of Italy's early approach to industrialism – alternating the pursuit of new models with the need to preserve old traditions, and experiencing the tension between conforming to a global standard and searching for distinction.[9]

Along the lines of De Amicis's report, which testified to the first impact of global industrialism on Italian culture through the products of the exposition, this investigation of the early evolution of the Italian industrial venture runs parallel to a study of the gradual creation of an aesthetic discourse around objects, as attested in their initial appearance in the Italian society, literature, and painting of the fin de siècle, in their relation to artisanal craft and the theory of decorative arts, and in the increasing inclusion of commercial items in the space of culture.

The Project of Italian Industrialism

For Italian industry 1878 was a key year. With the passage of the first protectionist law on domestic textile products, the Italian state began to openly support the industrial project.[10] From the economic perspective, the protectionist turn-round of 1878 – spurred by the effects of the agrarian crisis (since 1876) and triggered by the Italian success at the Paris exposition – which would be followed by the abolition of forced currency[11] in 1881, represented the founding act of Italy's industrialism. From the political perspective, the 1878 establishment of customs duties, leading Italy to a commercial rupture with France (as reflected in the diplomatic war over the possession of Tunisia, which became a French protectorate in 1881), was at the origin of Italy's 1882 decision to join the Triple Alliance with Germany and former enemy Austria-Hungary.

Before these developments the Italian economy (mainly based on agriculture) had faced a long period of stagnation (Bianchi 24), and, since unification, industry had been limited to textiles (55.4 per cent of the total industrial production in 1878; Carreras 202) and food manufacture (oil, tomato preserves, and pasta).[12] Italy's industrial growth had slowed after 1861, owing to the exigencies of political and social stability: the need to square the national debt after the wars of independence (a goal reached in 1875); the urgency to endow the new nation with infrastructures (railways, harbours, tunnels),[13] and the necessity to create a uniform productive system (laws, taxation, market) across the peninsula. Unlike England and Germany, which had rapidly expanded between the first Exposition of London in 1851 and the economic crisis of 1873, Italy had not developed a national industry, because of the limited size of its domestic marketplace and its reliance on foreign sources (for raw materials, technical and entrepreneurial skills, and imported goods). As in the pre-unification period, it continued instead to excel on international markets and expositions in the manufacture of luxury goods (glass, porcelain, coaches, and cabinets) and in the export of distinguished local products (e.g., silk from Como, marble from Carrara, straw hats from Carpi, majolica plates from Faenza, accordions from Castelfidardo, and citrus from Sicily; Mori, *Industrializzazione* 9–40).

In the aftermath of the 1878 Paris exposition the industrial project gradually expanded nationally, thanks to the government's involvement in the organization of congresses and expositions. In line with the early establishment of the Polytechnic University of Turin (1859) and the Polytechnic University of Milan (1863), and in connection with the push to spread literacy throughout the national territory (enforced by the passing of the Coppino law in 1877),[14] scientific congresses aspired to promote a new technological culture in Italy and to position the nation in the network of global research. Similarly, expositions sought to educate citizens in modernity (through the thrill of attractions, amusements, and reproductions of past monuments) and to stage an ideal representation of Italy through the best of its production. Within this context two events launched the project of Italian industrialism: the National Exposition of Turin in 1880 and the National Exposition of Milan in 1881. The Turin exposition, reported by De Amicis, envisioned industrialization as a way to realize the "piedmontization of Italy" (Della Coletta 19). The Milan exposition, offering a first review of the nation's production in the twentieth anniversary of its unification, promoted industrialism as a new age of light and progress (concomitant with the first application of urban illumination and the success of the ballet *Excelsior*). In subsequent years Italy hosted numerous other expositions, both domestic (Rome, 1883; Turin, 1884; Milan, 1886; and Palermo, 1891) and international, like the Esposizione dell'arte decorativa moderna (Exposition of Modern Decorative Art, Turin, 1902), the Esposizione del Sempione (Sempione Exposition, Milan, 1906), and the Esposizione del cinquantenario (Fiftieth Anniversary Exposition, Turin, 1911). Italy also participated in global expositions, promoting its distinctive "style" through the excellence of its products (as certified by the gold medals won by Martini & Rossi and Filppo Berio in Paris, 1878; Alinari in Paris, 1889; and Pasta De Cecco in Chicago, 1893) and through the architectural design of its pavilions (ranging from Giovanni Basile's *neocinquecentismo* of Paris in 1878 to Manfredo Manfredi's neo-Gothic of Paris in 1889, and Carlo Ceppi's pseudo-Venetian of Paris in 1900).[15] While supporting the effort to create an industrial culture in Italy, the state also started to implement industrial structures during the 1880s. After establishing the electricity plant of

Santa Radegonda in Milan in 1883 (following the excitement for urban illumination at the 1881 exposition), the various political administrations favoured the foundation of the Edison plant in 1884 (the first after New York), the adoption of electric tramway lines (beginning with the Fiesole-Florence line in 1890), and the empowerment of hydro-electric energy (Mortara 357–70). At the same time, parallel to the need to provide the country with an adequate military structure, the government directly invested in the production of steel (with the foundation of Acciaierie Terni in 1884), in the construction of locomotives and ships (after the Boselli law of 1885), in the expansion of the railway system (approximately 6,500 kilometres), and, last but not least, in the financial bailout of Terni in 1887 (thereby preventing its immediate collapse; Segreto 16).

Throughout these early developments the Italian press promoted the diffusion of an industrial market and mentality. Popular newspapers and magazines like *Illustrazione italiana* (1873), *Corriere della sera* (1876), and *La domenica del corriere* (1899) contributed to the expansion of industrialization, not only quantitatively through their growing diffusion (made possible by advancements in reproduction techniques like chromolithography or the linotype process) but also qualitatively through their capacity to absorb, showcase, and create a discourse around modernity's new products.[16] From the early 1880s, objects of any kind started to invade the pages of Italian journals, magazines, and newspapers, fashioning their editorial space into a virtual expositional space as had the item-filled pages of De Amicis's *Ricordi di Parigi*. The verbal and visual advertisements of the 1880s mediated to Italian readers both the tangible impression of industrial modernity as a chaotic multiplier of items, and an ambivalent approach to its new civilization. In a way, advertised objects represented the coveted image of global modernity, identified with its new values of technological progress (in telephones, light bulbs, phonographs, and ovens), health (in pills to combat cough, asthma, mucus, or fevers), hygiene (in perfumes and soaps), touring (in advertisements for hotels, bicycles, and travel packages), capitalism (in price lists of publishing houses, department stores, and auctions), and pursued novelty (in bizarre inventions like anti-polio water, anti-cholera armour, and the swinging hot tub). In another way, advertised objects proudly stated the difference between the Italian industry and the global context by visually representing the country's excellence (in the textile and food sectors) and the supposed authenticity of its products.

From Crisis to Industrial Boom

The first industrial boom of the 1880s produced many changes at both the social and the political levels. The establishment of industry initiated abandonment of the countryside and the formation of the urban proletariat. The absence of any regulation of the conditions, hours,[17] and salaries of factory workers (with the exception of a law, passed in 1886, that forbade the employment of children younger than nine) favoured strikes, public unrest, and the rapid spread in Italy of anarchism and socialism (which acquired parliamentary strength with the election of Andrea Costa as the first Socialist Party representative in 1882). Within this context of social instability, which was aggravated in 1887 by the outbreak of a global financial crisis and by the failure of the colonial project in Eritrea (after the massacre of Italian soldiers in Dogali), the state backed Italian industrialism with the passage of two laws: one extending protectionist measures on Italian grain, iron, and steel (1887); and the other legalizing emigration (1888). Despite the denunciation of the poor working conditions (e.g., by Verga and later Pirandello in relation to Sicilian sulphur mines),[18] the government assumed over the years an authoritarian stance against social turmoil, as a way to "protect" industrial production from the perceived dangers of Marxism (which had been observed with caution and fear since the Paris Commune of 1871). Notwithstanding the impact of De Amicis's polemical novel *Sull'oceano* (*On Blue Waters*, 1888), contesting the law of 1888 and foreseeing the massive exodus of Italians over the next decades, the state saw emigration as a valve against the persistent poverty of the Italian peasantry, a remedy against the spreading of anarchism, and an indirect way to acquire "colonies" in the Little Italies of North and South America (Choate 40–1). In the continued absence of a state policy on work (until the establishment of a specific ministry in 1901), the so-called *questione sociale* (social issue) was thus tackled mainly by the Catholic Church after the publication of Pope Leo XIII's encyclical *Rerum novarum* in 1891 and by the Partito dei Lavoratori (Workers' Party), founded by Filippo Turati in 1892. In the meantime the worsening of the financial crisis and the eruption of the Banca Romana scandal in 1893 led to the collapse of the bank system and to the creation of a centralized bank of Italy.[19] In response to this situation, foreign banks partnered with the Italian state to create new financial institutions and sustain the Italian industrial effort. The foundation in 1894 of the Banca Commerciale Italiana (out of a consortium of German, Austrian, and Swiss banks) provided immediate liquidity for the main industrial sectors (iron and steel, electrics, chemical, transportation, and textiles). The establishment of Credito Italiano in 1895 (now Unicredit), from the acquisition of two banks by the former Banca di Genova, granted coverage for the energetic need of the industry, as the bank acquired more than 30 per cent of Edison.[20]

Despite the political instability of the fin de siècle (marked by Italy's defeat in the Battle of Adwa in 1896, the subsequent fall of Prime Minister Francesco Crispi, the military repression of Milan's strikes in 1898, and the assassination of King Umberto in 1900), the years between 1896 and 1914 coincided with a "mini economic miracle" in the nation (Sparke, *Italian Design* 23). Over these years, during the political ascent of Giovanni Giolitti (minister of interior from 1901 and then prime minister from 1903 to 1914), Italy experienced an average annual increase in the gross domestic product (GDP) of 2.8 per cent (briefly interrupted in the 1907 financial crisis; Segreto 27). The industrial boom of the Giolitti era (which coincided with the peak of Italian emigration to the Americas) related to a large concentration of foreign capital and competences (Morandi 221), which lead to the rapid growth of the mechanical industry (climbing to 19.8 per cent of Italian GDP in 1911; Carreras 207) and to the reorganization of production according to the German regimental model (Pirelli, Falck, Krupp) or to American Taylorism (Fiat, Olivetti, Breda). The expansion of the early 1900s also related to the intervention of the state, which solidified financial structures, managed the monopoly of key industries (e.g., tobacco or train production), supported the main steel companies (Ilva, Ansaldo), and implemented hydro-electric energy (after the construction of the plants in Paderno d'Adda in 1898 and Vizzola sul Ticino in 1901). During these years, in light of a new industrial spirit and this peculiar "connection between state and big business" (Amatori 694), the Italian industry grew exponentially in all fields.[21] While developing a new form of "collusive" capitalism (Giannetti and Vasta 64) based on family ties and fiduciary bonds, it also constituted itself as a political reality, after the 1910 establishment of Confindustria (Confederation of Industries) in response to the foundation of Italy's first trade union, General Confederation of Work, four years earlier. The outbreak of the First World War propelled an "accelerated industrial revolution" (Segreto 40), which was concentrated over four years. The war provided a substantial incentive for the Italian mechanical industries by inspiring new design solutions for the engineering corps[22] and by increasing the production – and political leverage – of companies like Fiat (cars), Ansaldo (ships), Bianchi (bicycles),

Caproni (aeroplanes), Borletti (bomb fuses), and Olivetti (typewriting machinery), not to mention popular newspapers like *Corriere della sera*. The conflict marked the beginning of Italian mass industry, and considerably changed the Italian industrial structure, by virtue of both the state's direct investment in war production after the foundation of the Istituto per la Mobilitazione Industriale (Institute for Industrial Mobilization), and the collective entrance of women into the workforce (thereby replacing male labour and leading to definitive abandonment of the countryside).

Objects in Art

Starting from Italy's first industrial boom in the early 1880s, the display of a new cosmopolitan "universe of commodities" (Benjamin, *Arcades* 8) in advertisements and expositions coincided with the increasing appearance of things in contemporary literature. As they started to invade the space of art in the 1880s, objects enacted a form of containing the invasion of industrial products, compensating their perceived chaos and ephemerality with a deliberate aesthetic construction. At the same time, however, objects also carried a new experimental epistemology, as documented in the exemplary narratives of Matilde Serao, Gabriele D'Annunzio, and Marchesa Colombi.

In the initial scene of *La virtù di Checchina* (*Checchina's Virtue*, 1883) Serao introduces the meeting of Checchina with her friend Isolina by way of a singular object-based description:

> Isolina threw herself on the couch, made of yellow cotton fabric with red flowers and rendered stiff by a straight back-rest. She observed the living room *distractedly*. There were four little armchairs covered in a similar fabric [...] about a round table of white marble. [...] Then, six chairs of black wood, pale in colour, which seemed perpetually dusty; a shelf covered in grey marble, upon which there were six cups of white porcelain, the coffee-pot, the sugar bowl; two boxes for sugar-almonds, empty and old, one of pale green satin, the other of straw with tassels; a plate of artificial fruit, this too in marble, vividly painted – fig, apple, peach, pear, and cherries; a gaming table, covered in green fabric [...] upon the only window were little curtains of embroidered voile, transparent, thin, with cloth borders. Just in front of the couch there was a small rug. That was it. (Serao, *Virtù* 13–14; emphasis mine)[23]

While listing things in a rationalized order, Serao's item-filled page also stages a new aesthetic focus on objects, hinted in the nonchalance of Isolina's "distracted" gaze, precisely detailing their fabrics (cotton, satin, straw), colours (yellow, white, black, grey, green), materials (marble, wood, porcelain), decorations (with flowers, *à cretonne*, in tassels), shapes (artificial fruits, embroideries), and disposition. As in a still-life painting, the objects on the page turn into symbolic actors in the space of the house, revealing Checchina's temperament even before her appearance in the narration. Following the same technique, in the later novel *La conquista di Roma* (*The Conquest of Rome*, 1885), Serao describes the protagonist Sangiorgio by meticulously observing his apartment in Rome.[24] Sangiorgio's objects stage his expectancy and fear of Angelica (his femme fatale), projecting onto the theatre of the house's interior an atmosphere of soft and languid sensualism. At the same time, as they are "skilfully" arranged in the room (*Conquista* 206) in relation to a studied interplay of light and shade, or to the careful combination or variation of colours, and as they are sensorily experienced through touch (fabrics, curtains, and carpets) or smell (in the presence of a variety of flowers), Sangiorgio's objects transform from aestheticized pieces of furniture into the silent protagonists of the scene.

Serao's fictional focus on materials, textures, and flowery motifs matched the aesthetic posture of the contemporary decorative arts in their unique crafting of glass, fabrics, ceramics, iron, and wood, and her accurate gaze prefigured a new attitude in the planning of the objects' forms. In a way, Serao's literary descriptions mirrored the language of still-life painting (similarly rediscovered by French Impressionism and contemporary photography), both by their imitation of the baroque impulse to catalogue reality and by their witty search for variations in a precodified repertoire of themes (e.g., flowers, fruits, drapery, furniture, cloths, folds, glasses, ceramics, dust, food). In another way, the gaze of her characters also foresaw a new approach in the construction of objects, focused on their external decoration, on their three-dimensionality (in the expanded reference to touching, hearing, smelling, and tasting), and on their amalgamation with the surrounding space (thus transforming a *still life* into a *lifestyle*; Rinaldi 146).

Like Serao, D'Annunzio also detected in objects a defining element of his contemporary modernity and the experimental site of a new form of aesthetics. Objects played a significant role in his novel *Il piacere* (*Pleasure*), set in 1886–7 and published in 1889. As they are meticulously arranged in the "most perfect theatre" (17) of the protagonist's house (Andrea Sperelli), artisanal objects and collectable items appear as living actors, endowed with a symbolic sexual charge (as seen in their "latent aphrodisiac potentiality"; 18) and imbued with their owner's personality ("they had acquired for him something of his own sensibility [...] they were not just witnesses, but partners in his love stories, in his pleasures, in his sadness"; 17).[25] As valuable antiques, Sperelli's objects reveal an implicit element of resistance against industrial indistinction. As pieces of a collection, they offer a remedy against industrial ephemerality by carrying forever in their fragmentary nature a "magic encyclopedia" (Benjamin, *Illuminations* 205) of their time. As creative agents of a new mental space, they embody Sperelli's elaborate search for distinction and trendiness, thus becoming experimental sites for the fashioning of his new modern self.

Amidst the many objects listed in the novel, D'Annunzio deliberately focuses on fabrics, accessories, and pieces of clothing as objects of pleasure. Against the back-drop of D'Annunzio's reports of Rome's worldly life for the newspaper *La tribuna, Il piacere* portrays the characters' appearances meticulously, as seen in the detailed account of Elena Muti's dress upon her entrance into Andrea Sperelli's house, or of Sperelli's care in dressing before a dance at Palazzo Farnese.[26] Such aesthetic focus on the tactile and emotional impact of objects and such attentiveness to costumes (both social practices and dresses) informed D'Annunzio's own practice of fashion design – attested by his commissioning of the best tailors of his age to make clothes for himself and his lovers, as well as by his crafting of customized pieces, which were branded with the logo "Gabriel Nuntius Vestiarius Fecit" or "Made by Gabriele D'Annunzio, Tailor".[27] In this light, D'Annunzio's approach to objects in the aesthetic space of the novel was not only that of a collector and an art fetishist but also that of a designer creatively transforming commercial matter into art and of a style maker fashioning trends around himself and his lovers. In parallel with D'Annunzio (also known as the Vate), the space of art also became an experimental workshop of fashion design in the contemporary works of the painter Mariano Fortuny (who designed with the Vate a gown for the actress Eleonora Duse) and the portraitist Giovanni Boldini. Fortuny pioneered the evolution of the Italian fashion industry by transforming his Venice house into a laboratory for designing costumes (for Wagner's operas) and gowns (such as his famous Delphos gown of 1907). Boldini assembled, in his pictorial catalogue of the women of European high society, an experimental gallery of the fashion trends of his time with regard to accessories (purses, umbrellas, pearls, gloves, and shoes), fabrics (elaborate laces, draperies, and crochet), and cuts (of gowns, sleeves, and hairstyles) ("The Complete Works").

While Serao and D'Annunzio prefigured in their objects the sort of planning attitude that would later be central to Italian industrial design, the same objects – similarly deprived of their exchange value – appeared instead as enigmatic sites of reflection on modernity itself in the contemporary work of other Italian writers like Marchesa Colombi, Emilio De Marchi, Italo Svevo, Luigi Pirandello, and Guido Gozzano. As outmoded items, chaotically accumulated and covered with dust, their objects indicated the ephemerality of consumption and embedded an everlasting repository of memory to be nostalgically or ironically activated.[28] In the novel *Un matrimonio in provincia* (*A Small-Town Marriage*, 1885) Marchesa Colombi purposely contrasted with the opening of Serao's *La virtù di Checchina* by beginning her autobiographical story with a detailed list of the objects stored in her family house. Worn-out items provocatively stage the rapid aging of the present, such as the consumed elements of her parents' room ("two holy-water bottles in engraved silver, which time had *oxidized*, [...] a pile of *flaked*, *crumbled*, and *twisted* candles, [...] the daguerreotype portraits of mom and dad for their wedding, *almost entirely faded*, [...] a bunch of *yellowed* sheets, containing Dad's poems from his youth"; 7–8, emphasis mine).[29] At the same time, however, immobile junk crystallizes a nostalgic fragment of the past, setting in motion the author's personal quest for a unique memory in their link to life events (e.g., the gloves worn by the mother on her wedding day, the silver plate given as a gift on the protagonist's birth).

As in Colombi's book, outdated objects frequently appeared in late-nineteenth-century literature in conjunction with the topos of the *rovina* (ruin) following the sale of the family house (e.g., in Verga's *Malavoglia*, Svevo's *Una vita*, De Marchi's *Demetrio Pianelli*, and Pirandello's *Il fu Mattia Pascal*). Similarly, their abandonment and their dust signified both an irreparable loss and, as the past's sole remnant, a poetic anchor for fiction. Further to this imagery, junk would indeed become a site of non-conformist memory in the early twentieth century, in the works of the Crepuscularist poets Sergio Corazzini, Marino Moretti, and Corrado Govoni. In the poetry of Guido Gozzano it would even turn into a living source of inspiration, as seen in the poem "La signorina Felicita" ("Miss Felicita," 1911), which significantly indicated the "mousetraps, mattresses, vases / lamps, baskets, cabinets" accumulated in the attic of an old vacation house as the "junk / rejects so dear to my Muse" (*Opere* 194; verses 154–6;).[30]

From this early imagery, which accompanied the industrial boom of the 1880s, two concomitant models of object manufacture started to develop between the 1890s and the 1910s, later forming two creative sources of industrial design. First, during the contemporary rediscovery of Italy's crafts heritage and the early theorization of decorative arts, Italian artisans started to intentionally "apply" an artistic dimension to serial production as a way to "author" manufacture and extend it from local workshops to a larger (even industrial) scale. Second, with the appearance and increasing acceptance of new technologies in the social space, Italian intellectuals began to include industrial products or brands in their artworks, reconfiguring them from serialized artefacts into cultural symbols.

The Theory and Practice of Decorative Art

The gradual contact between early industrialism and the Italian crafts patrimony since the 1880s elicited the need for a more mature project to merge serial production and the arts. Its early seeds were manifested in the establishment of the Commission for the Teaching of Art and Industry after the exposition of Turin in 1884, and in the spreading influence in Italy of the British Arts and Crafts movement (launched in 1887 by John Ruskin and William Morris).[31] The cultural propeller for a new theory and practice of the "applied arts" came, however, from the journal *Arte italiana decorativa e industriale* (*Italian Decorative and Industrial Art*), founded by the Milanese architect and writer Camillo Boito in 1890, and published until 1911 under the aegis of Italy's ministry of agriculture, industry, and trade. Along the tradition of Carlo Cattaneo's journal *Il politecnico* (1839–44, 1859–69), the publication, addressed to collectors and artisans, advanced the idea of industrial culture as an original fusion of aesthetics and production – as confirmed in its stated aspiration "to be *beautiful*, but more than beautiful, *useful*" (no. 1, January 1890; emphasis mine).[32] *Arte Italiana decorativa e industriale* systematically presented faithful reproductions (drawings and photographs) of architectural details, decorations, and artefacts (doors, tables, chairs, ceilings, ironworks, jewellery, leather, embroidery, ceramics, fabrics) from the Renaissance to the present. Its aim was to provide artists and artisans with exact, reproducible prototypes (allowing them to start a larger-scale quality production, and overcome the risk of approximation) and industries with original models for developing a new style in the adornment of serial objects.[33] By grounding decorative arts in a reinvented practice of traditional crafts, Boito conceptualized over time a new idea of "industrial art" as a serialized form of artisanship. Boito's pioneering project constituted an important contribution to the evolution of an autonomous industrial culture, in parallel with the contemporary establishment of the art journal *Emporium* (1895–1964), the foundation of the Esposizione internazionale d'arte decorativa (Biennale) of Venice in 1895, and, most importantly, the Italian reception of art nouveau (reinvented as Liberty style) from the mid-1890s.

Presented at the International Exposition of Modern Decorative Art of Turin in 1902, the Italian Liberty style realized and divulged a new, accomplished model of decorative or industrial art. The 1902 event propelled the commercial success of the style and helped to define its peculiar tension to endow with an inimitable aura both industrial products (through the aesthetic decoration of their form) and their architectural space (as exemplified in the exposition's pavilions designed by architects Raimondo D'Aronco and Giuseppe Sommaruga, and in Turin's Casa Fenoglio-Lafleur designed by Pietro Fenoglio in 1902).[34] In the journal *L'arte decorativa moderna* (*Modern Decorative Art*, 1902), published during the exposition, the editor, artist, and critic Enrico Thovez defined Liberty as the accomplishment of a new Italian modern art: produced serially (industrially), with an aesthetic purpose, and targeting daily life in its practical dimension – "un'arte che porta in ogni angolo della casa, in ogni oggetto dell' esistenza famigliare e cittadina, un elemento di bellezza e di armonia" (quoted in Besio; an art that brings an element of beauty and harmony to every corner of the house, every object of familial and civic existence).

In parallel with Boito's and Thovez's theories, the Italian decorative industry, in the manufacture of ceramics, glass, metalwork, furniture, interiors, textiles, and costumes, enjoyed a period of extraordinary international acclaim at the turn of the twentieth century, thanks to the recognized excellence of its craftsmen and the long-lasting reputation of its traditions. At the same time, thanks to the increased demand of the bourgeoisie for individualized objects, the applied arts underwent a moment of commercial expansion and productive reconversion in their quest for manufacturing processes that mediated between artisanship and serialization (Sparke, *Italian Design* 29).

In the field of glass production Venetian artists like Duilio Cambellotti (in the art of stained glass) or, later, Vittorio Zecchin (in the production of vases) relaunched on a larger scale the glass-making tradition of Murano. In the field of ceramics two factories emerged in the wide panorama of Italian local workshops: Arte della Ceramica, founded by Galileo Chini and awarded the gold medal at the Exposition of Paris in 1900; and Richard-Ginori, established in 1735, which would be in the avant-garde of the design movement after hiring Gió Ponti in the 1920s. In the field of furniture Carlo Bugatti was the first to raise the working of ebony to a level of excellence, as confirmed by the gold medal won at the Exposition of Paris in 1900, by the later success of his apprentice Eugenio Quarti, and by his successful collaborations with the Liberty architect Giuseppe Sommaruga and the iron designer Alessandro Mazzucotelli in the planning of bars (like the Camparino in Milan), public spaces, and private houses (including those of Marinetti, Puccini, and Segantini). In the field of textiles and fashion the pioneering figure of Rosa Genoni freed Italian fashion from its French domination by retrieving its historical significance from the Humanist era and by applying Renaissance figurations to the design of modern-day dresses (as in the Primavera gown, inspired by Botticelli, which was presented at the International Exposition of Milan in 1906).[35]

The Culture of Industry

The plethora of technological innovations entering the market at the turn of the twentieth century generated new industrial forms of culture as well as a new culture of industry. New cultural spaces with a broader industrial reach opened in the fields of journalism, photography, advertising, and entertainment (in response to the growing enthusiasm for sound recording, cinema, and radio). Some intellectuals actively engaged in the new cultural industry by collaborating with newspapers (Carducci, De Amicis, D'Annunzio, Serao), illustrating journal covers or posters (Boccioni, Rubino, Dudovich, Metlicovitz), partnering with cinema (Gozzano, Verga, Pirandello), embracing the record industry (Caruso, Puccini, Toscanini), and working on commissions for other companies (e.g., D'Annunzio, who branded the Veglia alarm clocks and the department store La Rinascente). A few others intercepted the new dynamics of the culture industry and even turned into cultural entrepreneurs by promoting their work through advertising (as in Puccini's partnership with Leopoldo Metlicovitz in the design of posters for his operas); constructing their image through photography (as in D'Annunzio's collaboration with Mario Nunes Vais); enacting pioneering marketing strategies (as in

Marinetti's launch of Futurism in 1909 in *Le figaro* through the manifesto genre,[36] or in Giovanni Papini's deliberate pursuit of provocation in advertising *Lacerba,* 1913); or turning authorship into a signature label (as in D'Annunzio's branding of Pastrone's movie *Cabiria* of 1914). As serial cultural products (books, records, movies, photographs, illustrations) increasingly appeared in Italy's market, other intellectuals either rejected industry as such – as in the case of Benedetto Croce who discarded it as a degradation in light of his aesthetic purism (Sparke, *Italian Design* 26) – or reacted to the ongoing commodification of the arts with hopelessness and even despair. In opening the novel *Il fu Mattia Pascal* (*The Late Mattia Pascal,* 1904) with the vision of an abandoned library filled with random piles of books, Pirandello offered a symptomatic image of this disheartened resistance in front of the old world's decay, depicting both the crisis of an immutable idea of culture (exposed to consumption, as any other perishable item) and the suffocating paralysis (signified in dust) following the loss of its aura.

Along with their gradual acceptance in the market and society, some industrial products and brands started to break into the artistic space in a distinct, recognizable form, slowly acquiring new intellectual value. Technologies (radio, cars, aeroplanes, and typewriters) and brands (e.g., Manoli cigarettes and Tot detergent) openly entered the Futurist aesthetics, signifying a new dimension of art and the progressive modification of modern man's sensibility.[37] Albeit contained in the space of traditional literary language, cars and, above all, aeroplanes appeared in D'Annunzio's novel *Forse che sì forse che no* (*Maybe Yes Maybe No,* 1910) as symbols of man's new, limitless possession of reality. In Aldo Palazzeschi's poem "La passeggiata" ("The Walk," 1913), brands and local shops (e.g., "La pasticca di Re Sole," "Cinematografo Splendor," "Politeama Manzoni," and "F.lli Bocconi"; *Tutte le poesie* 583–7; verses 20, 64, 113, 122) invaded the space of poetry, conveying, in their analogical and rhymed succession, the excited whirl of modern life. After Picasso's shocking *Still Life with Chair Caning* (1912), daily objects also appeared in painting, indicating the intersection of eternal and ephemeral beauty in modern art, and shaping new techniques, like the creative collage of advertisements and newspaper clippings (as in Carlo Carrà's *Interventionist Demonstration* of 1914, or in Umberto Boccioni's *Charge of Lancers* in 1915) and the dynamic enactment of other sensorial realms (as in Ardengo Soffici's box of matches in *Still Life with Matches* of 1914).

As common industrial products like watches, bicycles, gramophones, cigarettes, toys, and photographs increasingly became the subjects of artistic representations, they gradually attained aesthetic relevance and new symbolic meanings. The elaboration of aesthetic codes upon these objects undoubtedly obeyed a commercial strategy (aimed at differentiating the Italian product through art), and their rigid stylization certainly contributed to a "rise of unanimity" in cultural production, induced by the industrial logic of infinite equivalence and substitutability (as theorized in Horkheimer and Adorno's *Dialectic of Enlightenment*, 36). Notwithstanding, the aesthetic reworking of these objects and the cultural investment in them by entrepreneurs like Senatore Borletti, Vittorio Alinari, Edoardo Bianchi, Giulio Ricordi, and Davide Campari also generated around them an independent and original culture that was able to set critical debates in motion, affect social and political spheres on various levels, and intermingle different worlds (art and industry) in a state of constant self-assessment.

The "aestheticization" of these selected objects then documents the slow metamorphosis of the arts from a humanistic endeavour into a large-scale business. In turn, their "culturalization" makes manifest the reconfiguration of industrial production into an authentic culture. Along this experimental process of mediation, the objects provide a traceable platform on which to reconstruct Italy's industrialism and to illuminate its a-systematic elaboration of an original way to industrial modernity.

Chapter Two

Timepieces and Italian Modern Times

In October 1883 scientists from all over the world convened in Rome for the seventh International Geodesic Conference to discuss ways to establish a common reference time for the globe. Twelve years earlier, in 1871, the first International Geographic Congress, in Antwerp, had launched the goal of extending the nautical prime meridian of Greenwich to all nations, and since that time Italy had actively promoted geodesic conferences. In 1875, in accord with the second International Geographic Congress, in Paris (which ratified the "Metre Convention" and defined the metric system as a new way to measure space), Rome had organized a conference, pushing for a diplomatic compromise between England and France, which led the former to accept decimal measurements and the latter to recognize Greenwich as the world's prime meridian (Thomas 24). In 1881 Venice hosted the third International Geographic Congress, revolving around the need to establish a zero point from which to measure the exact altitudes of the earth, and, for the first time, the congress actively included in its agenda the creation of a common meridian for timekeeping. Both Italian conferences, of 1875 and 1881, ended in diplomatic and scientific failure, postponing the timekeeping debate to an unspecified future.[1] Given this prior history, the 1883 conference was surrounded by high expectations.

The Rome conference took off from the issue (already debated in Venice) of determining a zero point from which to survey geographical altitudes and the earth's size and shape (geodesy). As scientists presented evidence that distances from the shore to ancient Roman harbours had changed, they decided to discard the sea as a valid reference for a universal level – because its variables (winds, waves, streams, and tides) alter the relationship between earth and water – and agreed to create artificial landmarks on the coast-lines as zero points. Their discussion on the unequal levels of the seas offered scientific support for Isaac Newton's hypothesis on the ellipsoidal shape of the earth (launched in his *Philosophiae naturalis principia mathematica* of 1687) and led to the related issue of the unequal length of day that sunlight produces on a revolving spheroid flattened at its poles. Moving from geodesy to contemporary society, the conference finally focused on the issue of having a common reference point for measuring time and approved some significant resolutions. During the sessions the participants agreed that "the unification of longitude and time was desirable" in the interests of science and global communications (resolution 1); they proposed the application of a sexagesimal system to hours and seconds (resolution 2), selected Greenwich as the prime meridian (resolution 3) for counting hours in one direction (resolution 4), and enhanced the need for a universal day (starting at noon, Greenwich Mean Time; resolution 6) and a universal time "to be used alongside local or national time, which must necessarily continue to be used in civil life" (resolution 5; Howse, "1884" 13–14).[2] Despite their consultative nature, the resolutions that were passed in Rome laid the groundwork for future political and diplomatic action, affecting the contemporary institution of American standard time (on 18 November 1883) and paving the way for the International Meridian Conference of Washington in 1884, which established Greenwich as the origin of modern time.

Italian newspapers extensively covered the International Geodesic Conference of Rome, hoping to capture the historical moment of a resolution on universal timekeeping, and presented the event as both a worldly celebration and a landmark in the nation's rebirth. At a cultural level the press constructed the conference as a favourable occasion on which to expose Italians to current scientific achievements, to assert Italy's leading role on the global scene, and to boast of the primacy of its civilizing culture (bonding the world together in the quest for a universal time standard). At a political level the conference showcased Italy's aspirations to emerge as a world power, a year after the signing of the Triple Alliance with Prussia and Austria-Hungary and after its first colonial penetration in Eritrea (with the purchase of the Bay of Assab from the Rubbattino Company).[3] Along with the Fine Arts Exposition (Esposizione di Belle Arti) taking place in Rome the same year, the International Geodesic Conference offered a new opportunity for a nation in search of an identity, a year after the death of the Risorgimento hero Giuseppe Garibaldi and the crushing, with the execution of the Triestine irredentist Guglielmo Oberdan, of Italy's territorial ambitions at the northeastern border. Lastly, while enacting the political agenda of promoting a new scientific and technological culture throughout the national territory, the International Geodesic Conference offered the press an occasion to reflect upon industrialism and its global needs, against the back-drop of Italy's dawning industrial venture, which had been launched two years before in the 1881 exposition of Milan. The press coverage of the Roman conference – focused on the urge to assert Italy's power among modern nations and to affirm the country's uniqueness – intentionally followed the rhetoric of the Milanese exposition, which had been presented by *Corriere della sera* as an occasion to showcase Italy's participation "in the stream of modern ideas and work" (4 May 1881), and as "a solemn feast in which Italy affirms its own existence and its own value, and, by displaying what it has done, offers a sure promise of what it will offer in the future" (4 May 1881).[4]

Even though the International Geodesic Conference of Rome in 1883 did not achieve an ultimate resolution on world standard time, it constituted a significant chapter in the global history of modern timekeeping (by virtue of its ground-breaking debates and resolutions), and a key moment in the expansion of the Italian industrial project. In contrast to other government-sponsored congresses and expositions that endorsed a new global modernity, the 1883 month-long debate on timekeeping captured a larger, more radical shift in Italian society, along with the process of industrialization and the later spreading of world standard time. Such debate turned into a larger cultural reflection both on watchmaking – conceived as crafts (in the humanistic tradition of the liberal arts) and a serial production (combined with the development of industrial time as a monetary or quantitative measure of labour) – and on the very nature of modern (and Italian) times, in which the old, circular order of a transcendent time coexisted with a new notion of immanent, linear, and accelerated times.

Over the following decades Italy's documented involvement with the global timekeeping debate not only testified to the historical impact of world standard time upon its society and culture but also contributed to the creation of an original idea of Italian modern time (and times) in industry, arts, and ultimately politics.

Italy and the Global Debate on Timekeeping

The global effort to establish a common reference time for modern industrial nations originated in the development of telegraphy, which was applied to transatlantic communications and, later, to railway systems. In transatlantic navigation the emergence of a new timekeeping method (the so-called Harvard Telegraph Time, presented at the 1851 London exposition), based on wired telegraphic beats, gradually replaced the old way of calculating the hour (and a ship's exact location) by using a chronograph set to Greenwich Observatory Time and subtracting the actual sun meridian. In North American railway systems the increased use of telegraphy led to new timekeeping models in the 1870s, outlined in two pamphlets by Charles Dowd (*System of National Time*, 1870), dividing the United States into four time zones and identifying Washington as a reference meridian, and by Sandford Fleming (*Uniform Non-local Time*, 1876), dividing Canada into three time zones and envisaging a twenty-four-hour universal time.[5]

The developments in transportation and the growth of industrial production increased the need to adopt a global reference for time-keeping in the early 1880s. Thc International Geodesic Conference of Rome in 1883 laid the scientific and cultural foundation for the development of modern timekeeping by way of its scientific and political recommendation of Greenwich as the world's prime meridian (Bartky 72).[6] The International Meridian Conference in Washington, organized in October 1884 by American president Chester A. Arthur and gathering delegates from twenty-five nations (ten from Europe, ten from Latin America, and one each from Hawaii, Liberia, Japan, the Ottoman Empire, and the United States), revolved around its political and economic implications. After ratifying the conclusions of the Rome conference – the need for a prime meridian, a universal twenty-four-hour day starting on the first meridian's midnight, and a universal time (which "shall not interfere with the use of local or standard time where desirable"; resolution 4; quoted in Dolan) – the Washington conference focused mainly on the diplomatic decision about the prime meridian, or rather about one nation dictating its own time to the planet. Among the proposed prime meridians of London, Paris, Washington, or Berlin (Italy proposed Jerusalem as a neutral solution; Pollina), the conference chose Greenwich because of the accuracy of its observatory, its pre-existing reference in nautical charts, and England's leading role among industrial societies. The resolution on "the Observatory of Greenwich as the initial meridian for longitude" (resolution 2; quoted in Dolan) was approved, with twenty-two votes for and one against (San Domingo), but it did not pass without conflict, as reflected in the abstention of Brazil and in the protest of France (which lobbied for decimalized time and which would continue to measure time according to Paris world time until 1911; Galison 156–220).

In Italy, despite the excitement raised by the Rome conference in the previous year, the press barely mentioned the Washington event, resentfully dismissing it as an unimportant ratification of a fictional universal time, vis-à-vis a true local time ("the universal day will not prevent any country from counting the true time that conforms to its respective longitude, but it will serve its purpose in international relations of all kinds"; *Corriere della sera*, 22 October 1884).[7] The nation ratified world standard time (WST) only in 1893, by adjusting the Rome meridian to Etna in order to conform to the European central time of its allies Austria and Germany.[8] This belated recognition, nine years after the Washington conference, did not diminish either Italian society's perception of international time as a human-based fiction (which would be repeated with the introduction in 1916 of daylight savings time from 3 June to 3 October),[9] or the importance of Monte Mario (established in 1870) as the *origine delle longitudini italiane* (origin of Italian longitudes), which was proudly indicated in maps until the 1960s.

Despite the acceptance of global timing, the prolonged adoption of an Italian prime meridian constituted "a significant indicator of the young Italian state's desire to stress its own uniqueness and peculiarity" (Ceen 100). Moreover, it also revealed the country's persistent resistance to standardized timekeeping, rooted – in urban areas – in the recent creation first of a common hour (agreed between Rome and Milan in 1866) and then of a national meridian (Monte Mario),[10] and – in rural areas – in the continued use of the medieval *ora italica* (Italic hours) or *tempo dell'Ave Maria* (Hail Mary time), which was calculated on a counterclockwise twenty-four-hour dial, starting from the hour of sunset or the last night prayer.[11] In this context, the issues debated in Rome (and the resolutions approved in Washington) thus divulged a radically new notion of timekeeping and, albeit unnoticeably, deeply affected the social and cultural self-awareness of Italians with relation to time in at least three ways.

First, the geodesic discussions of Venice and Rome spread the notion that, in light of the earth's ellipsoidal shape, sunlight no longer offered a reliable foundation for standardized timekeeping. Indeed, the resolutions in Rome indicated that, in order to create a universal day (uniforming the differences of daylight in the earth's diverse locations and seasons), the real sunlight – "the true time, indicated by solar quadrants" – had to be reduced to a fictitious length based on the calculation of a global average ("the average time based on an *artificial* sun"; Usioli, *Illustrazione italiana*, 27 September 1883, emphasis mine).[12] The declared unreliability of the sun thus moved time's foundation from a metaphysical source – the heavens (sun, stars, moon), a transcendent God (the Greek Kronos, the Creator in Catholic doctrine, the great Watchmaker in Protestant tradition), or an extra-temporal entity (the immobile engine in Aristotle, an a priori category in Kant, an empty structure in Newton) – to a new secular origin, rooted in the intra-temporal dimension of a political, imperial, and geodesic calculation. This silent yet radical shift, ratified by the Washington conference, had been captured in the same years by Nietzsche, who opened *Thus Spoke Zarathustra* (published between 1883 and 1885) with the non-casual hubris of his prophet towards the sun, right before his proclamation of the death of God: "He rose, stepped out before the sun, and spoke to it thus: Greetings, Great Star! What would your happiness be, were it not for those whom you illumine!" (9). Disconnected from the sun's circular motion and replaced by the development of electricity (propelling man to no longer equate day and sunlight), time detached from its eschatological origins, or its connection to meaning-bearing events (rituals, liturgies, anniversaries), and became linear as an infinitely divisible succession of instants, measuring human progress by a potentially endless algorithm.

Second, while the assigning of an arbitrary beginning for the earth's movement[13] confirmed the notion of a different source of time (in contrast to the traditional yet imperfect one coming from the heavens or a religious system), the election of the Greenwich astronomical observatory as new *umbilicus mundi* also ratified the construction of a new order of the cosmos,[14] based on the political power of imperial England and the needs of industrial capitalism. In Umberto Eco's words, if "for millennia the only reliable clock was the cock's crow and, in a predominantly agrarian economy, the only measures needed to regulate individual and social lives were sunrise and sunset, and the passing of seasons," the new "clock civilization" (11) imposed instead an urban, standardized time, as well as a new societal framework based on fixed, regular, and automatic control. The passage from countryside to urban life – or, in Hannah Arendt's terms, from *vita contemplativa* to *vita activa* – followed the development of industrialism, from England's "horological revolution" (Macey 17) of the seventeenth century[15] to the timekeeping revolution of world standard time. The establishment of a legal time around the first world clock constituted a powerful means to extend this order globally, and not by chance was the first recorded act of international terrorism in history (narrated in Joseph Conrad's 1907 *The Secret Agent* as the first attempt to dismantle the globe's engine) a bomb placed under the Greenwich observatory by a French anarchist (Martial Bourdin) in 1894.

Third, the acknowledged existence of a universal time, running independently only "for scientific need and for the service of communications, such as railways, maritime routes, telegraphy and post offices" without affecting the dimension of "local or national time that would continue to be used in civic life" (*Corriere della sera,* 2–3 November 1883),[16] divulged the awareness of a dual and concurring timing of reality. While nineteenth-century double-dial watches exclusively referred to the division into *ora francese* (*ore moderne*) and *ora italica* (or *all'antica*), the appearance in the Rome conference of two parallel "timepieces, with opposing quadrants, one indicating the local hour, and the other the universal hour" (2–3 November 1883),[17] envisioned a different idea of time's duality. The Rome and Washington conferences thus inaugurated a temporal, double citizenship of modern man, divided between the time of contingency – personal and local (as confirmed by the resistance of countries to giving up their authority on national times) – and a universal or fictional meta-time, connecting all individuals in a global network. The creation of such abstract time constituted not only a common currency, or a common language for the world (Zamenhof invented Esperanto in 1887), but also the first concrete step in the process of global unification.

Italian Timepieces in Contemporary Cultural Representations

In conjunction with the timekeeping debate during the last two decades of the nineteenth century, reflections on modern temporality emerged in contemporary Italian art and literature, either through indirect references or through the explicit symbolization of clocks and watches.

Two paintings of the period by Giovanni Segantini contain veiled allusions to the temporal shift engendered by the geodesic and meridian conferences: *Ave Maria a trasbordo* (*Hail Mary while on Board,* 1882, 1886) and *Mezzogiorno sulle Alpi* (*Noon on the Alps,* 1891). In the first canvas (whose two versions were composed a year before and three years after the Rome conference) Segantini alludes to the Italic hour by staging three shepherds (a man, a woman, and a child in the woman's arms) who hear the bells from a distant church tower as they return from work and recite the sunset prayer at the end of the day (which, according to the *ora dell'Ave Maria* or *ora italica,* represents its beginning). In the second canvas (plate 1), displaying a woman with her grazing sheep who is contemplating the radiance of an Alpine plateau at midday, the painter, perhaps unconsciously, hints at the contemporary timekeeping debate by representing the failed equivalence of the suggested noon (or the sun's real meridian) and the non-perpendicular shade produced by the sunlight on the ground (the fictional hour of the new clock civilization?).

Albeit indirectly, the two paintings convey a sense of unease at the contemporary transition to a new temporal model, assuming the forms of a nostalgic resistance and a latent disquiet. In first case, the reappearance of *ora italica* (vis-à-vis the establishment of world standard time) indicates a prolonged attachment to a past idea of time, related to the nostalgia for both the rural civilization and an everlasting metaphysical or social order (symbolized by the elements of prayer and the bell). The trope of longing for the past, seen as symbolically compensating for the spread of urban industrialism, continued in Italy until the turn of the century, as attested by Giovanni Pascoli's poems "L'ora di Barga" (in *Canti di Castelvecchio,* 1903; "The Hour of Barga") and "L'Avemaria" (in *Primi poemetti,* 1904; "The Hail Mary"), which evoke the Italic hour through the representation of the bell's sound and the last prayer at sunset.[18] In the second case, Segantini's discrepancy between "natural" and man-made time filters not only the need for a gap between legal and solar time as the condition for modern timekeeping but also a hidden element of disquiet, as the sun acquires an "asymmetrical" iconography. The title of his canvas, *Mezzogiorno sulle Alpi,* seemingly refers to the traditional imagery of noon, depicting the sun's zenith as immobile, radiant, and perfectly even (Perella 70–113). Such imagery would be manifested a few years later in Giosuè Carducci's poem "Mezzogiorno alpino" (in *Rime e ritmi,* 1898; "Alpine Noon,"), which staged the motionless serenity and symmetry of the sun's noon through the positioning of the word *mezzodì* (verse 4; midday) at the exact mid-point of the composition.[19] Notwithstanding the apparent similarities, the canvas displays an unprecedented image of an uneven sun, no longer seen in its eternalized zenith, but rather captured in its destabilizing transitoriness and almost irrelevant motion.

In conjunction with these indirect allusions, contemporary literature testifies directly to the cultural impact of the timekeeping debate on Italy and records symbolically the nation's unresolved tension towards modern times. The growing presence of clocks and watches in the literature of the 1880s, paralleling the years of Italy's early industrial push, defines the contemporary awareness of the present as a phantasmagoric and uncanny time of feverish acceleration. The parallel rediscovery of the baroque theme of the "dance of the hours" in music (in the famous tune of Amilcare Pomchielli's opera *La gioconda* of 1876; act 3, scene 6) or, later, in painting (in Gaetano Previati's canvas *La danza delle ore* of 1899) also confirms the association of the present with an inebriating yet destabilizing movement.[20] Against the back-drop of Italy's quest for the global hour, literary clocks and watches thus signify the exciting pursuit of the new and the anxiety about the fading of the old, as symbolically documented in the timepieces of two contemporary novels: *La virtù di Checchina* (*The Virtue of Checchina*, 1883) by Matilde Serao and *Il piacere* (*Pleasure*, 1889) by Gabriele D'Annunzio.

In the short novel *La virtù di Checchina*, published in instalments in *La domenica letteraria* between 25 November and 6 December 1883, right after the Rome conference, a wrist-watch appears as an idealized object for the protagonist Checchina, a married woman who is misunderstood and mistreated by her husband and aspires to (yet never realizes) an adulterous relationship with the nobleman Marchese d'Aragona. Checchina wishes for "un orologetto" (39; a small watch) and regrets not having one, especially after she learns from her friend Isolina that a golden wrist-watch constitutes not only a desirable fashion item for ladies but also an indispensable tool for extra-marital affairs: "You can't believe how terrible it is not to have a watch when you have a lover! You are always on the wrong time. You are too early: he is not there, and it's a slow death. You are late: a quarter of an hour has already passed, and for another quarter of an hour he pouts. Men are annoyed by waiting. If, once at his place, you ask him every five minutes, 'What time is it?' this question will irritate him. You are always late, on your return home, with such a bewildered look that it is a miracle that it doesn't betray you. My God, what would I do to have a watch!" (39)[21] After Isolina's advice, the idealized timepiece acquires an explicit sexual charge, serving as the missing tool for Checchina's escape with the Marchese d'Aragona and, above all, the symbol of a new possible freedom (expressed in the liberation of her erotic desires and in the emancipation from her pre-assigned gender role). In Isolina's words and Checchina's imagination the wished-for "orologetto" also fashions a new "watch temporality" for the characters, identified, in one way, with the seemingly indispensable capacity to register and co-ordinate time with precision, and, in another way, with the perception of an accelerated or intensified life.

While reflecting Checchina's new feverish state of energetic vitality or unsettling vertigo, of excitement for an illicit love and fear of discovery (leading her to abandon her adulterous plan at the end of the novel), the wrist-watch also mirrors contemporary attitudes of thrill and resistance towards the new dynamic, accelerated, and simultaneous temporality of industrial societies. By connecting the "orologetto" to fashion and adultery, Serao relates its temporality (whose characteristics mirror the new universal timing debated in Rome) both to an exciting novelty and to a destabilizing or even forbidden break from a centuries-long order of time. Moreover, in her association of the watch with a woman figure, Serao also refers to the two contemporary tropes of fascination and resistance, similarly outlining, in relation to a broader notion of modern times, the gendered vision of industrial modernity as an attractive yet ephemeral femme fatale, and, in relation to the market, of the male reaction to the growing wrist-watch trend as one of arrogance and fear towards an "irrational" female item. Such mental frames would mostly remain unaltered during the following decades,[22] at least until Carlo Carrà's shocking *Portrait of Marinetti* (1911), which reverses the codes and presents the Futurist founder wearing a wrist-watch as a masculine symbol of a trendy modernity and as an emblem of the new idea of rapid and instantaneous writing.

A few years after Serao, in D'Annunzio's *Il piacere* (1889), a singular jeweller-crafted clock, in the shape of an ebony skull, serves as an object of ambivalent attraction for Sperelli and his lover Elena in the auction episode. While releasing an "inexpressible appearance of life" (69) to her hands and "a *pleasure* never felt before" (68, emphasis mine) to him (via the vibration of her fingers), the timepiece also reminds Andrea of the vanity of his reawakened sexual desire by presenting the end of pleasure as a lugubrious omen of death (in line with the traditional coupling of Eros and Thanatos). Charged with a sensual and mystical tension, Sperelli's clock indicates an anti-scientific time, suspended, like the protagonist, in a prolonged state of decadence (as both bliss and decay) and mirroring, by analogy, the self-perceived disquiet of an age on the edge of vital blossoming (belle époque) and horror vacui (fin de siècle). D'Annunzio's uncanny timepiece, deliberately distanced from the commercial images of foreign industrial watches appearing in contemporary press, explicitly refers to the horologic imagery of baroque poetry (as in Giambattista Marino, Ciro Di Pers, Antonio Bruni, and Giovan Leone Sempronio) and still-life painting (as in Giovanna Garzoni and Bartolomeo Bettera). In seventeenth-century imagination, clocks staged an ambivalent "theatre of time" (Bonito 10) where old and new models of the cosmos could overlap in a prolonged suspension (as hinted in Leibniz's notion of *unruhe*, meaning both "pendulum" and "restlessness").[23] By their concurring temporalities – *praesentaneus* (fashion or taste) vis-à-vis crystallization (remoteness or death), or rather *krónos* (the moving and measurable exterior instant) vis-à-vis *kairós* (a stilled interior hour) – baroque clocks encapsulated the excitement of a time in transition (often equated to a wave or a dance), yet also mirrored the anxiety of a time without memory, decomposing in its atomic divisibility (as the sands of an hour-glass) and subduing mankind's life to its rigid mechanical dominion. Along with the diffusion of world standard time and the mass production of timekeepers, D'Annunzio's baroque clock similarly signals the trauma of a broken temporal order (and the end of the aristocracy) and a protest against the ongoing transformation of life from *contemplativa* to *activa* (D. Baroncini 121). In its juxtaposition of time of pleasure and time of vanity, D'Annunzio's clock also mirrors the novel's overall temporality, expressed as an attempt to forge a hybrid modernity (both ancient and à la mode) and enacted, in the book's opening pages, through the climactic overlapping of Sperelli's inner expectation of Elena's arrival at Palazzo Zuccari (*kairós*) with the intensified progression of external time indications (*krónos*) from the "clock of Trinità de' Monti" (D'Annunzio, *Piacere* 6).[24]

As the new trend of portable watches reached the market around the turn of the century, traditional clocks and watches increasingly appeared in literary representations as outmoded, antique, or broken. In opposition to the affirmation of a new pragmatic temporality, this imagery of timekeepers displayed the reassuring nostalgia of a purely decorative time of old, seen as full of memory and meaning. In the novel *Demetrio Pianelli* (1893), Emilio De Marchi equates an out-of-date watch with the warmth of the past, as the protagonist (Demetrio) sells his dear *cipollone* (pocket-watch) to repay his brother's debts. The old pocket-watch, turned into rubbish ("a second-hand dealer [...] purchased it as out-of-use junk, not as a timepiece"; 216), mirrors its owner's life, as Demetrio "compared himself to his old watch from Vienna and realized that he, too, was an out-of-use object" (216).[25] In the object De Marchi represents the gap separating the dear time of his memories, full of warmth and meaning, from the awareness of a present ruled by cold monetary exchange, pricing the "life lived in another world" as a "useless time" (216) worth little money.[26]

Moving into the twentieth century, the figure of a stuck clock – or, in the words of Guido Gozzano's poem "La signorina Felicita," of a "ticking broken clock" (verse 65, *Opere* 191; tic-tac dell'orologio guasto) – would become a dominant metaphor, indicating a double-edged temporality (seen both as an internal depository of experience and memory and as an external force of mechanical consumption), as well as the paradox of poetry itself, turned into junk in modern industrial societies.[27] Giovanni Papini's short story "L'orologio fermo alle sette" (in *Il pilota cieco*, 1907) presents a variation of the theme in the image of "an old clock stuck for many years at 7 a.m./p.m." (189). In reference to time-keeping, the stuck clock hints at a possible balance of local and universal time by suggesting the opposition or coincidence of a time of uselessness (in its stillness) and a time of revelation (in the realization that twice a day, at 7:00 a.m. and at 7:00 p.m., it expresses a real movement in sync with the world). In reference to the poet's intellectual life, the clock symbolizes the contrast between an external time of meaninglessness, boredom, and mechanical repetition ("the majority of the time my life is empty, ordinary, habitual, and filled with unexpressed boredom"; 190–1), and an internal time of value, intensity, and harmony with the universe ("in these instants, so fleeting, I live many more things than in all of the time between one passage and the other"; 192).[28] Along the same lines, Aldo Palazzeschi uses the image of an old stuck clock in the poem "L'orologio" (1910; "The Clock"), representing the paradoxical metaphor for a hybrid time (both stilled and in motion) and a way to distinguish the poet's uniqueness from the "statistical uniformity" (Arendt 43) of modern societies. In his interpretation of the theme Palazzeschi presents himself as breaking the clock's mechanism (which registers and reminds him of the hour of his death) and as comically enacting a mock suicide not just to reaffirm his own control over time ("now I give the orders / I will give the hour to you, Now!"; verses 125–6; *Tutte le poesie* 268) but also to react against the still widely perceived standardization produced by the quantification of time.[29]

Producing and Marketing Time

At the turn of the twentieth century the process of standardization and universalization of time, according to the needs of industrial capitalism, coincided with its growing mass consumption in the form of portable watches. The affirmation of wrist-watches, in the decades following the Washington meridian conference, contributed to the success of Greenwich Mean Time and was instrumental in defining a new pragmatic notion of time, as opposed to its decorative idea. The success of wrist-watches over pocket-watches followed their slow transformation from luxury feminine items (as the first wrist-watch, commissioned in 1868 by Patek Philippe for the Hungarian countess Koscowicz)[30] into reliable masculine items, which were produced on an industrial scale for an affordable price after the introduction of the Roskopf mechanism and Omega's first series of wrist-watches, launched at the 1900 Universal Exposition of Paris. Initially discarded because of their frivolousness, their lack of precision, their connection to female emancipation, and the supposed greater masculinity attributed to pocket-watches (which lacked the feminine element of the bracelet), wrist-watches started appealing to males in the early twentieth century, in connection with sports, science, and war. They became popular in relation to sports through the Brazilian aviator Alberto Santos Dumont, who flew over Europe (in 1901 with a dirigible and in 1906 with an aeroplane) and achieved in 1906 the record of 220 metres of height in twenty-one seconds while wearing a customized wrist-watch designed for him by the watchmaker Louis Cartier, which allowed him to check time in flight without taking his hands off the control panel (Strazzi 18). Wrist-watches' supposed lack of scientific accuracy was overcome in 1914 when the Greenwich Observatory approved the exactitude and precision of a model produced by Rolex. Lastly, wrist-watches acquired a masculine connotation in relationship to war, as anticipated in 1879 by Emperor Wilhelm's commission to the watchmaker Perregaux of two thousand gold wrist-watches (with a special protective grid) for the officers of the German navy (Strazzi 12), and in 1901 by the Goldsmiths Company's production of leather straps as a wrist support for watches used in co-ordinating military operations during the Anglo-Boer War. It was in the First World War, however, that the need to synchronize military actions made pocket-watches impractical and spurred the massive distribution of trench watches, which were more reliable against dust, moisture, and shrapnels (thanks to their metal grids). As soldiers wore them on the front line, wrist-watches turned from objects of convenience into objects of survival (on whose precision depended a soldier's life or death), and, as soldiers returned from the front, the watches were converted from war items into popular masculine products.

As wrist-watches made the experience and consumption of time more accessible, moving from the dominion of a few (kings, queens, aristocrats, jewellers, sailors, and scientists) to the common individual, they also carried to European societies (Italian included) their embedded tradition and unaltered functioning mechanism (Uglietti 9). From a technical perspective, they transmitted a centuries-long process of adjustment and miniaturization, dating back to the fourteenth century – when the first mechanical clocks appeared in Milan (Saint Eustorgius, 1309), Padua (1344), Genoa (1353), Bologna (1356), and Ferrara (1362)[31] – and extending throughout the sixteenth and seventeenth centuries, when Italian *maestri di bottega* (Lodovico Manelli, Giovan Battista Pirazzoli, and Domenico Fornasini) excelled in European courts for their craftsmanship in *orologeria minuta* (small watches embedded in rings or jewels). From a cultural perspective, wrist-watches intrinsically reflected the multiple values at the core of the modern age, for example in their association of watchmaking with the Protestant work ethics of precision and exactitude[32] (indicating man's new relationship to God, described as a great Watchmaker who no longer intervened in human history but impassibly set the universe in motion and oversaw its order). While reflecting the spirit of the scientific revolution (in its search for reliable time measurements), clocks also represented, after Huygens's application of Galileo's pendulum notion to horology (1657), "thc first industry to put into practice the theoretical findings of physics and mechanics" (Cipolla 58–9) as well as the first items to be produced in industrial series. With the invention of Harrison's chronographer (1735), clocks became indispensable tools for orienting and locating ships in transatlantic navigation. In the philosophical context of the Enlightenment, they represented at last both a Cartesian metaphor of the rationalization of time, and the materialistic ideal of the automaton (as expressed in Julien Offray de La Mettrie's idea of the man-machine; *L'homme machine,* 1748).[33]

As wrist-watches conveyed this extensive history to the masses, they also fashioned a new industrial experience of time as global, artificial, immanent, synchronous, and, above all, portable. While foreign brands were progressively invading the Italian market at the end of the nineteenth century,[34] watches exemplified Italy's rush to catch up with the new world time, in the context of the nation's dawning scientific (positivism) and industrial revolution (capitalism). Despite this fact, the country delayed developing a national industry (until the foundation of Borletti in 1896) in light of a persistent resistance to the temporal and productive order introduced by industrialism.

Instead, Italy retained its cultural attachment to past models, which related clockmaking to a humanist endeavour (as confirmed by its location in the liberal arts pavilion at the 1881 Milan exposition), to artisanal craft (in the ongoing production of old furniture clocks), and to small businesses (as in the initial workshop of Panerai in Florence). Despite its previous leadership in European watchmaking until the seventeenth century, Italy indeed "did not develop any great centre of clock- and watch-making in the modern age" (Cipolla 53–4), because of the inability of its *botteghe* (family-oriented businesses focused on timekeeping furniture) to cope with the new Northern European models of industrial production (based on serialization and miniaturization; Simoni 36) and because of the persistence in the peninsula of the outmoded Italic hour (which had been replaced elsewhere by the French hour that calculated the day as starting at midnight).

A second element of resistance to "precision horology" (as later defined in Aldo Palazzeschi's poem "La passeggiata," 1913, verse 23, *Tutte le poesie* 583; "The Walk") lay in its identification of modernity with capitalism, in its scientific and industrial quantification of time as a unit of measurement, and in its transformation of time into a currency for calculating the increasing rate of an amount of money (finance, no longer seen as usury) or the exchange value of a product (work). The Italian resistance to industrialism, politically motivated by the fear of a Marxist uprising, moved socially and culturally from a general opposition to the domination of a mechanical time order, based on repetition and division of labour and ultimately expressing, in its assemblage of productive gears, an alienating relationship between the individual and society.[35] In accordance with the reflection on the *questione operaia* (the issue of workers) in the 1890s, the emergence in contemporary literature of figures of unfit *impiegati* (white-collar workers) like Demetrio Pianelli (De Marchi) or Alfonso Nitti (Svevo) suggests indeed that the resistance to industry was ultimately a reaction against industrial clockwork and the alienation produced by the mechanization of life – later epitomized in Piero Jahier's figure of Gino Bianchi (1915), the model of the "administrative Italic man" (56; uomo amministrativo italico), who regulated his empty life merely on a working or monetary schedule.[36]

The Factories of Modern Italian Times

While the belated development of the Italian horologic industry documented a persistent *epoché* in the nation's relationship to industrialization, it also constituted the creative soil for its expansion. Two companies starting their production in Italy at the end of the nineteenth century, Bulgari and Borletti, not only broke into a market monopolized by Swiss and German products but also experimented with a new Italian notion of timekeeping (and temporality), bridging crafts and seriality, arts and production, and tradition and innovation. The Greek silversmith Sotiris Voulgaris (who escaped from the Greek-Turkish conflict in Epirus and

Italianized as Sotirio Bulgari) opened his first atelier on Via Sistina in Rome in 1884. The entrepreneur Senatore Borletti (son of Romualdo, owner of Linificio e Canapificio Nazionale, one of Italy's first textile industries) established in Milan the first Italian factory of watches with the Roskopf system in 1896. Even though they were expressing two different entrepreneurial philosophies and productive models – the one rooted in luxury and the long Italian tradition of the *bottega*, and the other based on mass consumption and industrial serialization – the two companies shared a common effort in the cultural making of Italian modern times. As timepieces, allegorizing "a world of cultural value that could not otherwise be represented except by means of oblique allusive adornment" (Fumerton 22), their respective products carried out a complex philosophical *kósmos*, embedding – as in the trifold meaning of the Greek word – in a *jewel* (or ornament) not just an intrinsic social *order* but also a new structure of the *universe* itself.

Sotirio Bulgari linked the Italian tradition of *orologeria minuta* (miniature watchmaking) to the craft of jewellery (in the choice of gold and silver as his materials). At the same time, while designing modern timekeepers, Bulgari deliberately constructed them with reference to antiques (as reflected in the Dickensian name of his first store, "Old Curiosity Shop"), to classical or "neo-Hellenic" taste (imported from Greece and mirrored in Rome), and to the aesthetic scenario of the eternal city (in the choice of Via de' Condotti for the main store, beginning in 1905).[37] By targeting the market of foreign aristocrats (mainly British and American tourists) and collectors (like Andrea Sperelli of D'Annunzio's *Il piacere*), Bulgari thus positioned watches in the category of luxury and reconverted them from mass-produced items into aesthetic symbols (once destined for royalty) of a unique way of being. By virtue of their precious materials, aesthetic design, artisanal craft, and classical taste, Bulgari aimed to overcome the association of industrial timing with mechanical repetition, an approved standard, and commercial ephemerality and aspired instead to reinvent past crafts and styles in a new modern key. Bulgari deliberately constructed modern watches, in contrast with other contemporary products (which were focused on the qualities of precision and exactitude), as everlasting pieces of art charged with an explicit cultural and aesthetic value.

Over the years, Bulgari's business ideal (which was emulated by the fashion industry in the years after the Second World War) shaped a new hybrid model of watchmaking, expanding the *bottega* into a small industry, yet maintaining a limited production in order to construct its exclusivity. By turning Italy's watchmaking delay into a creative resource for developing innovative solutions, Bulgari crafted a new aesthetic dimension to the manufacturing of products and constructed luxury as a culturalized model of production, bypassing the intrinsic industrial aspects of replicability, fetishization, and perishability. While experimenting with the notion of decorative art (which had been theorized by Camillo Boito since the early 1890s) and applying design and craft to objects of common use, Bulgari reconfigured fashion not only as a new tradition within modernity but also as a modern culture-maker. As the Bulgari shop gradually became an intersection of aristocracy and the arts, and as the photographs of its façade and interior appeared in the Treccani encyclopedia to illustrate the entry "negozio" (store), Bulgari's products acquired a new aura, thus turning commercial products into actors of a modern museum.

Unlike Bulgari, Fratelli Borletti (born as Industrie Femminili Lombarde) was the first and only Italian large-scale manufacturer of timekeepers, at least until the founding of Breil and Morellato in the 1930s. While acquiring a significant portion of the Italian market in the early 1900s (especially after the 1905 agreement with the state railway system; Introna 54) and expanding its production to alarm clocks and to chronographs for automobiles, the company, driven by its multifaceted owner Senatore Borletti, developed a singular business model, explicitly referring industrial work to a cultural endeavour. As emblematized in a cover advertisement for *La rivista mensile del Touring Club Italiano* of August 1912 (plate 2), the "first Italian watchmaking factory" (prima fabbrica italiana di orologeria) explicitly combined industrial achievements (the quality certification at the Brussels exposition, as well as a stated daily production of 1,400 Roskopf-system watches and 700 alarm clocks) with past culture in the clear reference to the classical statue of the discobolus. As confirmed in the juxtaposition of the factory's exact orthogonal projection with the body's muscular tension (reflected in the contrast between the precision of a watch and the energy of an industrial fire or smoke), Borletti's Italian factory staged a hybrid model of watchmaking, in-between serial production and cultural commitment, and located Italian modern time at the intersection of technological progress and humanistic tradition. As the factory becomes one of the engines of Italian industrialization, Borletti's vision of a culturalized industry will deeply affect Italian society, politics, and culture in the years to come.

Senatore Borletti enlarged his business in 1917 by investing the capital accumulated during the war (when the factory produced bomb fuses instead of watches) in an additional line of precise-measurement instruments (such as voltmeters and speedometers), as well as in organized distribution with the purchase of Milan's department store Aux Villes d'Italie (modelled on the Parisian Le Bon Marché and owned by the Bocconi brothers since 1889). Borletti culturalized his industrial plan by commissioning the poet Gabriele D'Annunzio to brand the two enterprises as, respectively, Veglia (literally "vigil," yet evoking *sveglia,* the Italian for "alarm clock")[38] and La Rinascente (the resurgent). The Milanese store introduced an entrepreneurial formula, inspired by Borletti's reading of Émile Zola's *Au bonheur des dames* (*The Ladies' Delight,* 1883), expanding from the retail of ready-made suits to everything else. After a fire destroyed it on Christmas night in 1918, the store assumed an even greater cultural resonance. Mirroring the recent story of Italy's rebirth in war, from the Caporetto defeat in 1917 to the Piave victory in 1918, the re-opening of La Rinascente on 21 March 1921 (Turani) and its new expanded idea of fashion became a collective metaphor for the nation's post-war resurgence, as suggested in the polyvalent motto crafted for it by D'Annunzio: "Italia nova impressa in ogni foggia" (Introna 35; The new Italy impressed in every style).[39] In the 1920s the Rinascente business acquired an even deeper aesthetic dimension after Borletti hired the graphic designer Marcello Dudovich for its advertising series, and the innovative architects Gió Ponti and Emilio Lancia for the design of a new furniture line, Domus Nova (launched in 1927).

Thanks to Borletti's collaboration with D'Annunzio, his impact on Italian industry went beyond his watchmaking and retail businesses and soon applied to politics and culture. Borletti conspicuously financed D'Annunzio's self-proclaimed occupation of Fiume (1919–20),[40] made in opposition to the Treaty of Versailles and the Italian government, and in September of 1920 he even paid the ransom for the ocean liner *Cogne,* which had been held by the Legionari (the occupants of Fiume), after being diverted to the Istrian harbour, as a way to meet the city's financial needs (Ledeen 194–6). After the *impresa fiumana* (Fiume feat) Borletti became involved in the Italian culture industry by hiring, in 1921, Arnoldo Mondadori, a relatively unknown typographer from Verona, and acquiring, in 1923, the newspaper *Il secolo.* Borletti entrusted Mondadori with the task of converting *Il secolo* "into an authoritative pro-Fascist voice capable of competing with Italy's best independent paper, *Corriere della sera*" (Bonsaver 43). In 1926 Borletti negotiated with D'Annunzio regarding Mondadori's acquisition (over the competitor Treves) of the rights to publish his collected works, thus establishing the foundation of the publisher's later and long-lasting leadership in the Italian editorial market. As these developments suggest, Borletti's watchmaking model ultimately pursued a much larger cultural and political vision[41] related to the establishment of Fascism as the nation's new temporal order, based on industrial time (as in Mussolini's effort to run Italian trains on time) and originally constructed as a hybrid Italian temporality (both reinventing the past in a modern key and grounding progress within a longer tradition).

The Enigma of the Italian Hour

As timepieces appeared in literature, the visual arts, and industry at the turn of the twentieth century, they made visible and registered the slow processing of a profound epistemological shift – from "Time" to "Times" (Galison 13) – engendered first by the establishment of world standard time and then by the theory of relativity. As they turned into cultural sites, clocks and watches also enacted an independent space of aesthetic investigation, which tied their representations (as stuck clocks, dysfunctional solar timekeepers, and bearers of enigmatic hours) first to the contemporary European debates on the nature of time, then to an overt critique of old epistemological models, and lastly to the experimentation of new hybrid temporal solutions.

In their multifaceted exploration of the relationship between inner and mechanical time, the recurring images of stuck clocks (e.g., by Gozzano, Papini, and Palazzeschi) reveal an intimate connection with the European debate on time, opposing Henri Bergson's philosophical idea of interior time to Albert Einstein's physical understanding of mechanical temporality.[42] In Bergson's reading, as outlined in the *Essai sur le données de la conscience* (*Time and Free Will: An Essay on the Immediate Data of Consciousness,* 1889), time corresponds to an internal, qualitative, and heterogeneous *durée* (duration), composed of the uninterrupted interpenetration of conscience and memory in the self.[43] In the 1905 publication of the pioneering article "On the Electrodynamics of Moving Bodies," Einstein theorized that time represented a relative notion, defined as a "clock-coordinating procedure" (Galison 13), and conceived, in mere physical terms, as an external, homogeneous, and quantifiable succession of simultaneous instants. In this context the paradoxical image of Italian stuck clocks reveals not simply a nostalgic desire to revert or even block modern timing but rather the first poetic attempt to capture the point of intersection between physics and *durée* and then reunite the dual temporality (e.g., interior and exterior, local and universal) of modern man.

The increasing representations in the Italian culture of the early twentieth century of solar-based timekeepers (e.g., broken sun-dials, neglected church bells), seen as no longer functional or useful mechanisms, express the contemporary "disenchantment of eternity" (Archetti 133) and the definitive abandonment of the old metaphysical temporality, as attested in two contemporary novellas by Luigi Pirandello. In the first novella, "Leonora addio!" ("Farewell, Leonora!" 1910), Pirandello depicts the figure of Mommina, the neglected wife of the protagonist (the opera lover Rico Verri), who hears the deserted church bell of the town sounding the hours while she is contemplating the starry sky, and she asks, "Why did that clock measure time? For whom did it sound its hours? Everything was dead and vain" (1079).[44] Moving away from Segantini's and Pascoli's attempts to retain rural and metaphysical time, Mommina's (or rather Pirandello's) unanswered questions to the clock indicate the pointlessness of rural time (vis-à-vis the advent of urban civilization) and of metaphysics itself, as the church bells fail not just to provide a reliable source of timekeeping but, most importantly, to endow time with an ultimate significance. In the second novella, "Pallottoline!" ("Marbles," 1902), Pirandello presents the paradoxical figure of an astronomer and philosopher (Jacopo Maraventano) who rejects the sun as unfit for modern timekeeping ("the sun [...] is not even good enough to regulate watches!"; 996) and contraposes his watch to its mendacious hours: "his watch, indeed, upon whose quadrant was written in red ink: *Solis mendaces arguit horas* [*it discloses the sun's deceitful hours*], was not regulated by solar time" (996).[45] In the figure of Maraventano then, Pirandello blatantly ratifies the crisis of solar time, making manifest a phenomenon that had found many indirect references in the post–world standard time culture. In a way, Nietzsche first weakened the sun's cultural association with light, order, and stability by shifting its traditional figuration from noon to twilight – both at dusk (as later hinted in the Italian poetics of *crepuscolo*) and at dawn (as later outlined in the myth of the *sol dell'avvenire*, which would become the theme of Giuseppe Pellizza da Volpedo's painting *The Rising Sun*, 1904). In another way, the post–world standard time awareness of a "wrong"[46] sun also went along with its portrayal as a slow-moving star, surpassed by electricity (as in Gino Severini's depiction of a lamp replacing the fading light of the sunset, in *Via di Porta Pinciana al tramonto* [*Porta Pinciana Road at Sunset*], 1903) or by the accelerated speed of modern life (as symbolized in Giacomo Balla's later metaphor of the quicker *Mercury Passing before the Sun* in the series *Mercurio passa davanti al sole,* 1915–16).[47]

Against this background, Giorgio De Chirico's paradoxical image of a clock, stilled yet working, in the famous painting *The Enigma of the Hour* (1911, plate 3) represented a singular attempt to investigate the nature of present time (seen, during the industrial boom of Italy and the jubilee of its unification, as an *ora inquietante*),[48] as well as a platform on which to forge an experimental notion of a modern Italian temporality.

In the context of an empty piazza, De Chirico's canvas revolves around a central timepiece, which registers an ambiguous hour, expressing in relation to the train station a linear and divisible time of business (as opposed to the circular, slower time expressed in the fountain) and in relation to three unmoving human figures on the piazza a time of loss, melancholy, and grief. As in other contemporary clock paintings by De Chirico – *The Delights of the Poet* (1913), *The Philosopher's Conquest* (1914), and *Gare Montparnasse* (1914) – the timepiece suggests the paradox of an "eternal present," located on the edge of a-temporal revelation and moving temporality, and enacts the enigma of its nature.

De Chirico's time set at 2:55 p.m. replicates the precisely quantified hour (*krónos*) of the train station, which visibly differs from the hour of the sun projecting longer shades on the piazza. As documented, the representation of the unco-ordinated solar hour hints at the "immanentization" and secularization of time, which was propelled by the industrial and urban civilization and ratified by the establishment of world standard time. Moreover, De Chirico's 2:55 p.m. also represents a frozen instant or a supreme hour (*kairós*) that is eternally stuck on a clock. While juxtaposing the countable hours of modern time (*krónos*) with a supreme hour (*kairós*) – lost (like metaphysics) yet everlasting (in the eternity of art) – De Chirico's clock does not reproduce two parallel temporal tracks but rather forges a mixed temporality, meant as a new secular transcendence at the intersection of industrial (yet no longer eternal) timekeeping and metaphysical (yet no longer religious) timelessness. De Chirico's pursuit of a new hybrid *ora-Ora* (meaning both "hour" and "now") encapsulates the enigma of an old, yet new nation searching for a temporal identity in the modern world, and its broader attempt to construct Italian modern times as a mixture of secular eternity and transcendental contingence. In this sense, the *Enigma of the Hour* visually synthesizes the multilayered historical attempts (from Risorgimento to Fascism) to articulate the oxymoronic notion of modern and Italian times from a combination of the movement of progress and the everlastingness of Italy's glorious past.

In light of De Chirico's concomitant hour of eternalized contingence and secularized metaphysics, it is possible to read the Risorgimento's ideological construction of the nation as a mystical entity, and ecclesiastical Modernism's endeavour to reduce religion to immanence, as two sides of a common cultural tension to construct Italian modernity as an oxymoronic transcendent present.

During the Risorgimento, Italian intellectuals had tried to anchor the nation's progress to a mystical aura of everlastingness as a way to remedy its identity-deficit and foster its spiritual unity. Vincenzo Gioberti in particular had shaped this narration of uniqueness in *Del primato morale e civile degli italiani* (*On the Moral and Civil Primacy of the Italian Race*, 1848), locating Italy's eternal spirit or unchanging truth in its cultural primacy and in the universal mission of the Catholic Church. After the unification his primacy template informed Italy's need to assert its uniqueness amidst modern nations at various levels: in Bertrando Spaventa's scholarly re-evaluation of Italy's philosophical heritage (Telesio, Campanella, Bruno, Galileo, and Vico) as the foundation of modern thought; in the country's proud affirmation of its civilizing mission in organizing congresses and expositions; and in the acknowledgment of the nation's lost superiority, later evolving into the political discourse of nationalism.[49] Paralleling the attempt to create a political metaphysics of secular times, an opposite (yet concomitant) tension towards the "secularization of religion" (Del Noce 15) was manifested in the early 1900s in the Modernist debate within the Catholic Church – which was heavily covered in Italian newspapers, especially after the scandal following the inclusion of Antonio Fogazzaro's novel *Il santo* (*The Saint*, 1905) in the list of forbidden books. The Modernist attempt at a democratic reformation of the Catholic Church[50] was condemned by Pope Pius X, in the encyclical *Pascendi dominici gregis* (1907), which openly rejected the movement's push to secularize revelation as well as its reduction of faith to mystical immanence, a philosophy, or a historical phenomenon, and of Catholicism to "a secular religion of enlightenment, surcharged with positivism and nationalism" (Misner 69–70).

Along with these phenomena, De Chirico's enigmatic exploration of a possible fusion of times (transcendent and immanent, mechanical and interior, Apollonian and Dionysian)[51] into a modern Italian time not only corresponded with the contemporary reseach of the philosopher Giovanni Gentile but also established an indirect template for the political construction of Fascist temporality.

In parallel with the affirmation of an immanent notion of time through Greenwich Mean Time (leading to the establishment of a new industrial or mechanical time order), the alleged end of metaphysics, and the growth in Italy of atheist philosophies of history, evolution, progress, praxis, and action, Giovanni Gentile started his philosophical career with two related monographs – *Rosmini and Gioberti* (1898) and *The Philosophy of Marx* (1899) – in which he elaborated a new model for constructing Italian modernity, via the reconciliation of Catholic reformism (Rosmini) and Universalism (Gioberti) with Marx's idea of historical progress. Gentile's bold endeavour to meld the post-Hegelian need to explain the formation of a specifically Italian incarnation of the Spirit with the need to anchor the nation to its everlasting *primato* (primacy) anticipates the development of his philosophy of actualism and his theory of education (which will later inform the creation of Fascism as an unprecedented political form). In line with Gentile's *Sommario di pedagogia come scienza filosofica* (*Summary of Pedagogy as a Philosophical Science,* 1912), actualism constituted a deliberate instrument of collective education, offering the nation "its own spiritual identity, a strong conscience of its own tradition and goals" (Colletti 20). In line with his attempt to merge Hegelian-Marxism and Catholic reformism, actualism aimed to realize a new sacralized metaphysics of culture, simultaneously located in the times of immanence (industrial progress) and everlastingness (the past and religious tradition). Gentile's secular pedagogy and religion will find actualization in the immanence of history, in his reform of the Italian school system (as minister of education from 1922 to 1924), and in his elaboration of the first section ("Idee fondamentali") of the *Doctrine of Fascism* (*Dottrina del Fascismo*).[52]

Against the back-drop of Gentile's philosophy, Fascism similarly aimed at realizing a political model that could bypass the immobility of past forms or the instability of modern ways by deliberately fusing "regressive and progressive elements, modernizing projects and Romanizing throwbacks" (Spackman, "Fascist Puerility" 16). In its concomitant "sacralization of politics" (E. Gentile, *Sacralization*) and immanentization of religion, Fascism aspired to forge modern Italian time as a hybrid of eternity and progress, rootedness and movement. The elaboration of modern Italian time will find its ultimate realization in the establishment of a new Fascist calendar, which will define 1922 as year one for the regime, and the March on Rome (later portrayed as a revolution in Alessandro Blasetti's 1935 movie, *Vecchia guardia*) as the mystical founding act of a new, modern era.

Chapter Three

Industrial Photographs and the Fictional Vision

On 26 May 1889 in the auditorium of the Florentine Regio Istituto di Studi Superiori the anthropologist Paolo Mantegazza, amateur photographer and president-elect, saluted the birth of the Società Fotografica Italiana (established under the honorary guidance of Prince Victor Emmanuel III). In his inaugural speech, delivered on the fiftieth anniversary of François Arago's presentation of the daguerreotype in front of the French Academy of Sciences in 1839, the Italian positivist praised the qualities of photography: its objectivity, its capacity to capture the instant, and its "democratic" nature.[1] At the same time, in front of an audience of renowned Florentine photographers and intellectuals (including Giorgio Roster, Vittorio Alinari, Carlo Brogi, Ugo Bettini, and Yorick), he purposely launched the project of an "Italian" photographic club.

The foundation of a photographic society in Florence was certainly not coincidental, given the city's leading role in Italian photography. Florence had hosted a plethora of local and foreign photographers during its tenure as Italy's capital (1865–70) and had also held both the first display of photographs in Italy (in the chemical industry section of the National Exposition of 1861) and the first Italian exposition of photography (in June 1887). Thanks to its history, the city also represented an ideal location for photography, enhancing a visual national identification[2] and displaying a recognized international flair – as certified by the gold medal awarded to the Alinari firm at the Universal Exposition of Paris (which opened on 6 May 1889).

The launch of the Società Fotografica Italiana (SFI) coincided with a time of global growth in the photographic industry. The year before, in fact, George Eastman had released the first portable camera (Kodak no. 1) and a new model of photo-development, based on his recently invented photographic film and advertised in the famous campaign "You press the button. We do the rest." As these innovations reconfigured cameras as accessible tools for entertainment, and photographs as industrial items, photography gradually moved from the practice of a few into "a mass retail market of goods and services" (R. Hirsch 173). As of 1889, Italy lacked an industry of photographic materials and equipment and relied heavily on imports. Given the global appeal of its art, and the pioneering activity of the Florentine worskhops Alinari (since 1852) and Brogi (since 1864), the Italian photographic "industry" coincided instead with local craft, mainly focused on the production of photographic images to export to the international editorial market. In the decades following its establishment and before its closing in 1915, the SFI used its initiatives and monthly magazine, *Bullettino della Società Fotografica Italiana,* to slowly fashion a new market for Italian industrial photography and build a new photographic culture in Italy. Envisaged as a photographic comradeship, the club organized social events for its members – like the *passeggiate fotografiche* (photographic walks) and the *serate di proiezione* (screening nights) – and openly engaged in political activism, such as opposing the tax on photographs of state-owned monuments (a law as of 25 February 1892) or favouring the protection of photography under the copyright law (Puorto 69–74). Thanks to its monthly magazine, the club advertised new studios and cameras (Lamberti e Garbagnati, Koristka, and Zeiss) and introduced amateurs to photographic best practices (with technical suggestions on poses, lighting, and exposures). Its publication also provided a privileged vehicle for the diffusion of photographic culture throughout Italy, by offering updates on current debates (copyright, chronophotography, and pictorialism), fostering the application of photography to other fields (e.g., geodesy, medicine, and criminal justice), and promoting photographic expositions (as the one in 1899, led by Vittorio Alinari and Mario Nunes Vais).

Within this context, the founding of SFI also inadvertently coincided with a significant moment of transition in the history of Italian photography. During the year after its birth, in fact, Romualdo and Giuseppe Alinari (co-founders of the Alinari firm with their brother Leopoldo) suddenly died, and Leopoldo's son Vittorio (an active member of SFI and a friend of Carducci) took over the business in 1892. Over the following decades, Vittorio's leadership would give the company a qualitative leap, transforming it from a family-size workshop into a world-renowned industry (Maffioli, "Fratelli Alinari" 40). Vittorio Alinari defined the new space of industrial photography in relation to publishing by steering the firm's production from decontextualized collections of images (of ethnicities, costumes, flowers, fruits, and monuments) to more sophisticated editorial projects (following the foundation of the Alinari publishing house in 1885). Linking photography to documentation, travel, and art, he produced illustrated volumes on Florence (*Firenze e contorni,* 1891), Italian Renaissance artwork, and Italian landscapes, and, beginning in 1903, he published the journal *Miscellanea d'arte* (later titled *Rivista d'arte*) edited by the prominent art critic Igino B. Supino. Vittorio reconfigured his company vis-à-vis the growth of photography from an exclusive dominion of experts (atelier photographers, professional painters, and scientists) to a mass social practice (Quintavalle 457–80).

In collaboration with the SFI, he also prompted two main changes in the photographic business. First, he fashioned his Florentine atelier into a global industry of photographic images by establishing recognized compositional standards for its pictures (e.g., frontal perspective and large plates), assembling an archive of art photographs for retail, and directing his publications to the foreign market.[3] Second, he purposely configured industrial photography as a modern form of art by crafting his photo books as decorative objects (in their limited run and careful composition) and using the photographic lens as an aesthetic tool to reinvent literary and pictorial masterpieces.[4] In 1900 the Alinari reproductions of artwork at the Uffizi Gallery obtained the "grand prix" at the Universal Exposition of Paris, an achievement that would lead in subsequent years to other European museums (e.g., Athens, London, Dresden, Paris) commissioning several photographic works. In 1902 Vittorio Alinari launched the project of a photographic book on Dante's *Inferno* in celebration of the firm's fiftieth anniversary, featuring works by master photographers like Adolfo De Carolis, and Giovanni Costetti. Inspired by the success of the initiative, he later invested in photographic illustrations of Boccaccio's *Decameron* (1909–15) and of the sites in Dante's *Divine Comedy* (*Il paesaggio italico nella "Divina Commedia"*) after the sale of the company to a group of Florentine nobles, intellectuals, and entrepreneurs in 1921 ("The Chronology").

The transformation of photography from a pioneering technology into an industry, and the increased production and consumption of photographs (in publications or in amateur practice), profoundly influenced Italian society in the late nineteenth century, affecting literature, painting, and the later evolution of an independent photographic art. In a way, the reinvention of photography as an industry had an impact on the formation of Italy's autonomous photographic culture, especially in the cities of Florence and Turin. In another way, the serialization of industrial photography also led to the parallel evolution of hybrid visual languages, verbalizing, aestheticizing, fictionalizing, or setting in motion the photographic image.

Photography before the Industry

A brief overview of the historical development of photography will illuminate the ways in which industrialization radically shifted its cultural meaning. The invention of the daguerreotype in 1839 was the result of a long evolutionary process, combining optical studies (dating back to the discovery of perspective)[5] and chemical research (dating back to Johann Schulze's discovery of photo-impressionable substances). Following this invention, the first decades of photography, an experimental phase, were marked by the research and development of new image-fixing techniques (e.g., heliotype, Talbotype, collodion, albumen, ferrotype). The process of realizing a photographic image went through a radical simplification in the 1870s, thanks to Richard L. Maddox's invention of gelatin (1871) and the start of the serial production of dry-plates. The new system, which shortened and uniformed the procedure to obtain a photograph, gradually replaced earlier techniques, which had entailed long exposure time, heavy equipment, and specific development and image-fixing skills. From its pioneering phase to its early industrialization, photography slowly moved from a handful of chemists and professionals to a larger pool of anthropologists, explorers, scientists, and humanists.[6] Stirred by their "frantic urge to see" (or *folie du voir*, in Buci-Glucksmann's words), the photographic instrument, still endowed with an aura of magic,[7] also fashioned a new "visual thought" (Zannier, *Leggere* 15). Since the mid-nineteenth century, photography indeed applied to geography, science, anthropology, and historical memory, defining man's extended dominion over reality and its "anaesthetized" knowledge (Sontag 20).

Driven by the impulse to chart and collect the unknown, photographs represented a visual exploration of the earth's geography and inaugurated a "new age of discovery comparable to that of the explorers who charted the globe from the fifteenth to seventeenth centuries" (Marien 159). Photographers expanded the space of human (Western) influence by unveiling a vision of the world's oldest cities (as in the photographic surveys of Rome),[8] its most remote locations (as in Felix Beato's catalogues of Athens, Constantinople, Calcutta, and Japan), and its most impervious sites (as in the alpine pictures by Quintino Sella, a Piedmontese politician and founder of the Club alpino italiano in 1863).[9] Against the back-drop of the global urge to track the earth's geography, photographs of Italy singularly enacted a process of territorial self-ascertainment, as expressed in the habit of assembling *vedute* (scenic views) of Italian cities, landmarks, and sites into an "ideal miniature museum" (Mignemi 57). As confirmed by their clichéd representations (populated by picturesque elements like clouds, maritime pines, and olive trees), by their theatrical panoramas, and by their deliberate exclusion of human figures from the illustration of monuments, these photographs served to fix the stereotypical image of Italy as *il bel paese*, visually matching Antonio Stoppani's famous book of 1875.

In relation to science and anthropology the enhanced, "objective," and reproducible vision of photography offered a valuable tool in various fields for the rationalization and diffusion of scientific knowledge. In Paolo Mantegazza's Darwinian book *Atlante della espressione del dolore*, published in 1876 in collaboration with Giacomo Brogi, photographs formed a systematic "atlas of the expression of pain." In Cesare Lombroso's collections of local inhabitants and social categories, photographs allowed the immediate facial recognition of potential criminals and different ethnicities. Following his research, photographs became valid instruments for police (as tools of identification, record-keeping, and judiciary evidence)[10] and for the classification of diverse ethnographic groups (thanks to their visual patterns).[11] Despite their easily recognizable reconstructions of regional differences, photographs also imposed a paternalistic and stereotypical view of "Italians," as documented in Carlo Naya's Venetian fishmongers and Neapolitan *lazzaroni* (scoundrels), replacing the actual condition of the photographed subjects with theatrical re-enactments by "real-life 'actors'" (Zannier, *Venice* 23). With regard to Italy's colonial possessions in Africa, photographs visualized different races, local costumes, and social roles by framing them in the self-contained space of a catalogue or within a narrative of sexual and political dominion (as implied in the accepted use of female nudity; Campassi and Sega 55–61).

Thanks to their capacity to visualize past and present, photographs also defined a new space of collective memory, once again exerting over it a strong cultural and political control. The photographic survey of Italy's ruins and monuments constituted a way to "occupy" the past and perspectively frame it in its nicest, most suggestive façade (Manodori 8). At the same time, the emphasis on ruins, imposed by the technical need of prolonged exposure, defined the paradoxical ontology of photography itself as a fragmentary "vestige of what has been" (Dubois 46). In the work of the German-Italian photographer Giorgio Sommer the ruins of Rome and Pompeii became metaphors of the photographic present by their capacity to identify both the *duration* of an ever-changing observation, for example in the vision of the archaeological unearthing of the Coliseum, and a *cut* of "Absolute Particular" (Barthes, *Camera lucida* 4), as documented in the vision of the human corpses that were petrified by the AD 79 eruption of Vesuvius.[12] Contemporary photographer Michele Amodio's exemplary image of a gentleman camouflaged against the ruins (at the bottom right of the arch) as he crosses the ancient street of Mercury in *Pompeii* (*La Strada di Mercurio a Pompei*, ca. 1870; fig. 3.1) captures the ambiguity of a moving, yet petrified, photographic present in the coincidence of the subject's walking act (mirroring photography's capacity to report a live contingence) and its "medusification" into an eternally stilled presentness.

In light of its dual capacity to visualize the present and transform it into a ruin, photography fashioned Italy's recent present – people and events – accordingly by capturing it at its making and immediately turning it into a monument.

With regard to the vision of contemporary Italians, photographic portraiture offered a quicker and seemingly more democratic way of self-representation, as well as an effective tool of control over the bodies of others. By its rigid visual standards – the three-quarters profile for upper classes and the flattened frontal poses for lower classes – photography indeed ratified and perpetuated an immobile social hierarchy. Moreover, while feeding the bourgeoisie's "need for self-glorification" (Freund 9) as a new ruling class, it also enacted in portraiture a ritual practice (compared by Bazin to the Egyptian habit of embalming; 11), mummifying the passages of life (birth, first communion, engagement, military service, wedding) within a fixed performance code. As pictures were assembled in personal albums, ethnic repertoires, and self-celebratory collections, they shaped a controlled social narrative. In the case of Alessandro Pavia's *Album fotografico dei Mille* (*Photographic Album of the Thousands*, 1862), a photographic montage of portraits of Garibaldi's soldiers in uniform, they even constructed the monumental vision of a unified national body (Marcenaro 71–80).

With regard to current events, the photographic view of the Risorgimento followed an analogous pattern of controlled visualization and sudden monumentalization of the present. The recurring pictures of rubble and "contemporary ruins" (Pizzo 16) – after the fall of the

Figure 3.1 Michele Amodio, *La Strada di Mercurio a Pompei,* ca. 1870. Raccolte Museali Fratelli Alinari.

Roman republic in 1849 (taken by Stefano Lecchi), after the arrival of the Mille in Palermo in 1860 (by Eugène Sevaistre), and after the seizing of Porta Pia in 1870 (by Gioacchino Altobelli) – depicted not the actual events but rather a present just passed and immediately stilled. Photographic images staged an anaesthetized version of recent facts (e.g., the wars of independence) by eliminating the disturbing vision of combat (because photographing war was prohibited) and any explicit sign of violence or death (with the exception of Luigi Sacchi's picture of a soldiers' cemetery after the battle of Magenta in 1859). At the same time, they defined a "correct" version of the facts for future reference by prefiguring the events (as in the visual yet unrealized "conquest" of Mentana through its *veduta* before the 1867 battle), by manipulating the evidence (as in the case of the first fake photograph of Garibaldi with a wounded leg in 1862), and by staging the events a posteriori, as in the photographs of dead Neapolitan soldiers arranged on the ground after the siege of Gaeta (1861) or of Altobelli's soldiers posing on the ruins of Porta Pia on 21 September 1870, the day after the *breccia* (breach). As these photographs "became, de facto, the iconographic form of Risorgimento" (Pizzo 33), they extended the vision of the present to future generations, yet also flattened and fetishized its memory.[13]

The Birth of the Photographic Industry

At the turn of the twentieth century the increasing emergence of photographic clubs, the rapid spread of cameras, and the growing demand for specialized magazines and informative handbooks[14] across Italy quickly shifted photography from an exclusively professional to an amateur practice. Thanks to their lightness, clarity, and instantaneity, portable cameras became trendy objects for the bourgeoisie, related to youth, fashion, leisure, and travel. With the parallel success of the new development technique, which separated the act of taking photographs from the process of making photographs, the demand for photographic images grew exponentially, as attested by the doubling of Italian professional photographers (from 1,749 in 1881 to 3,502 in 1901; Rocchetti 43) and by the sudden rise in the number of amateurs. Since the late 1880s, in conjunction with the invention of linotype and phototype techniques, photographs began to appear alongside drawings in Italian publications and periodicals, boosting sales by virtue of their unprecedented immediacy in representing events, scoops, and celebrities. During the 1890s, following the growth of tourism, photographs started to be produced as separate artefacts in the form of postcards, thus launching a new format of visual and verbal communication and a new industry.

Following the increased facility of the photographic experience and the massive circulation of photographic images, the industry of serialized photography also became a powerful culture-maker, as it both extended its visual knowledge to a larger number of people and fashioned new trends and styles. At the same time, however, it also imposed uniform visual standards, including deferring the need for a unique individual vision to ready-made images (celebrity portraits, postcards) or subordinating its aesthetic ends to mere practical uses (solemnizing family events, certifying a vacation).[15] In response to the perceived fixity and homogeneous format of photography, the industrial proliferation of images also led to an ever-growing endeavour to apply photographic vision to art, as seen in contemporary painting and literature.

The industrialization of photography had an impact on the evolution of new pictorial languages. As French painting transitioned from *impressionnisme* to *pointillisme* (or *néo-impressionnisme*), the impact of the photographic practice on Italian art similarly evolved from the first experiments of the Macchiaioli painters (after their encounter with Leopoldo Alinari in 1861) to the later elaboration of *divisionismo* (first presented at the inaugural Triennale of Milan in 1891; Paulicelli, "Art" 246–8). Since the mid-nineteenth century, photography had informed a new notion of realism in the paintings of Giovanni Fattori, Silvestro Lega, and Telemaco Signorini (in their depiction of outdoor scenes and daily life), and blurred photographs had inspired the formation of their abstract, anti-imitative, and anti-academic technique of the *macchia* (colour patch), deliberately capturing in their lighting effects the "mood" of forms (Boime 193–4).[16] In the later work of divisionist painters (Giovanni Segantini, Gaetano Previati, Angelo Malerba, Plinio Nomellini, and Giuseppe Pellizza da Volpedo), the photographic vision prompted the creation of an optical effect of brightness (chromoluminarism) obtained by the deliberate use of pure (or "divided") colours and the optical interaction of individual colour dots. The new divisionist technique would apply over the years to distinct pictorial styles and themes – ranging from Giovanni Segantini's naturalism (expressed in *Due madri* [*Two Mothers,* 1889], first exhibited in 1891), to Gaetano Previati's symbolism (as in *Maternità* [*Motherhood,* 1891]); from Angelo Morbelli's specific focus on social concerns (as in his *Giorno di festa al Pio Albergo Trivulzio* [*Feast Day at the Trivulzio Nursing Home,* 1892]) to Giuseppe Pellizza da Volpedo's explicit political dimension (as in the famous painting *Quarto stato* [*Fourth Estate,* 1901]), informed by his socialist ideology. In subsequent years the work of divisionist painters influenced the pictorial research of Giacomo Balla (beginning with his move to Rome in 1895) and offered an explicit model for Futurist painting, as recognized in the "Manifesto of Futurist Painting" and in the "Technical Manifesto on Futurist Painting" of 1910 (co-signed by Umberto Boccioni, Carlo Carrà, Luigi Russolo, Giacomo Balla, and Gino Severini; Avanzi 87–8).[17]

The increased availability of cameras and photographs also influenced contemporary literature. The intense activity of Luigi Capuana and Giovanni Verga as photographers[18] certainly had an impact in their theorization of verismo as an impersonal, self-made, and objective form of writing.[19] The photographic narrative of Capuana's pictures of his hometown Mineo or of Verga's *novelle* offered a seemingly impartial record of the life in Sicily. At the same time, however, as sites where the invisible became visible, their photographic activities also coincided with the deliberate opening of a parallel imaginative space. Along with the work of the positivist Cesare Lombroso and with the appearance of the first X-ray images (1895), Capuana's numerous photographs of ectoplasms, which seemingly recorded ghosts and spirits, revealed not just his scientific motivation to trace objective proof of the supernatural but also his explicit ties with occultism (which also manifested later in the photographic work of Anton Giulio Bragaglia). In a similar way, despite the "photographic" intention of his novel *I Malavoglia* (*The House by the Medlar Tree*, 1881), Verga never explicitly relied on or mentioned any identifiable photographic documentation;[20] instead, he constructed his narration from a distance as an "imagined photograph" (Lolla 37), or a fictional illustration of Sicilian life following the method of contemporary magazines, overlapping drawn reconstructions with the lack of photographs in the reports of embarrassing events (like the colonial defeats of Dogali in 1887 or of Adwa in 1896).[21]

The mass increase of photographs also coincided with an attempt to aestheticize the photographic act and language itself. Photography was not recognized as an aesthetic form per se, but rather it constituted a technical process, a product for the market, or, in other words, a "diminished art." Along with their commercial growth, photographs gradually acquired a cultural status at the turn of the twentieth century, as documented in the experience of contemporary photographers (like Giuseppe Primoli and Mario Nunes Vais), in the formation of new entrepreneurial models (around the figure of Vittorio Alinari), and later on in the evolution of the pictorialist style.

The experimental work by Giuseppe Primoli (1851–1927) and Mario Nunes Vais (1856–1932) sought to define a new photographic aesthetic in the heavily codified fields of photojournalism and portraiture. In their laboratory both photographers transformed the photographic image from a serial reproduction into a unique object of art, and photography itself from a mechanical language into an intellectual gaze that could, in Baudelaire's words, "extract from fashion whatever element it may contain of poetry within history, to distill the eternal from the transitory" (*The Painter of Modern Life* 12). The Roman and Parisian aristocrat Giuseppe Primoli changed photojournalism from an expression of unskilled excitement and clichéd instantaneity (equivalent to that of amateur photography) to an artistic tool for investigating modernity. As a flâneur in Rome's streets, a reporter of the aristocracy's mundanity, and a "byzantine" intellectual (working with D'Annunzio, Nencioni, and Martini for *La Capitale*, *Il fanfulla della domenica*, *La domenica letteraria*, and *Cronaca bizantina*), Primoli captured the essence of modern city life in its ephemereal instances (the *passeggiate* on the Corso, aristocratic races, parties, and dances) and its current events (e.g., the visit of Kaiser Wilhelm in 1888 and the funeral of King Umberto in 1900). In the key figures of his time he visualized significant cultural issues: the decay of aristocracy (the Roman *nobildonne*), the emergence of feminism (Matidle Serao), Italy's intellectual subservience to France (D'Annunzio reading *Le Gaulois*), and the tension between state and Church (a priest reading *La voce* under Bernini's colonnade).[22] In another field the Florentine photographer Mario Nunes Vais (who worked for Vittorio Alinari) changed portraiture from an act of identity validation, or a fixed series of codes (of space, posture, and dress), into a language of subject exploration. In his visual galleria (or, in his own words, photographic pantheon) of Italy's most famous writers, artists, composers, actors, and politicians of his time, Nunes Vais (whom Pelizzari defines as "the Italian equivalent of Nadar"; 73) shaped portraiture as an artistic form, penetrating faces as written texts (by comparing the subject and his or her signature) and seeking in them the liminal moment of an unrepeatable yet emblematic expression (O. Ferrari 9).[23]

During those years, the photographer and entrepreneur Vittorio Alinari steered his company in the same direction by defining not just a new business model for photography (in its deliberate association with art publications) but also a new photographic style, moving from static documentary reproductions of monuments to their dynamic and aesthetic fictionalizations. The image *Scala della torre di Arnolfo in Palazzo Vecchio* (*Stairs of the Arnolfo Tower in Palazzo Vecchio*, 1900–5; fig. 3.2) exemplifies this shift by capturing a gentleman climbing the stairs of Palazzo Vecchio's tower (reflecting the new dynamism of photography) and by framing the view of the duomo opposite him (mirroring a new phantasmagoric experience of seeing). While staging a rediscovered beauty in a renowned monument – by way of its unprecedented perspective, its well-studied framing, and its contrast of light and shade – the picture also suggests another space of unknown wonderment and creative imagination, symbolically represented by the mysterious door in the forefront and the stairs' continuation beyond the frame. The construction of such fictional space upon photographs marks the beginning of their distancing from ready-made models and of their evolution into a mature art language.

Figure 3.2 *Scala della torre di Arnolfo in Palazzo Vecchio,* Florence, 1900–5. Fratelli Alinari.

The Art of Photography in Turin

Like Florence, Turin also emerged in the fin de siècle as an important centre of Italian photography. Along with the city's industrial growth, photography expanded as a commercial entreprise and evolved as a malleable product, engendering new languages and forms of expression.[24]

The foundation of the growth of Turin's photographic industry and culture can be traced to the period from April to October 1898. During that time the city hosted the Italian General Exposition, commemorating the fiftieth anniversary of the Statuto Albertino (the then Italian constitution, promulgated in 1848 by King Carlo Alberto of Savoy) and featuring the first Italian Photographic Congress. The congress overlapped with the special Exhibition of Sacred Art, which celebrated the four-hundredth anniversary of Turin's cathedral.

The organizers of the sacred art exhibition successfully obtained from King Umberto the permission to display and photograph the Holy Shroud (legally owned by him) as a way to attract visitors and revitalize the veneration of the relic (conserved in the cathedral). On 25 May 1898 the official photographer, the Piedmontese Secondo Pia, took the first photographic image of the shroud, which surprisingly revealed the negative contours of a suffering man (Christ himself?) impressed on the veil, and launched the unsolved *querelle* about the effigy. By virtue of such discovery, Pia's photograph suddenly turned from a "reproducible" industrial artefact (promoting an industrial exhibit) into a paradoxical icon, endowed with a mystical aura, different layers of meanings, and an inherent critical apparatus. Over the next years, Pia continued to conceive photography as a reproducible yet unique form of vision in his pioneering work on both "architectural photography" (Falzone del Barbarò 20) and colour photography (after the patenting of the Lumière's autochrome process in 1907). In his photographic study of the less popular monuments of Aosta Valley, Piedmont, and Liguria (excluded from the Alinari repertoires), Pia used photography as a many-eyed discovery tool, revealing the details of the buildings' architecture (column capitals, windows, bas-reliefs on portals). Likewise, in his technical experiments with inanimate objects (motivated by the need for long exposure times in the autochrome process) he shaped photography as an aesthetic language of observation and revelation (of colours, imperfections, and variations) and – in dialogue with pictorial still life – as a symbolic way of seeing, turning the exposed objects (fruits or flowers) into enigmatic presences or absences.[25]

The rise of public interest in the shroud photograph also brought attention to the first Italian Photographic Congress, hosted in the General Exposition of Turin between 19 and 22 October 1898. The organizers (the photographic committee of the General Exposition) saw the event as an occasion to exhibit the works of the main Italian photographers (Luigi Primoli, Guido Rey, Mario Nunes Vais, Giuseppe Incorpora, Secondo Pia), trace the status of national photography, and advance the need to create an Italian school. Throughout its six sessions (on copyright and artistic property, telephotography, photography schools, photography and the graphic arts, the conditions of photographers in Italy, and photographic material) the conference rejected the positivist myth of photography's mechanical reproduction and launched instead a new idea of photographing as a poetic or pictorial act, an idea that would have great influence in the following years.

As an immediate outcome of the congress, the former mayor of Turin, Edoardo di Sambuy, established the Società Fotografica Subalpina in 1899 (granting honorary membership to Queen Elena) and organized the first international exposition of photography in 1902 (in the context of Turin's International Exposition of Modern Decorative Art; Costantini, "L'esposizione" 95–105). The 1902 exposition, which propelled the spread of the Liberty style in Italy, also marked a significant watershed in Italian photography – first, because of its theorization as an independent aesthetic expression, in alignment with the other decorative arts, and, second, because of the contact established between Turin's photographers (Pia, Rey, Sella, Schiapparelli, Sambuy) and the international Modernist movement, which would launch the project to develop an Italian equivalent to Stieglitz and Steichen's journal *Camera Work.* This project came to fruition in December 1904, when, under the sponsorship of the Società Fotografica Subalpina, Annibale Cominetti founded *Fotografia artistica.* Published in Italian and French until 1917, the journal launched the school of pictorialism and developed a critical discourse on photography. The publication showcased pictorialist works, organized expositions (in 1906 and 1911), and delineated the independent characters of artistic photography – identified in representations *en plein air* (often of scenic landscapes), in the portrayal of poetic instances in daily life, and in the artistic enactment of tableaux vivants. *Fotografia artistica* also accompanied the evolution of pictorialism by creating a critical debate around it, as reflected, for example, in the *querelle* of *flouisti* and *nettisti* (opposing advocates of blurriness or sharpness, respectively, in photography; Zannier, "Pittorialismo" 10–11). The journal not only fashioned a new photographic language by way of its numerous renowned contributors (including Alinari, Brogi, Gioppi, Namias, Lumière, and Nadar) but also articulated a new photographic culture by fostering a debate on authorship, launching a campaign to safeguard the archival patrimony (after the 1904 burning of Turin's national library), and introducing new repertoires (e.g., Bragaglia's *fotodinamismo* in 1913).[26]

The ongoing transformation of photography from a serial language into a *kunstindustrie* ("decorative art" or "art industry"; Payne 46–7), and of photographs from depicting immobile ruins into being imaginative platforms, finds contemporary literary correspondence in a 1907 poem, "L'Amica di nonna Speranza" ("Grandma Hope's Friend"; *Opere* 208–15), by the Turin-born Guido Gozzano, which stages an old photograph as a unique site of both nostalgic resistance to industrialization and creative exploration of new expressive forms. First published in *La via del rifugio* (*The Path to Refuge,* 1907), and later in *I colloqui* (*The Colloquies,* 1911), the text describes the poet's discovery of an 1850 photograph of his then- young grandmother Speranza with her friend Carlotta, in the midst of her accumulated junk. Surrounded by "good things of awful taste" (verse 2; buone cose di pessimo gusto), the photograph represents a fragment or a quotation of the past, enabling the vision of its ephemeral contingence, yet also distancing it in an outmoded stillness. Standing out among the other objects, the photograph opens a peculiar space of imaginative suspension and wonderment, which leads the poet to start his poetic fiction. Enclosed by Gozzano's initial and final reflections of 1907 (in sections 1 and 5), the poem thus coincides with a live re-enactment of the 1850 picture in an imaginary three-episode fiction (in sections 2, 3, and 4). In the first scene (section 2) the young *nonna* and her friend Carlotta arrive in a vacation home on Lake Maggiore and sing together while imagining their Prince Charming. In the second scene (section 3) Speranza's parents introduce Carlotta to her pro-Austrian uncle and pro-Piedmont aunt, conversing about current events (Verdi's new opera, Radetzky's armistice, King Victor Emmanuel II's ascent). In the third scene (section 4), outside by the lake, Carlotta tells Speranza of her love for an unknown poet, ignited by his gift of Goethe's *Werther.*[27] Identifying himself with Carlotta's young poet, Gozzano prolongs and returns her love, in an impossible call to her from his own present – "where are you, / the only one whom, perhaps, I could love with real love?" (verses 109–10; ove sei / o sola che, forse, potrei amare d'amore?) – thus changing the photograph from a singular object into a new creative space, endowed with a unique epistemology.

The photograph object reflects the contradiction of a new technology pervaded by a nostalgic striving for the past (mirrored in the poet's melancholy for the past appearance of photography as the "newest thing"; verse 108)[28] and the ambivalence of the photographic present, depicted as a time and space of marvel and melancholy, fixity and movement, interiority and exteriority. The two girls of the poem reflect its concurrent and contrasting temporal dimensions. The aging of Grandma Speranza represents the duration and continuity of *krónos,* ideally leading to the time of the poem's author, and identifying the decay of everything into pastness. The everlasting beauty of Carlotta, stilled in the photographic "cut," instead portrays the discontinuity of a *kairós,* signifying her ever-blossoming vitality and the poet's nostalgic desire to freeze time in order to avoid her absence.[29] In such an inward and outward theatrical staging, Gozzano's photograph object also delineates, however, a specific epistemology in the visual exchange (or love dialogue) between Carlotta's forward gaze and Gozzano's retrospective look, which takes the form of a tragic dialectic of necessity and freedom. Gozzano's theatrical drama ultimately plays out in his "ti fisso nell'albo con tanta tristezza" (verse 103, emphasis mine; I *stare* at you in the album with much sadness), which expresses the coincidence of two gestures: that of literally *fixating* on the girl's romantic posture, her youth, and her eternal love for the poet in the immobility of the photographic object; and that of continuously *staring* at her (in her ephemeral movement and in his imaginary fiction). Analogous to Pirandello's famous dialogue about the "tragedy of Orestes in a marionette theatre" (136), opposing the characters of Meis and Paleari in the contemporary novel *Il fu Mattia Pascal* (*The Late Mattia Pascal,* 1904), the photographic space thus represents a tragic interstice between the necessity of stillness (contained, as is the ancient tragedy's enclosed space) and the drama of movement (opened, as in modern tragedy, by the break of "a little hole torn in the paper sky of the scenery"; *Il fu Mattia Pascal* 136); or rather between an objectified, inward plot (the myth of Orestes) and an outward vision, turning characters into Hamletish figures ("in other words, Orestes would become Hamlet"; *Il fu Mattia Pascal* 136).[30]

In the paradox of this "stillness in motion" (Hill and Minghelli) Gozzano then metaphorically stages an attempt to break the visual conformity produced by the serialization of photography and the inherent "sclerosis" of the photographic image. At the same time, he analogically reflects in it the endeavour to overcome the standardization generated by the commodification of culture (e.g., in the contemporary branding of *dannunzianesimo*)[31] or by the academic crystallization of literature into a monumental collection of antiques. Gozzano's photographic gaze designs a new, experimental poetic language in-between, forged as an analogical visual-verbal communication and imagined as a metalinguistic space, made of allusion and recognition,[32] ironically reassembling objects, fragments, and poems in a creative puzzle (Baldissone 16). In the interstice between literary tradition and the new media industry Gozzano uses his photograph as an experimental platform for developing a new allegorical vision, conceived as an enigmatic rebus, where "the written word tends toward the visual" and the image transforms into a "rune" (Benjamin, *Origin* 182). In his photographic "mutoscopio" ("silentscope," in his later definition in the short story "I sandali della diva" [154; "The Diva's Sandals"]) Gozzano also locates the laboratory for the evolution of a crystallized image into a moving fiction, silently combining the slowness of words and the rapidity of the image's deitic evidence in a new cinematic narration.

The New Cinematic Language of Photography

Against the back-drop of the industrialization of photographic images, Gozzano's poem reimagines photography as a malleable interface of fixity and motion, standard and variation, and reproducibility and creativity. Gozzano's vision would mirror the analogous attempt of other contemporary photographers to break the boundaries of photography (stillness and bidimensionality) and to develop it as a cinematic space. Professional photographers worked on the re-enactment of the photographic image, as seen in Futurist photography or in the evolution of tableaux vivants into motion pictures. In a similar way, amateur photographers turned photographs into "moving screens" or "acting pieces" of personal narration, as later documented in the photographic memory of the First World War.

In Futurist art, photography assumed an ambivalent role (Lista 133–89), both as a serial marketing tool (in the well-studied group portraits or effigies of the artists that were used to promote Futurism in the press) and as an experimental tool of aesthetic investigation. However, in parallel with the pictorial pursuit of dynamism (as expressed, for example, in the photographic vision of Giacomo Balla's 1912 painting *Dynamism of a Dog on a Leash*), it also embodied a contrast with its intrinsic crystallization of movement (as openly denounced by Boccioni).[33] Against this back-drop, the practice of photodynamism by Anton Giulio Bragaglia and the visual experimentation on photographic portraiture by Fortunato

Depero constitute two parallel attempts to develop photography into a cinematic language. The (aspiring) Futurist photographer Bragaglia experimented with the creation of motioned images in his *fotodinamismi*, rendering the movement of contingence through the technique of overexposure, in order to overcome the immobility of pictorial representations and reflect the movement of modern life. The Futurist painter Depero violently attacked the pre-established codes of portraiture in his three famous self-representations of 1915 *with a fist, with a grimace*, and *with a cigarette*. In his mimicked breaking of the lens, his defiant pose, and the image's deliberate overflowing into a pictorial space, the artist expressed the desire to crack the bidimensionality of photo-portraiture and open a new alternative space of meaning in it.

Along with these attempts, the work of the Turin photographer Arturo Ambrosio (1870–1960) proved particularly relevant to the transformation of photography into an aesthetic and dynamic representational form. Ambrosio practised photography in Turin during its photographic boom at the turn of the century; he opened a shop (selling his own camera, "Ambrosio"), a photography studio, and a laboratory and even became the official photographer of the royal family, at the request of Queen Margherita. After his travels to France and England, and his encounter with the photographer Giovanni Vitrotti, Ambrosio started experimenting with the production of tableaux vivants and short cinematographic documentaries in 1904, anticipating the growing success of cinema. Influenced by the closeness to France and following the establishment of the first Italian film company, Cines (Rome, 1905), he started his own cinematographic business in 1906, which became one of the leading producers of silent movies and laid the foundation for Turin's golden age in early cinema (Armenante). After the founding of Ambrosio Films, other major production houses started in Turin – Aquila Film (1907), Rossi & C. (1907), and Itala Film (1908) – and the city rapidly established itself as the capital of Italian early cinematography, as evidenced by its growing number of movie theatres and film productions (107 titles in 1907, versus the 40 of Rome; Sangiovanni 90) as well as by its establishment in 1907 of Italy's first specialized magazine on cinema – *Rivista fono-cinematografica e degli automatici, istrumenti pneumatici e affini* (*Journal of Phonography, Cinematography, Automatic and Pneumatic Instruments, and Similar*), directed by Gualtiero Fabbri.

The foundation of Ambrosio Films symbolically marked the end of cinema's itinerant period and the beginning of its new industrial phase, focused on configuring a more structured market and articulating an independent aesthetic language, beyond the existing model of a "twenty-minute phantasmagoria" (as defined by Giovanni Papini in the 1907 article "La filosofia del cinematografo," published for the Turin newspaper *La stampa*). In this context Ambrosio's experience with photography influenced the forging of cinema as both an industry and an art. Ambrosio applied the entrepreneurial model of his previous photographic activity onto the new cinematographic business, envisioning a production cycle that handled the entire film manufacturing in house (project, realization, distribution, and retail). In light of this model and his network, he developed professional connections with first-rate photographers (Roberto Omegna, Giovanni Vitrotti, and Luca Comerio), authors (like Gozzano, or D'Annunzio whom he first approached in 1908),[34] actors (like Eleonora Duse), and theatre or opera stage designers (starting with his own cinematographic version of Ruggero Leoncavallo's *Pagliacci* in 1907). Moreover, Ambrosio built cinema around the same repertoire as that of photography, connecting it to both Italy's monumental past and its literary masterpieces, in an endeavour to appeal to the national and international markets. Inspired by nineteenth-century archaeological photographs and paintings (Herbert 586), he produced historical films, starting with Omegna's *The Last Days of Pompeii* (1908), which would inspire Mario Camerini's famous version (Ambrosio Films, 1913) and the genre's later success, emblematized by Enrico Guazzoni's *Quo Vadis* (Cines, 1912) and by Giovanni Pastrone's colossal *Cabiria* (Itala Films, 1914). Along with its effort to "cinematize" Italy's ruins, Ambrosio Films bet on cinematic adaptations of theatrical and literary works as a way to compete with French cinema and endow the new industrial art with a pronounced cultural dimension.[35] Like Vittorio Alinari, Ambrosio moved cinema from his local atelier into a global business, by standardizing its language (in the repetition of winning formulas), creating its ideal archive (or cinematheque), and turning it into "a magical means of locomotion for all kinds of trips into the country's history and its artistic and literary past" (Brunetta 29). He similarly shaped cinema as a hybrid industrial art, refashioning the masterpieces of other arts and inventing new expressive forms. In the push to break the bidimensional space of the reproduced image, or the traditional visual standards of theatrical and operatic set design, Ambrosio and other houses successfully applied the laws of Renaissance perspective to cinema (Brunetta 36) and developed new ways of orchestrating the movements of large crowds in historical films. Parallel to photography, cinema thus intrinsically fashioned a new ambivalent cultural space of

Italian industrial modernity, by grounding contemporary progess in Italy's past tradition (monuments, imagination, iconography) and reinventing Italy's antiquity through its most innovative technology.[36]

These dialectics of serialization and "aestheticization" of the photographic image would find their apex in the visual narration of the First World War.

The Photographic Fiction of the Great War

In the context of the First World War, dominated by an unprecedented mass mediation of images and ready-made visual narrations, the official, and, more significantly, amateur photographic materials represented an ambivalent industrial language, both standardizing the conflict in its documentary version and latently opening up a malleable space of its fictional re-elaboration.[37]

Serialized photographs, on postcards and magazines, displayed war's immediacy and satisfied the growing demand to "see" the conflict. Their codified visual standard aimed at sugar-coating the war, staging the front line as a theatre filled with breath-taking sceneries or heroic adventures, and territorial Italy as a compact internal front joining the war effort by the production of machinery, weapons, and uniforms. As mandated by the Army Supreme Command,[38] official photographic documentation controlled the conflict's imagery by offering a reassuring guide to trench war (through captions, providing its "correct" interpretation) and by invalidating the frightful reports of soldiers coming from the *zona di guerra* (through the deliberate removal of any unpleasant sight of violence and death). Official war photography thus documented and presented events in accordance with a pre-packaged "cultural template" (Ashplant 34). In a way, it reiterated nineteenth-century visual patterns: positivistic classifications of soldiers in uniform (encouraged by the military command as a form of surveillance against desertions and a symbolic instance of national uniformation),[39] or explicit representations of the war's modern ruins in the recurring images of destroyed buildings and deserted rubble (which contributed to the spread of hatred against the enemy and to the mummification of war in its making). In another way, in its new mass-mediated presence, repeating to saturation a deliberately premade pose of the events,[40] photography also imposed on them a "general amnesia" (McQuire 130). By serializing its visual and narrative templates (the portrait of a man in uniform, the views of modern ruins, the pre-conflict myths of redeeming violence and of social order against chaos), industrial photography thus subtly imposed its visual knowledge of the war onto the reality of war, erecting a neutralized façade and a memory of marble, which safely contained trauma under the cover of a frame, a monument (the *sacrari*), a tombstone (the unknown soldier at the Vittoriano), a stoic attitude, or an unchangeable myth.

In addition to the officially controlled images the war produced a rich repertoire of private visual materials. Thanks to the dispersion of cameras among officers and middle-class soldiers, amateur photographs mediated the servicemen's desire to communicate with relatives and to collect a somewhat coherent narration of war in photo-diaries ("no differently than they would on the occasion of a long adventure-filled trip"; Fabi 57). Located in the "intermediate area" (Bollati 5) between the muted narrative space of the individual and the mass-mediated rhetoric of the global war, private photographs "spoke" a commodified language. Albeit individual, the portraits of soldiers in uniform repeated a common cliché, presenting the servicemen in a singular space (lacking any real "infra-knowledge"; Barthes, *Camera lucida* 30) and in a standard pose (serene, with feet crossed, hands relaxed sideways or leaning on a support, against the background of gardens and seemingly domestic furniture). Likewise, the photo-diaries from the front followed a fixed pattern, depicting the "normal" war of the *retrovia* (rearguard) as an immobile hiatus between the military and the civil, as a therapeutic space of leisure, separated from the daily vicinity of death, and as a mixed narration of personal deeds (such as jokes, feats, encounters with local women, eating, and shaving) and inserted postcard images (e.g., the explosion of a bombshell, a speech, the adoption of an orphan by the brigade). In spite of their highly codified representations, however, private photographs also filtered latent aspects of the conflict by their visual variation and intrinsic "call" to imaginative association. These pictures performed their uniqueness by displacing the distinct facial effigy or individual gaze of the otherwise "muted beings" (in Camillo Sbarbaro's words; *Trucioli* 67; viventi ammutoliti). In their slightly varied backgrounds, poses, and expressions, they certified a soldier's presence to relatives and friends (as it happened with postcards, which were often left blank, bearing only the date, a signature, and the stamp "verified for censorship") and codified an implicit space of signification (or a particular situation or condition of life at the front) to be reconstructed. As evidence of a "buried life" (Fussell 326), these pictures also embedded an elliptical space, engendering the fictional reprocessing of the war's trauma, as Aldo Palazzeschi ascertained in his war memoir *Due imperi mancati* (*Two Failed Empires*, 1920). Recalling his assignment in the section of "military leftovers" (118), Palazzeschi represented himself in the act of reviewing the collected remnants of deceased soldiers and imaginatively fabricating about them "a compassionate lie" (119; una pietosa bugia) that would offer their relatives a brief appeasement. The portraits of deceased soldiers in uniform, often the only remaining fragment of their lives, engendered the same kind of fictional imagination in a public space: as the ceramic ovals for tombstones (Favaro) or as the covers of private booklets, published by relatives in celebration of the lives of their beloved ones.[41] The success in Italy of the post-war genre of biographical necrologies coincided with the need to elaborate, upon the screen of a photographic image, the a posteriori fiction of the conflict, compensating for a soldier's missing funeral by embellishing his death (often meaningless and brutal) and transforming a private grief into a public or political exposition.

In general, the photographic sources of the First World War ultimately dwelt on an ambiguous "presente!" (as symbolically carved on the marble stairs of the Redipuglia war shrine), suspending the vision and memory of the ongoing conflict between the bright, repeated façade of a camera lucida (monumentalizing an objectified contingence) and the latent development of a camera obscura (reprocessing a poetic or fictional reinvention of the events). In a way, the photographic image's imprinted documentary evidence replicated the trauma ad infinitum, creating, by its compulsion to repeat an "authentic" façade, a protective shield for memory. In this sense, the exposed presence of photography worked to cathartically remove war's uncanniness by deflecting trauma onto a distant myth, containing pain within its frame, and banalizing it through repetition in order "to domesticate it, normalize it, and, in time, absolve it" (McQuire 153). In another way, photographs also turned, for later observers, into unconscious screens of fantasy and imagination, in which a poetic development of memory (mythopoiesis) and a fictional reprocessing of trauma could constantly take place. Similarly, oral and choral tradition, war poetry, and few other novels of invention enacted, in their latent non-absolute language, a Hamletic and malleable space of memory.

As in Gozzano's dialectics of standard and invention in "L'amica di nonna Speranza," the photographic "medusification" of the myth of the Great War thus coincided with the latent elaboration of a hybrid language of visual and verbal fiction and, thereby, of an alternative narration of the facts. In opposition to the mechanism of cultural self-defence erected by serialized photographs, the opening of an allegorical and associative breach in them thus envisions an ultimate space not just of resistance to conformation but also of cultural challenge, always inviting a renewed interpretation of collective memory.

Chapter Four

Bicycles and the Moving Body of the Nation

On 7 July 1893 an unusual letter exchange appeared on the pages of *Corriere della sera.* Luigi Masetti, a young cyclist competing in the Milan-Turin race that was recently organized by the Milanese newspaper, offered a daring challenge to its founder and director Eugenio Torelli Viollier: "Give me a 500 lire ticket or dry up the ocean, and I'll show you the practical utility of the bicycle [*bicicletto*] by taking a round trip from Milan to the great World's Fair of Chicago in about two months."[1] Italy was in the middle of an unprecented financial and social crisis, following the scandal of the Banca Romana and the uprising of the Fasci dei Lavoratori (Workers' League) in Sicily (which would lead to the political collapse of Giovanni Giolitti's government and to the appointment of Francesco Crispi as the new prime minister).[2] Yet such a bold and bizarre wager, followed by the cyclist's promise to send a weekly report of his long trip, immediately captured the imagination of the journalist ("We love feats endowed with boldness and eccentricity; we accept") and, over the months to come, the imaginations of the Italian readership.

After renaming his *bicicletto* Eolo (Aeolus), Masetti inaugurated the "viaggissimo" (as defined by *Corriere della sera*; the trip) just one week later, on 13 July. His seven-thousand-kilometre journey took him from Europe (through Zurich, Bruges, Calais, and Oxford) to America (on board the steamship *Chester*), then from Niagara Falls to the Chicago World's Fair (on 13 September), and, on his return, from Washington (where he met U.S. president Grover Cleveland at the White House) to Milan in triumph. The reports of his adventures, published each Monday in *Corriere della sera*, became a live version of Jules Verne's *Le tour de monde in quatre-vingts jours* (*Around the World in Eighty Days,* 1873), feeding the curiosity of the nation about the world, its ambition to emerge in the global culture of universal expositions, and its thirst to know more about America (against the back-drop of the social debates on emigration, and the cultural phenomenon of *americanismo*).[3] The summer story of the "anarchist on two wheels" (in Torelli Viollier's words),[4] who was riding on a trendy item and emerging as a global hero in the midst of the economic recession, represented an important watershed not only in the Italian bicycle market but also in the contemporary debate on cycling.

At the time, velocipedes were spreading all over Italy, and cycling was gradually transforming from a fashionable practice into a recognized habit, as confirmed by *Illustrazione italiana* in its commentary on the victory of Luigi Airaldi in the Milan-Turin race: "This is the moment of velocipedes. They own the roads, the newspapers, and the public interest. Will this trend pass? The fanatism will, but the delightful, healthy, and above all useful habit will remain" (9 July 1893).[5] Against the many detractors of the "imbeciles on two wheels" (*Il resto del carlino*, 4 June 1893), the public debate on cycling had gained momentum a few months earlier on the occasion of the first congress of Italian velocipedists, organized in Milan on 28 May 1893. By bringing together the Veloce clubs of Milan and Turin for the first time, the conference provided a new public platform for reflection on the social value of bicycles. Two cycling philosophies collided in Milan: *biciclettismo*, which promoted bicycles as recreational and democratic vehicles of leisure and travel, and *velocipedismo*, which supported velocipedes as vehicles for professional racing and competition. Against this back-drop, Luigi Masetti's adventures extended the contents of the congress to the public (as implied in Masetti's goal to show the "practical utility" of cycling, in the letter to Torelli Viollier) and effectively contributed to the spread of a broader understanding of bicycles as cultural objects for travelling and knowing, as well as binding and regenerating the national body. At the same time, in addition to the popularity of the report by Masetti, his double profile as a racer and a world explorer, narrating a story and competing for a prize, embodied a visual synthesis of *biciclettismo* and *velocipedismo*, informing, in different ways, their parallel development in the subsequent years.

On the one hand, in the wake of Masetti's adventure to Chicago (and his later trip, in 1900, touching Ceuta, Cape North, Moscow, and Constantinople),[6] Federico Johnson and Vittorio Bertarelli founded the Touring Club Ciclistico Italiano in 1894 (known as the Touring Club Italiano after 1900) as a further elaboration of *biciclettismo*. The Milanese club envisioned travelling as a pedagogical and practical tool for providing a living knowledge of Italy, and it invested in cycling as a way to shape a collective imagination, create "unified fields of exchange" (Anderson 44), and establish a bond of comradeship among Italians.

On the other hand, more focused on Masetti's participation in cycling races, the Northern Italian bourgeoisie saw in bicycles and in the culture of *velocipedismo* the favourable ground for developing a new modern sport. *Velocipedismo* represented a unique phenomenon in the peninsula thanks to its success in growing a specialized press and in organizing events. The rising number of bicycle races boosted sales and generated an ever-greater coverage of cycling, as documented by the establishment of weekly and monthly publications for fans, like *La bicicletta (The Bicycle)*, *Illustrazione ciclistica (Illustrated Cycling)*, *Italia ciclistica* (*Cycling Italy*), and *Il ciclo* (*Cycles*) in Milan; *Rivista velocipedistica* (*Journal of Velocipedes*) in Turin; and *La pista* (*The Race Track*) in Verona.[7] The cycling market was launched by the foundation of *La gazzetta dello sport* (*The Sports Gazette*) in 1896 – out of the fusion of *La tripletta* (*The Triple*), founded by Eugenio Camillo Costamagna in 1894, and *Il ciclista* (*The Cyclist*), founded by Eliso Rivera in 1895 – which led to the invention of longer races, like the Giro di Lombardia (from 1905) and the Giro d'Italia (from 1909); these in turn had a great impact on the construction of a national imagination.

Within this context, and throughout the 1890s and 1900s, cycling constituted a significant moment of technological experimentation and innovation, laying the groundwork for the growth of the Italian mechanical industry (in the production of motorcycles, automobiles, and aeroplanes). Along with the development of the independent cultures of tourism and sport, cycling also acquired more complex social, political, and cultural dimensions, as bicycles gradually turned into myth-making tools supporting the industrial bourgeoisie's project to endow Italy with a shared national imagination. As they appeared in literature and painting, bicycles ultimately enacted an experimental space of aesthetic negotiation and investigation into industrial modernity, the rediscovered value of corporeity, and the new "moving" dimension of culture.

The Italian Bicycle Debate and the Birth of Italy's Cycling Industry

Throughout their technical development and industrial diffusion from the 1880s to the 1910s, bicycles polarized fervent political, social, and cultural debates in Italy. As the new cycling fashion gained ground in the country, many sectors of society reacted with fear and unease. Governing bodies tried to contain the new trend, as documented by the high taxes imposed on velocipedes by the government, the vetoes against bicycles ordered by several municipalities, and the cycling prohibition decreed by General Bava Beccaris during the 1898 turmoils in Milan (Schirinzi 40). The scientific community also fed the Italian *ciclofobia*, as bicycles were accused of deforming both the body and morality. Physicians argued that "the new machine overdeveloped leg muscles, harmed joints, neck muscles, and the head's vascular system, and deformed hands and feet" (Thompson 30), and positivist scientists like Paolo Mantegazza and Cesare Lombroso warned against cycling as a cause of licentious behaviour by women and as a vehicle of transgression, offering criminals a rapid escape from any infraction (Boatti 16).[8] At the level of public morality, bicycles (which began to be referred to in the feminine, *bicicletta*) were negatively related to a frightening and irrational femininity, as they represented an occasion for women to wear less modest clothing and expressed a destabilizing instance of liberation (of women and of sexual impulses) in their double connection to the emancipation movement and to the explicit eroticism and nudism in contemporary advertisements and publications.[9] In addition to being associated with moral deformation and feminine irrationality, bicycles catalysed, for Catholics and socialists, a broader critique of industrial modernity. Although they embraced sports culture, Catholics rejected bicycles, as being symbols of mundanity and luxury – to such an extent that priests were forbidden to ride them (in certain dioceses, with the threat of suspension *a divinis*) because of "the modernistic 'whirl' they assumed and suggested" (Pivato, "The Bicycle" 182).[10] While considering cycling a frivolous distraction from the political struggle, socialists identified and rejected bicycles as symbols of bourgeois classism and exploitation (as confirmed, in 1912, by Benito Mussolini's journalistic campaign in *Avanti!* to cover the streets of Giro d'Italia with nails; Pivato, *Bicicletta* 224).

In contrast, bicycles magnetized the ambitions and dreams of the emerging industrial bourgeoisie, which foresaw in them the embodiment of its values and the accomplishment of a sought-after new modernity. Fuelling the excitement for this new tool of self-transportation, the Northern middle class channelled the cycle mania into a cultural discourse in support of industrialism, which slowly grew into a real cycling ideology. In the late-nineteenth-century images of elegant gentlemen pedalling in Milan's Piazza Duomo (Vergani, *Uomo a due ruote* 45–7), bicycles appeared as symbols of economic prosperity, cultural exclusiveness, and social distinction. Associated with trendiness, motion, and rapidity (in contrast to the slowness of pedestrians), bicycles also visualized new modern concepts, like independence of movement, the regeneration of the body, and the acceleration of progress. In the early 1890s, in the wake of the growing success of cycling races and Veloce clubs[11] bicycles turned from being eccentric luxury toys for the wealthy into being cultural symbols of unprecedented appeal. As a result of the deliberate effort by the Northern industrial bourgeoisie to promote their social utility, the market rapidly grew,[12] and, with the expansion of *biciclettismo* and *velocipedismo*, cycling became the fertile ground for the creation of a cultural and political project, aimed at rejuvenating the country by means of a new ideology of movement that fostered health, scientific progress, and well-being.

As bicycles entered Italy's cultural space, symbolizing an exciting or frightening modernity, the Italian production of "the machinery" (as they were referred to in contemporary publications) began to develop. Until the 1880s only a few mechanics (Carlo Michel in Alessandria, Raimondo Vellani in Modena, Bartolomeo Balbiani in Milan, and Giuseppe Garolla in Padua) and small workshops (Luigi Figini, 1874; Turri e Porro, 1875; and Carlo Ciocca, 1881) had been producing cycles (replicating the *michaudine* model that had been presented at the Paris exposition of 1867). Indeed, bicycles were still considered unconventional means of transportation in Italy, in rapid and constant evolution,[13] or trendy items, mostly imported from France, Germany, or England (Boatti 9). The turning point of the Italian (and global) cycle industry came in 1885 when, after the establishment of his workshop in Milan, Edoardo Bianchi launched the revolutionary "safety bicycle." Thanks to his technical innovations – a pair of matching wheels with reduced diameter, and lowered pedals – Bianchi stabilized the *bicicletto*, thus creating the prototype of modern-day bicycles. The immediate success of his model boosted Italian cycle production and, with it, the country's mechanic industry; over the years many other Italian factories started production – Olympia (1893), Velo (1894), Maino e Dei (1896), Frera (1897), Fiat (1899), Lygie (1905), Legnano (1906), Atala (1907), Torpado (1908), and Ganna (1910) – rapidly transforming bicycles from luxury items into affordable objects (priced from 600 lire in 1896 to 150 lire in 1906; Bozzini 34).[14]

Bianchi's progressive affirmation as "one of the leaders of Italy's industrial system" (A. Gentile 8) heavily relied on the skilful combination of technological innovation and the pursuit of a larger cultural horizon for his products. On the one hand, Bianchi transformed cycling into an experimental platform for new industrial developments. The company was the first to apply air-filled rubber tyres to bicycles, after John Boyd Dunlop's invention of the pneumatic in 1888, and to add engines to cycles in 1897; it rebranded itself as Fabbrica Automobili e Velocipedi Edoardo Bianchi in 1899; and it diversified its production during the expansion of the automobile industry (manufacturing 45,000 bicycles, 1,500 motorcycles, and 1,000 cars by 1914). On the other hand, Bianchi associated its products with cultural, social, and political discourses.[15] After teaching Queen Margherita the art of cycling and designing for her the first lady's bike in 1895, Edoardo Bianchi became the official supplier by appointment to the royal court. Paralleling its interest in tourism (confirmed by its early partnership with the Touring Club Italiano) with its pioneering investment in advertising, the industry also engaged in global competitions, sponsoring racers like Giovanni Tommaselli (winner of the Paris Grand Prix de la Ville in 1899), winning a prize at the 1901 Exposition International du Salon de Cycles in Paris, and supporting the organization of cycling races. At the outbreak of the First World War, the Bianchi company (which suddenly found itself on the receiving end of a flow of orders from the government) actively campaigned in favour of Italy's intervention and purposefully transferred the imagery of its sports successes to the realm of war (as noted in its war-time slogan "Bianchi, which reigned in peaceful sport competitions, now valiantly contributes to the glorious advance of the Italian forces").[16]

Over the years Bianchi thus shaped the Italian cycling industry into a hybrid experimental ground in between technology and culture, generating innovative partnerships and elaborating new promotional languages. On the one hand, in light of its agreement with Pirelli in 1890 – ratified by the latter's patenting of the first bicycle and car tyres (Flexus in 1897, and Ercole in 1901) – Bianchi contributed to the Italian rubber company's shift in production from telegraphic and electric cables to air-filled pneumatics (Zamagni 100) and also laid the groundwork for the expansion of the nation's entire mechanical sector, in terms of collaborative assembling, technological research, and raw material development. On the other hand, in light of its active involvement with Touring Club Italiano (and later Giro d'Italia), Bianchi created a cultural imagery around its industrial products, slowly fashioning bicycles into literary symbols, visual icons, and collective myths. As visualized in an early promotional poster from 1896 (fig. 4.1), Bianchi deliberately paired the brand with an independent cultural discourse, relating bicycles to the concepts of flight (above Milan), light (around the cycle),

Figure 4.1 Anonymous, poster advertisement for Bianchi, 1896. www.museodelmarchioitaliano.it.

lightness (the angel holding the bicycle with a finger), divinity (the angels of heaven), deity (the triangular shape traced by the angel's cords supporting the bicycle), royalty (the mention of the "Real Casa"), and race (the winged wheel).

Even before the company commissioned promotional works by high-calibre artists like Umberto Boccioni (in 1908) or Fortunato Depero (in 1924), Bianchi's 1896 advertisement not only captured the social, literary, and visual imagery emerging around bicycles during the boom years from 1893 to 1909 but also provided early documentation of the company's core strategy to culturalize cycling and build an aesthetics of modern life around the new phenomena of tourism and sports.

Cicloturismo as a New Model of Knowledge

An important agent in the development of the cycling industry and in the cultural construction of bicycles was the Touring Club Ciclistico Italiano. After its foundation in 1894, which capitalized on a moment of expansion in the bicycle market, the Touring Club Italiano (its name since 1900) significantly contributed to building not just a narrative around cycling but also an experimental platform of collaboration between industry, art, and cultural politics. As it rapidly grew from 784 members in 1894, to 20,915 in 1900, and to 157,897 in 1915, the TCI constituted a privileged showcase for the Northern bourgeoisie.[17] Effective in projecting the new industrial class onto an imagined national community, the TCI also represented a powerful nation-making tool.

In the vision of the club's founders and funders, bicycles embodied a tangible suture between the aspirations of the capitalist bourgeoisie and society as a whole, seen both as a market to be conquered and as a mass to be shaped. Through social initiatives and publications the TCI hence developed the idea of *cicloturismo,* conceived as an organized culture of leisure (born out of the economic surplus of industrial modernity, and the increased availability of leisure time) and centred around a new idea of travelling – intended as a novel way of relating to extra-urban space (as in the increasingly popular habit of the *gita fuori porta*) and, beyond that, as a new model of self-knowledge. In a nation where, before unification, travelling was impeded by political fragmentation and, after 1861, was impelled by the need to emigrate, the cultural affirmation of *cicloturismo* coincided with the end of an older, idealized model of self-representation (the Grand Tour) and marked the beginning of a different way of touring and knowing Italy, conceived as a "*petit tour*" by and for the Italians and organized around the idea of discovering the nation's real landscape in short stages. This new form of travelling deeply affected Italian society, politics, and art by shaping the epistemology of a new culture in movement, which was focused on wandering exploration in space, and by aiming at the active revitalization of Italy's monumental representations.

A striking image of the TCI's ideal of setting in motion the nation's immobile individual and social bodies is provided by the extensively covered *passeggiata ciclistica* (cycling trip) from Milan to Rome, which was organized by the Club in 1895 and concluded with the visit of seventy members to Queen Margherita (who became a member in 1902). In the drawing and photograph of the event published in *Illustrazione italiana* (Vota 30) the two pedalling founders of the club, Vittorio Bertarelli and Federico Johnson, lead the group with their trendy and innovative bicycles, cutting through the crowds of Milan and the ruins of Rome, in a testament to the bourgeoisie's will to vivify the country's immobile mass. After this first *passeggiata* the club promoted similar educational cycling trips, aimed at bonding its members into a common body and at offering, in the knowledge *de visu* of Italy's historic sites and landscapes, a lively new image of the country – no longer relying on static, bookish, or picturesque representations.[18] As thousands of amateurs (and cheering crowds along the way) attended these *passeggiate ciclistiche,* bicycles turned from recreational objects into pedagogical tools for incorporating individual bodies into a collective movement across the country's unknown geographical space (as in the first alpine trip to Monte Rosa, in 1898, or even in the archaeological trip to Libya in 1914 after the colonization of Tripolitania). Through the *passeggiate* the club aimed to revitalize the patrimony of the nation's recent past while bonding citizens in a collective memory, as attested by the pilgrimage to Vittorio Emanuele's tomb (1901); the Sardinian Tour, following in the footsteps of Garibaldi (1901); the itinerary to the sites of the independence wars (Curtatone, Montanara, Custoza; 1902); the commemoration of the *spedizione dei mille* (1910); and the trip to Rome's Vittoriano for the fiftieth anniversary of unification (1911).

The progressive diffusion of this culture of travel was certainly accompanied by a sense of omnipotence and exaltation at the new power of conquest and movement in space guaranteed by bicycles. At the same time, however, against the back-drop of a persistent social unease with regard to cycling, the new bicycle experience manifested a latent sense of anxiety and insecurity, generated by the en route discovery of Italy's unforeseen realities (criminality, poverty) and its dire lack of infrastructures. As a remedy to this the club supplied its members with maps and guides,[19] providing cyclists with an exact and detailed measurement of the national territory (altitudes, mileages) and with useful information regarding spare parts, first aid kits, restaurants, hotels, and passports. While signalling the ambition to control and manage the unknown, these publications also reveal the bourgeoisie's ambivalent attitude towards *cicloturismo* in their disconcerting representation of a nation yet to be built (in the need of streets and more lodging)[20] and yet to be conquered in its entirety (as it appears with the inclusion of Dalmatia, Istria, Trentino, Corsica, and Nice in the national map, or with the translation into Italian of foreign city names like Rovereto for Rofreit, Trieste for Terst, and Bressanone for Brixen). In conjunction with its publications, aimed at supporting the recognized social value of cycling, the club promoted the safety of its recreational lifestyle by launching (and actively bringing to success) political campaigns for lowering the *tassa sul moto* (circulation tax) imposed on cyclists (from twelve lire in 1895 to six lire in 1909), approving a code of rules for velocipedists (in 1897), constructing new roads, and propagating school tourism.

The club's privileged instrument for promoting an aesthetic imagery of modernity in *cicloturismo* – as a new civilization of movement, health, and wealth – was its monthly magazine, *La rivista mensile del T.C.I.* (*The T.C.I. Monthly Journal*), which was offered free to its members and conceived as a point of intersection between industry (marketing, advertising, and customer acquisition), society (debates and event reports), and art (illustrated covers and travel books). Thanks to its capacity to create a fruitful collaboration between bicycle-related brands (Bianchi, Pirelli, Frera, and Fiat) and artists of the calibre of Umberto Boccioni, Gian Emilio Malerba, and Plinio Codognato (or, later, Leonetto Cappiello and Marcello Nizzoli), *La rivista mensile del T.C.I.* can be considered a pioneer in Italian industrial art. In the hands of great artists its covers transformed bicycles from mere products, mediating a new lifestyle, into social, political, and aesthetic symbols endowed with an independent iconography and cultural meaning.

The magazine launched a new graphic format in 1908, with covers featuring illustrated, full-colour advertisements. Mirroring the club's dynamic philosophy of modern living and its political commitment to building infrastructure, the road appeared as a privileged subject in the magazine's early colour publications, as attested in the cover advertisements of the first and last issues of 1908, designed by Umberto Boccioni and Gian Emilio Malerba, respectively.[21]

The front cover of the January issue, featuring Boccioni's advertisement for Gola & Conelli, società anonima per la costruzione e manutenzione delle strade ordinarie (anonymous company for the building and maintenance of ordinary roads; plate 4), showed a steam-roller paving a road in the countryside, against the back-drop of a factory's chimneys (contrasting with the trees) and a small crowd of people (mostly composed of children pointing or observing the novelty). While the advertisement reflected the club's cultural and political emphasis on the Italian state's need to construct new roads, Boccioni's image of both men and the machine gaining ground over nature presented the vision of modernity as a muscular conquest of space, an attractive marvel, and an energetic force of transformation, moving from the city to the countryside – a mirror of the peasants walking from a rural context to the urban industries in Boccioni's contemporary painting *Officine a Porta Romana* (*Factories at Porta Romana*, 1908).

The picturing of the road, similarly portrayed through a diagonal perspective, was also the cover subject of the December 1908 issue, featuring Gian Emilio Malerba's advertisement for Pirelli (plate 5). Displaying a wintry image of a cyclist pedalling on an icy road, the cover not only enhanced the visual connection between partnering brands (Pirelli and Bianchi) but also imposed upon them an aesthetic lifestyle, which reflected the club's values, in its emphasis on outdoor living (in the bicycle's overcoming of the adversity of winter) and the conquest of space (the road).[22]

While the magazine visually staged the connections between cycling, the club's ideals, and the mechanical industry, its investment in a high-calibre painter like Boccioni also suggested a deliberate attempt to use cover advertisements to fashion a recognizable artistic iconography of bicycles. After the first issue of January (and perhaps the February advertisement for La motosacoche, featuring an elegant man effortlessly climbing a mountain on his bicycle while smoking a cigarette), Boccioni designed two bicycle advertisements for Bianchi and Frera on the covers of the March and April 1908 issues.

Boccioni's endeavour to create a cycling aesthetics (along the lines of his patron Bianchi's cultural strategy) is manifested in his deliberate reference to the French contemporary painters and illustrators Jules Chéret and Henri de Toulouse-Lautrec. Boccioni's cover advertisement for Bianchi in March 1908 (plate 6), modelled on Chéret's advertisement for Hummes cycles, depicts bicycles in relation to fashion, femininity, and seduction. Staged in a pavilion and surrounded by bright lights and colours, the bicycle appears as a theatrical marvel, the focus of a high-society gala, and seemingly drawing the protagonist (a lady à la mode in the forefront) into the party. Connected to feminine elegance and the woman's gesture of looking (suggested by the spectacles in her hand), the bicycle is transfigured from a technological wonder into an alluring symbol of modernity and an object of seduction, admiration, and awe – also enforced by the slogan "La bicicletta Bianchi è sempre ammirata e preferita" (Bianchi bicycles are always admired and preferred). Modelled on Henri de Toulouse-Lautrec's advertisement for La Châine Simpson, Boccioni's cover advertisement for Frera in the April 1908 issue (plate 7) instead relates bicycles to movement and racing. Against the background of the Simplon Gate in Milan (recognizable in the sketch of Napoleon's Arch of Peace), the poster stages a tight race between a bicycle and a car. In contrast to the immobility of the other figures – the indifferent pedestrians in the background and the curious child hanging onto a street post in the foreground – the bicycle visualizes a new energy in motion, yet simultaneously captures an intermediate means of transportation, in between human forces (the passers-by) and mechanical forces (the car). As they appear on these two covers, Boccioni's cycles emblematize an attractive new way of living that is associated not only with industrial work (in the advertisement for a brand) but also with fashion and racing. They also give visible form to a new aesthetics of leisure, intended as a seductive time of wealth and mundanity, movement and entertainment, happiness and freedom.

Italian Cycling Literature

As in the visual arts, bicycles also inspired the imagination of contemporary writers. With the increasing popularity of bicycles an abundant corpus of literary texts developed around cycling, rapidly moving from a set of shared themes into its own genre. Against the back-drop of contemporary debates and advertising, cycling prose and poetry transfigured bicycles into symbols of exploration, modernity, or liberation, and cycling itself into a recurring metaphor both of a new intellectual attitude, seen as wandering activity, and of a new narrative style, crafted in a state of endless motion.

In prose works – by Emilio Salgari (*Al polo australe in velocipede,* 1895), Alfredo Oriani (*La bicicletta,* 1902), and Alfredo Panzini (*La lanterna di Diogene,* 1907) – cycling represented the ambition to explore the world and travel the national territory. Inspired by Masetti's trip to Chicago in 1893 and by Raffaele Gatti's bicycle conquest of the Arctic Circle in

1895, Salgari's *Al polo australe in velocipede* (*To the South Pole by Velocipede*) portrays velocipedes as the tools for man's conquest of the geographical space, in the staged bet between a velocipedist (the American businessman Wilkye) and a boat owner (the Englishman Liebermann) on who will first reach the South Pole. As Wilkye wins the challenge, his cycling voyage of exploration of the earth's space (across the sea and the Antarctic continent) matches the self-confident tale of man's progress in his struggle against nature: in the passing of Cape Horn,[23] in the fight against natural adversities on the way to the Pole, and in the race against time on the way back to the headquarters. In contrast to Salgari's fictional adventure, Oriani's *La bicicletta* (*The Bicycle*) approaches cycling through a threefold systematic perspective, including a philosophical essay on the liberating power of cycles ("bicycles are us winning over space and time," 87), a fictional part in four short stories, and a live narration of a round trip by bicycle from Faenza to Tuscany ("Sul pedale" ["On Pedals"]). The narration is filled with adventures, stories, and humorous anecdotes written while the author is pedalling around the country without plot or destination.[24] Along the lines of Oriani's last section, and following the model of Lawrence Sterne's *Sentimental Journey*, Panzini's *La lanterna di Diogene* (*Diogenes' Lantern*) narrates a bicycle ride from Milan to the Adriatic Sea, across rural Romagna, as a metaphor for individual and national rejuvenation. Like Oriani, Panzini, as a modern humanist, enacts a new kind of writing, no longer monumental or rhetorical but rather in motion upon the immediacy of cycling, in its mix of moralizing maxims, classical quotations, and local episodes (Dellanoce 19–26).[25]

Bicycles in poetry, by contrast, represented some distinguishing traits of a modern man: his lightness (associated with the images of flight and descent), his rejuvenated body (in relation to the themes of youth, spring, and rebirth), and his novel contact with nature (in connection with the ecstasy of love, the inebriation of scents, or idyllic visions).[26] The first bicycle poet is considered to be Luigi Graziani from Romagna, who wrote two hymns in Latin hexameters: "In bicycula" ("On a Bicycle," 1899), celebrating the progress from the first velocipede forged by Vulcan to modern cycling races; and "In re cyclistica Satan" ("Satan in Cycling," 1902), condemning cycles as objects of perdition, in the Faustian figure of a priest selling his soul to a *bicycula*. The first accomplished poetic synthesis of cycling imagery, however, was offered in the "Canto dei ciclisti" ("Song of Cyclists") by the poet Vittorio Betteloni, which was published on 10 June 1900 by *Illustrazione italiana*. In

Betteloni's "Canto," bicycles relate to a precipitous flight across space ("rapidly / precipitating in flight / we devour space"; verses 1–3), a palpitating inebriation ("impetuously palpitates / the heart filled with elation"; verses 5–6), a new possession of nature ("to run, run, run / the endless plains / to climb mountains and descend / immersed in divine sun / to admire a hundred different / spectacles of nature"; verses 17–22), and a powerful rejuvenated body ("muscles turn into steel / in the bold exercise"; verses 25–6).[27] The same imagery also informs the poetry of Lorenzo Stecchetti (the alias of the politician and journalist Olindo Guerrini, who also disguised himself as a female poet under the pseudonym of Argia Sbolenfi), who, in his *Rime* (*Rhymes,* 1902), portrays cycling as an idyll of rebirth, a blossoming spring, and a vital rejuvenation of the self ("I run across the solitary land [...] and I live and I fly!").[28]

As bicycles increasingly appeared in literary representations and acquired a cultural status, some intellectuals clamoured for their broader recognition – like Oriani, who challenged Italian intellectuals in *La bicicletta* (1902), "Who will be the Italian poet who will soon write the ode to bicycles?" (43; Come si chiamerà dunque il poeta italiano che fra non molto scriverà l'ode alla bicicletta?).[29] Despite the social success of bicycles, the Italian literary establishment had received them with indifference (even disgust), and the idea associated with them (of cycling, as a new form of literature in motion) with suspicion and shame. Some writers had discarded bicycles as too popular and eccentric; the journalist Matilde Serao disavowed them ("Every time I see a man squatting on this fleeting machine on the city streets, I can't refrain from fear and disgust"; *La bicicletta,* 12–13 May 1894), and the poet Giosuè Carducci defined cyclists as "knife grinders gone mad" (*La bicicletta,* 23 October 1894).[30] Others, who gave in to the allure of cycling, nonetheless opposed the cycling genre, either discarding it as an ephemeral trend or marginalizing it as "para-literature." Against this widespread intellectual reluctance to turn bicycles (objects of fascination and shame) into literary themes, two poems of the early 1900s by Giovanni Pascoli and Guido Gozzano stand out as counter-cultural representations, refashioning the marginalized repertoire of cycling poetry into suitable material for reflecting on a new idea of modern culture, intended as *in* and *of* motion (Bosi Maramotti 122–4).

In the poem "La bicicletta" ("The Bicycle"; in *Canti di Castelvecchio,* 1903) Pascoli offers the impressionistic sketch of a rapid "encounter" in rural Romagna between a still person (the poet) and a moving person (a cyclist). The sudden passage of the bicycle ("one instant [...] one pulse [...] one heartbeat"; verses 3, 7, 11) breaks the pastoral silence (as indicated by its onomatopoeic "dlin ... dlin ..." repeated at the end of each of the three stanzas) and elicits a radical and dramatic question from the poet: "my land, my transient road / it is you moving or I?" (verses 27–8). Pascoli's question, which underlines the juxtaposition of the one's fleetingness with the other's rootedness, enforces the symbolism of the bicycle as the watershed between the rapidity of a new world – flying with "spread wings" (verse 31) in the "elation of day" (verse 32; *Tutte le poesie* 584–5) – and the slowness of an old world, suspended, before night, in the nostalgia for an everlasting nest (*nido*).[31] Such imagery ultimately provides a variation on a broader theme in contemporary Italian culture – namely, the investigation of the relationship between moving technological progress and the immutable eternity of cultural or rural traditions – which in poetry (Carducci) and painting (De Nittis) is most commonly referred to with the symbol of the train.[32]

Building on Pascoli's imagery, Gozzano similarly located bicycles on the edge of quickness and slowness (or modernity and tradition) in the poem "Le due strade" ("The Two Roads," 1907; *Opere* 150–5), further enriching this contrast with a correlation between cycling and femininity. The text describes the poet's slow walk on a "beautiful alpine road" (verse 2) in the company of a married woman (Grazia), which is interrupted by the sudden appearance of a young female cyclist (Graziella). As Graziella stops and walks with them, the poet offers to carry her bicycle up to the top of the hill, mentally comparing along the way the seductive power of her adolescence and the other woman's older age. While projecting his erotic imaginings onto the bicycle, he associates it with both regeneration – in his own "ascent," "on fire" with sensual desire for the girl (underlined by the paronomasia of *ascesa* and *accesa*) – and decay, in the awareness of her imminent descent into nothingness ("you'll descend to Nothingness [...] these things I was thinking, while guiding the *ascent* of the bicycle, *on fire* with a grand bouquet of roses"; verses 57, 61–2, emphasis mine).[33] Having reached the top of the climb, the girl then throws herself into precipitous flight down the hill, causing the poet to ponder: "vola – dove? – la bicicletta ..." (verse 89; the bicycle flies ... where?). In addition to staging the split between a still world and a rapidly accelerating civilization, in the contrast (already envisioned by Pascoli) of an immobile spectator and a moving cyclist, Gozzano creates a deliberate sexual tension (of temptation and shame) towards the bicycle (and Graziella), which, combined with his incidental, almost embarrassed "Where?," points to a deeper intellectual anxiety with regard to cycling. Gozzano's tie between bicycles and femininity (seen as alluring and destabilizing) is, of course, a recurring element in the genre, previously outlined by Oriani ("women and bicycles look alike"; 60) in his assertion of their parallel seduction ("bicycles are more seductive than women; their speed becomes a caress impossible to resist"; 58) and volubility ("the point of a pin is enough to deflate the armour of a bicycle, and a word to empty the heart of a woman"; 60).[34] Yet, against the back-drop of the positivists' misogyny and the contemporary gender transformation of the noun *bicicletto* into the feminine *bicicletta*,[35] Gozzano's fear and fascination of bicycles also relates to the hidden repertoire of pornography, recurringly associated with cycling, both in the visual arts (photography and advertising) and in poetry, as documented in the work of the crepuscular poet Corrado Govoni[36] or in Lorenzo Stecchetti's poetic camouflage as the female poet Argia Sbolenfi (allowing him to contain the outburst of male pornographic desire through the use of a feminine voice).[37]

In light of these observations Gozzano's gendering of bicycles clearly indicates his intellectual attitude towards cycling, split between the recognition of an irresistible allure and the anxious desire to contain its chaotic power. Two contemporary short stories – "Mondo di carta" ("Paper World," 1909) by Luigi Pirandello and "La tentazione della bicicletta ("The Temptation of the Bicycle," 1906) by Edmondo De Amicis – reflect such intellectual fluctuations of fascination and fear towards cycling by explicitly attempting to marginalize or tame the irrational attraction coming, in the one case, from the culture of movement (brought forth by tourism or the industrial market) and, in the other, from the culture of the body (brought forth by sports).

Along the lines of Pascoli and Gozzano, Pirandello represents cycling in "Mondo di carta" (first published in *Corriere della sera* on 4 October 1909) with reference to the gap opened by the industrial age between two concepts of knowledge. In the staged contrast between Valeriano Balicci (old, blind, immobile, and living in a world of paper) and Tilde Pagliocchini, a young girl always in motion who "used to fly, run, run, by train, cars, on the railways, *by bicycle*, on steamers!" (486, emphasis mine), Pirandello indeed depicts the split between a culture intended as a still, a-temporal erudition and its new form as a moving, worldly endeavour. On the one hand, Balicci – who hires Tilde to shelve, read, and bring his own books back to life – is the symbol of a static intellectualism and a bibliolatrous conception of culture, entrenched in immobility and in the non-acceptance of a changing reality. On the other hand, Tilde is a moving traveller, a tourist, and a cyclist, stirred by "perpetually restless perplexity" (484), bored by the constraints of the old knowledge ("she felt suffocated in that world of paper"; 486), and daring to challenge it with her different, first-hand experience of the world (e.g., in rebutting a travel book about Norway, which lacked any real information: "I've been there, ya know? And I can tell you it's not like what it says here!"; 486). Balicci eventually dismisses her, supposedly because of her poor reading ("I don't want be offensive, but you colour everything differently, you know? And I need nothing to be altered; everything to remain as it is"; 484). In rejecting her impudence and her first-hand knowledge ("I don't give a damn that you've been there! It's like the book says, and that's it! It must be like that!"; 486), he denies any compromise with the real world, ultimately reaffirming the need for a protected and self-contained space ("His own world. His own paper world. His whole world") where "nothing had to be touched" (486).[38]

De Amicis' short story "La tentazione della bicicletta" (perhaps influencing Gozzano's "Le due strade") presents a similar tension in cycling, opposing the author's sexual desire towards bicycles ("every time I could thoroughly examine, unseen, a bicycle leaning against a wall, I felt forced, as if attracted by a forbidden fruit, to grasp it, caress it, and set it upright and in motion"; 76–7)[39] with his moral effort (driven by his socialist ideology) to resist their empty allure. De Amicis takes Pascoli's or Gozzano's opposition to the cyclists' exciting rapidity and his own "bitter sense of envy and regret" (84) as a slow pedestrian and explicitly refers to it as not just a contrast between two ways of writing – the one in motion[40] and the other in stillness – but also a splitting into two concepts of intellectual activity, either a bodily form of thinking or an a-corporeal experience. Cycling (as a metaphor for a new way of bodily writing in movement) is thus equated to an ambivalent feminine temptation, which the author, on the one hand, rejects as a destabilizing fascination (or as a form of moral degeneration) and, on the other, embraces in secret in the protected narrative area of a forbidden dream ("I became a cyclist on my pillow. During my sleep [...] I had the full and live illusion of the feel of a race. Ah, at last! Did you need this much time to decide!"; 83).[41] Along the lines of the author's previous story "Amore e Ginnastica" ("Love and Gymnastics," 1891), paralleling the protagonist's discovery of sports to his fatal infatuation with his gymnastics teacher, "La tentazione della bicicletta" ultimately stages a similar space of fictional negotiation and containment of the new culture of the body, vis-à-vis the contemporary evolution of sports, the Olympic Games, and cycling races.

Italian Sports and the Aesthetics of Industrial Competition

The establishment of the Giro d'Italia in 1909 marked the culmination of the process (set in motion by the foundation of *La gazzetta dello sport* in 1896) that transformed *velocipedismo* from a racing practice into a structured sports culture. In the early 1900s *La gazzetta dello sport* had been promoting cycling competitions that took place over one day or multiple days (like the Milan-Sanremo in 1907 or the Giro di Lombardia since 1905), in conjunction with partnering industries, as a way to spur technical innovation, expand the market, and advertise brands (via the athletes' bodies and feats; Thompson 16). With the creation and success of Giro d'Italia, *La gazzetta* redefined bicycle races in a mythopoeic space of collective imagination (Bassetti 23), shifting cycling from its previous association with bourgeois entertainment to its new self-construction as a national sport, offering both a base structure for capitalism (in the joint collaboration between media and the industrial bourgeoisie) and a powerful nation-making tool.

The massive coverage of the first giro documents the race's distinctive mingling of competition, spectacle, and business, as well as the effort to aestheticize the country's industrial take-off during the Giolittian era through the creation of an independent sports culture. The competition (divided into eight stages of three hundred to four hundred kilometres each) touched upon the great Italian cities (Milan, Padua, Bologna, Chieti, Naples, Rome, Florence, Genoa, Turin) and was extensively covered by *La gazzetta dello sport* (which had organized it in collaboration with the TCI), *Corriere della sera*, and *Illustrazione italiana*, from its start (at 2:53 a.m. on 13 May 1909) to its arrival in Milan on 30 May. *La gazzetta* covered the giro as an epic narration ("the great battle among the giants of the road," "an epic spectacle," "a titanic struggle"; 12, 24, and 26 May),[42] with a general attitude of self-praise and with a tone balancing traditional literary echoes with modern sports words borrowed from English and French (e.g., *équipe, outsider, forfait, défaillance, sprinter, exploit*). *Corriere* followed the competition in a scientific and descriptive way by offering (like the TCI guidebooks) maps, altimetries, and information related to each town or city visited by the cyclists and by listing all of the detailed facts and obstacles of the race (herds of animals, dust on the street, heat). *Illustrazione,* in the reporting of Mario Morasso, covered the giro as "the most marvellous of journeys" (23 May) ever taken across the country ("the greatest cycling feat not only in Italy but in Europe"; 16 May),[43] as the tale of a whole nation, and as a powerful force joining racers and spectators in a common regeneration.

While portraying the "extraordinary apotheosis of the bicycle" ("an innumerable, endless, majestic triumph of the small machine, of the humble steel device"; *Illustrazione italiana,* 6 June),[44] the media coverage of this competition and trip around Italy shaped the giro into a powerful narrative incorporating the individual into the national body, and as a vital moving force for a static country in search of a dynamic modernity. When one observes the photographic and pictorial images of the race in *Illustrazione italiana* (fig. 4.2), the recurring elements of the street, the crowds, the group of cyclists, and the hero Luigi Ganna visually suggest, in parallel with *cicloturismo,* the idea of a national body set in lively motion by bicycles.

The Giro d'Italia, in fact, stages the exploration of the country's body, both as a geographic space (through the measurements of its territory, the creation of a mental map of its streets, and the traversing of the areas between cities) and as a living social entity (through the emotional connection established between the corporeity of the racers and that of the masses). By invading urban spaces and participating in racers' feats, the crowds thus become part of Italy's wandering self-discovery, as unanimously stressed in *La gazzetta* ("in that supremely beautiful spectacle of thousands and thousands of people united in one common sentiment, the quivering soul of our generous people vibrated"; 1 June); in the *Corriere* ("crowds were everywhere [...] the road swarmed with cyclists in an interminable procession. [...] An avalanche of men and machines invaded the road"; 31 May); and in *Illustrazione italiana* ("One could see in Bologna, Chieti, Naples, Rome, entire cities frenetic at the arrivals [...] flow like torrents to the finish line to hold the winner tight, to scream *bravo* or *evviva* in his ear"; 23 May).[45] As crowds gathered around the cyclists' bodies in streets across the Italian territory, the giro created the collective awareness of a national community, born out of the synthesis of romantic idealism – in the "struggle against space" (Augé 30) or the titanic challenge "to dominate things" (Barthes, *What Is Sport?* 41) – and Darwinian positivism – in the selection brought about by the race (only eight cyclists out of eighty-one arrived at the end, as its "survivor heroes"; *La gazzetta dello sport,* 1 June). The winner, standing out as the fittest survivor and as a male romantic hero (in a moment in which the very idea of masculinity was in crisis), catalysed and intersected in his symbolic body, both admirable and measurable, the dreams of greatness of the bourgeoisie and the aspirations of freedom of the lower classes (Cardoza 357). For the industrial entrepreneurs, the winning cyclist embodied not just the ideal of extreme competition[46] as a driving force for progress but also a new dimension of corporeity,

L'ILLUSTRAZIONE ITALIANA

Anno XXXVI. - N. 22. - 30 Maggio 1909.

Centesimi 75 il Numero (Estero, Cent. 95).

Per tutti gli articoli e i disegni è riservata la proprietà artistica e letteraria, secondo le leggi e i trattati internazionali.

Published in Milan, May 30th, 1909. Privilege of copyright in the United States reserved under the Act approved March 3rd, 1905, by Fratelli Treves.

IL GIRO D'ITALIA IN BICICLETTA.

Ganna dopo l'arrivo a Roma.

L'arrivo di Ganna al traguardo di Roma. (Ganna è in testa; 2.° è Oriani) (fot. D. Paolocci).

Figure 4.2 The winner of the Giro d'Italia, Luigi Ganna, framed above the group and the crowd. *Illustrazione italiana,* 30 May 1909, front cover. BiASA.

promoted in relation to health and preventive medicine and objectified as a moving advertisement for brands of bicycles and pneumatics. For the workers, the body of Luigi Ganna, the proletarian hero from Varese, symbolized, in its union with the bicycle's "democratic mechanism" (Morasso, *Illustrazione Italiana*, 6 June 1909), the classless nature of records and the struggle of the lower classes to emerge in a more democratic society.

As described, the Giro d'Italia excitingly provided Italians with a new awareness of their country's geography and with a sense of collective participation in a truly national experience; however, in its first version and in the years to come, it also put forth (as did the TCI maps and guides) an ambivalent image of the nation, suspended (like the industrial bourgeoisie) between the feverish exaltation of a great new modernity and the hidden anxiety over an existing and perceived backwardness. As reflected in the image of a truncated Italy (south-less), in the map tracing the itinerary of the 1909 giro (fig. 4.3; *Illustrazione italiana*, 23 May 1909, p. 531), cycling not only projected an idealized country, conceived as an extension of the industrialized north over the south, but also staged the "older bourgeois anxieties and insecurities about the national community" (Cardoza 363). The exclusion of the islands and the southern regions from the program, due to their lack of roads and infrastructures, transformed the giro into a double-edged event. The race was indeed charged with an immense power of regeneration, yet it was also pervaded with ashamed anxiety over the acknowledgment of the South's backwardness, making visible, with the passage of the racers, the real country, which was still trapped in immobility. Over the subsequent years, up until the war, this anxiety of the bourgeoisie towards the country's unsuccessful image transformed into an explicit feeling of embarrassment at the incapacity of Italian cyclists to win any competitions outside of Italy.

In spite of these difficulties, the success of the 1909 Giro d'Italia had a great impact on the creation of a new narrative of sports and fashioned a new aesthetics of industrial competition. Along with rising interest in the Olympic Games, following Dorando Pietri's victory and disqualification at the marathon of the 1908 London Olympic Games (celebrated by Arthur Conan Doyle),[47] the Giro d'Italia captured the attention of the masses and profoundly contributed to a reversal of the country's widespread indifference and general resistance towards sports in the late-nineteenth century. At that time, in fact, Italians considered organized sports to be an imported British fashion or a form of

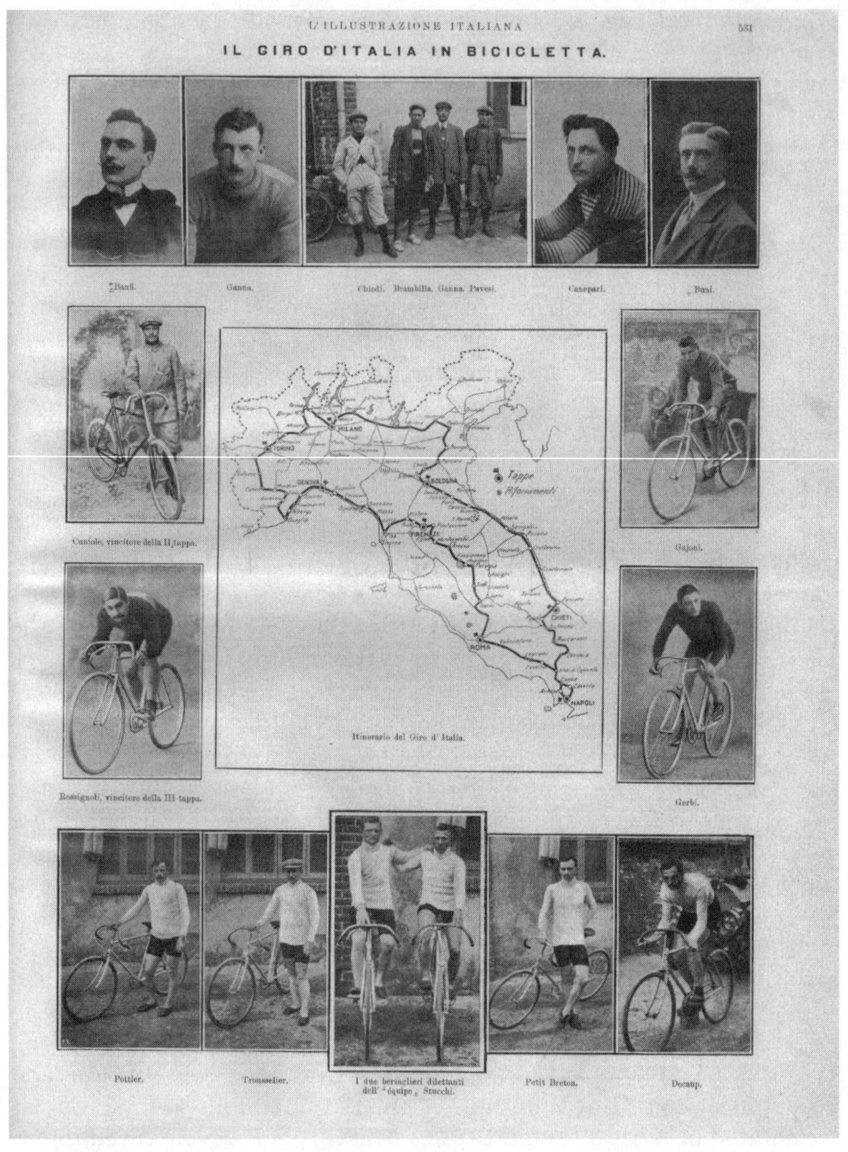

L'ILLUSTRAZIONE ITALIANA 531

IL GIRO D'ITALIA IN BICICLETTA.

Banfi. Ganna. Chiodi. Brambilla. Ganna. Pavesi. Canepari. Bani.

Cuniolo, vincitore della II tappa.

Itinerario del Giro d'Italia.

Gajoni.

Rossignoli, vincitore della III tappa.

Gerbi.

Pottier. Trousselier. I due bersaglieri dilettanti dell' "équipe" Stucchi. Petit Breton. Decaup.

Figure 4.3 Map of the 1909 Giro d'Italia. *Illustrazione italiana,* 23 May 1909. BiASA.

bourgeois amusement, as seen in Gabriele D'Annunzio's rubric of current trends "Sport e altro" in *La tribuna* (in which Il Duca minimo wrote about hunting, dog and horse races, and fencing from 1881 to 1891) and as later confirmed by the scarce press coverage of the Olympic Games prior to 1908. Morever, Italians viewed sports with some reluctance, as either the overt manifestation of global industrialism or the expression of a new spreading culture of the body.

In a way, modern sports were seen as by-products of British industrialism. Starting with the invention of rugby and soccer, they had been conceived as an organized time of leisure in contrast to work and as an educational method in schools, summed up by the motto "Mens sana in corpore sano" (Ickringill 36–42; healthy mind in healthy body). As they had been promoted by the Victorian ideology of athleticism, sports represented a way to aestheticize the values of capitalism (competitiveness, teamwork, spirit of improvement, great achievements, and record measurements; Bassetti 28–34) and forge a synthesis of the two nineteenth-century ideologies of romantic idealism and rationalistic positivism. Their link with industrialism was also made clear by the creation of the modern Olympic Games, envisioned by the anglophile Pierre de Coubertin as a platform for exporting the British model of capitalism on a global scale (Pivato, *Bicicletta* 19; Mandell 196–218), and organized not by chance, after Athens 1896, as attractions within the universal expositions of Paris 1900, Saint Louis 1904, London 1908, Stockholm 1912, and Antwerp 1920.[48]

In another way, sports also related to a new culture of the body, which had started to spread in Italy after the establishment of the National Gymnastics Federation in 1869. The passage of a law by Francesco De Sanctis (literary critic and, at the time, minister of education), imposing the mandatory introduction of physical education into Italian schools in 1878, signalled not just the interest of the young Italian state in promoting a culture of movement as a powerful force to "bind the subject in the state" (Stewart-Steinberg 141), but also the start of a fierce debate on different ways to build such a co-ordinated social body. There were two opposing models of physical education in the early 1880s: one, promoted by Emilio Baumann, was motivated by the German authoritarian or militaristic approach, and the other, championed by Angelo Mosso, promoted the Swedish competitive or democratic method. Reflecting the nation's divergent alliance strategies in these two schools, the gymnastics debate filtered a larger political divide, contrasting the authoritarian models of Prussia and Austria-Hungary to the democratic models of

England and France. In addition, as Catholics started to run their own gymnastics societies, the issue of physical education aggravated the already tense relationship between the state and the Church.[49]

In this context and paralleling the expansion of racing culture around *La gazzetta dello sport* in the early 1900s, two intellectuals – Gabriele D'Annunzio and Mario Morasso – first captured the ties between sports and modernity. D'Annunzio embodied the link between the sports ideology, fashion, and the new culture of performance both in his own persona (as a swimmer, a cyclist, a driver, and later an aviator) and in his fiction (for example, in the car-racing scene or in the flying contest for the altitude record in the later novel *Forse che sì forse che no*). His athleticism and the cult of his own body (as seen in his photographic nude portraits) fashioned him not just as "the prototype of the sportsman" (Vergani, *Piacere* 14) but also as an idealized and inimitable new man, modelled on Nietzsche's superhuman and conceived as a creature in a constant state of self-overcoming. Along with D'Annunzio, the philosopher Mario Morasso (who would report on the first Giro d'Italia for *Illustrazione italiana*)[50] first theorized the connection between the sports mindset and industrial civilization, in his book *La nuova arma: La macchina* (*The New Weapon: Machines,* 1905). Before the Futurists, Morasso saw in sports a metaphor of modern man's "hallucinated thirst for omnipotence" (94), and in racing, the synthesis of a new culture, of corporeity and movement, shaped around the ideal of "running in order to run ever faster" (*La nuova arma* 4).[51] In listing bicycles among the key machines of the industrial age, he glorified them as tools allowing modern men and women the capacity to transport themselves, move beyond the limits imposed by nature (time and space) or society (in reference to female emancipation), explore reality *in loco,* and know the world through first-hand experience.[52] In parallel with other machines, bicycles ultimately were seen as the forging sites of new human beings, characterized by an unprecedented power, and proudly described as hybrid creatures "of an unknown species, centaurs of flesh and iron, wheels and muscles" (54).[53]

A visual synthesis of D'Annunzio's and Morasso's early imagery and ideology of sports is offered by the illustrator Plinio Codognato in the 1910 poster advertisement for Cicli Fiat (plate 8). The advertisement revolves around a naked male figure (seen from behind), caught in the act of tightly tying wings to his bicycle's pedals. The cyclist's body is the centre of multiple staged contrasts: of man and machine, nudity and technology, light and darkness, and muscular strength and winged lightness. At the same time, the cyclist's body enacts a space of imaginative connections, as offering a metaphor for the mechanical industry (in the joint branding of Pirelli tyres and a Fiat bicycle) and the vision of a new, hybrid creature, in a state of motion and tension, overcoming the human limit in his symbolic tie (or knot) of iron and flesh.[54]

Culture as Body, Culture in Motion

As the Giro d'Italia captured the attention of an ever-growing number of Italians from all classes, bicycles turned into hybrid objects, entailing a dual commercial and cultural dimension. As intellectuals charged them with new symbolic meanings, they also became the site of a new aesthetic reflection on culture itself in its relation with corporeity and movement.

In 1913, perhaps in conjunction with a famous Pirelli advertisement designed by Ballie for *La rivista del T.C.I.* (Lopez 71), representing a cyclist dynamically breaking through a dark space followed by a trailing group, the Italian poet Dino Campana depicted the Giro d'Italia through a similar image of one excited cyclist, energizing the other racers and the crowds with his frenetic energy. In his poem "Giro d'Italia in bicicletta (1° arrivato al traguardo di Marradi)" ("Giro d'Italia by Bicycle: The First to Arrive at the Marradi Finish Line")[55] Campana relates bicycles to a dazzling state of motion, symbolized both in the cyclist's body and in the excited corporeity of the cyclists' group, fused into one by the vital energy and moving lightness of their vertiginous flight and descent on the precipice of an imminent catastrophe. The "whirl" (verse 2) of the cyclists' chaotic force reflects the vitalism and agitation of modern life itself, as also signalled by the phonic association of the word *turbine* (whirl) with the ambivalent *turba* or *turbe* (meaning both crowds and disquiet). In one sense, this whirling relates to the explosive mass of the group and the crowd, "loosening and tying" (verse 12) in a common magnetism, and, in another sense, to a new inebriating and primordial life ("vita primeva"; verses 10–11), clearly identified in the final repetition ("Dionisos Dionisos Dionisos"; verse 13) with Nietzsche's Dionysian spirit.

In the same year the Futurist painter Umberto Boccioni powerfully depicted the propelling force of industrial modernity in the canvas *Dinamismo di un ciclista* (*Dynamism of a Cyclist*; plate 9) through the image of a cyclist excitedly pedalling and dynamically breaking through the pictorial space. Along the lines of Campana's poem, Boccioni's painting portrayed the modern world as a whirlwind (enacting the energetic fusion of a new man and machine and the accelerated speed of a new time and space in motion). Along the lines of Morasso's thought, Boccioni pictured a new emblematic figure of modernity in the cyclist's excited and metamorphic body.

As Boccioni moved from cover advertisements to abstract language, the *Dinamismo* visualized his radical shift from the earlier depiction of the bicycle as an actual object with contingent qualities to its symbolic representation in reference to the cyclist's body and movement. Under the influence of the avant-garde (and the general excitement for the Giro d'Italia) he emphasized the cyclist's body as a site of explosive muscular tension (in the calves, thighs, and arms) and an experimental space for the chaotic blending of forces and perspectives (corporeity versus the surrounding space, human versus machine). The cyclist's body then became a platform on which to experiment with the technique of compenetration of levels and with the theoretical quest for a plastic dynamism where "moving objects constantly multiply themselves, change shape, succeeding one another, like rapid vibrations, in the space which they traverse" (quoted in Rainey 64; "Futurist Painting: Technical Manifesto"). In addition, Boccioni focused his attention on the bicycle's motion, as emphasized in the cyclist's wedging into the reassuring bidimensionality of the pictorial space (typical of Futurist symbology) and the highlighted spinning of the wheels (in the rapid brush strokes of the spokes). In mixing power and lightness, enthusiasm and anxiety, Boccioni's canvas thus reconfigured the bicycle-cyclist iconography into an experimental site of exploration (of the abstract concept of *dinamismo* as a key factor of industrial modernity) and investigation (of the subtle edge where the vitality and chaos of this movement coincide or collide). During its transformation from a commercial item into an aesthetic icon, Boccioni's Futurist bicycle ultimately reveals the image of an idealized yet somewhat troubling modernity, as evidenced first by its chosen object of representation and second by the gradual evolution of his representational style.

As of 1913, at the cusp of the commercial success of the bicycle, yet also at the beginning of its decay, Boccioni's choice of this machine as a pictorial subject conveys a double-edged message. In a way, bicycles represented such Futurist ideals as the celebration of movement and ephemerality, the success of the machine, and the inclusion of everyday items or contingent objects into art. In another way, however, bicycles also represented passéist objects, surpassed in speed by the fully mechanized cars and aeroplanes, harking back to a still fundamentally human-centred vision of the machine, and somewhat hampering the pursuit of speed, as emblematized in the narration of the founding "Manifesto of Futurism" in which a car-driving Marinetti veers off the road to avoid two cyclists.

In a similar way, Boccioni's stylistic evolution from the advertisements for *La rivista del T.C.I.* to the *Dynamism of a Cyclist* signals an uncertain attitude towards bicycles, as evident in the varied picturing of the actual object and its surroundings. As for the object, Boccioni distanced himself from the somewhat static style of his previous cycling images, adopting instead a more dynamic view – perhaps under the influence of Luca Comerio's pioneering cinematographic documentary *Secondo giro ciclistico d'Italia* (*Second Cycling Tour of Italy,* 1910).[56] By embracing the racing aspect of cycling (foreseen in the Frera advertisement of April 1908), Boccioni not only privileged the connection of bicycles to dynamism and speed over their concomitant relation to elegance and fashion (outlined in the Bianchi advertisement of March 1908) but also shifted his focus from a destabilizing "feminine" association (now marginalized and contained) to a more "masculine" imagery, visible in the added emphasis on the muscular body (along the lines of Codognato's 1910 poster for Cicli Fiat). As for the surroundings, Boccioni's *Dinamismo* certainly related the bicycle to racing – in parallel with the contemporary collage on canvas by Jean Metzinger, *Au Vélodrome* (*At the Cycle-Race Track,* 1912), which staged the Cubo-Futurist vision of a cyclist and the crowd at the velodrome – yet also moved away from the depiction of an actual competition by deliberately omitting both the spectators and the track. As anticipated in the preparatory sketch for the canvas (published in *Lacerba* on 15 October 1913 under the title "Dinamica di un ciclista"), in which the image of the moving cyclist stands alone in an uncontextualized blank space, Boccioni referred the bicycle to an abstract or mental race, seen as an idealized symbol of modernity, along the lines of his 1914 theoretical manifesto "Contro il paesaggio e la vecchia estetica": "Tracks, athletic contests, races exalt us! The finish line is for us the marvellous symbol of modernity" ("Against Landscape and the Old Aesthetics"; quoted in Salaris 106).[57] In its imagined race and racing, the canvas thus turns into an ambiguous site, exploring a "unique form of continuity in space" (in the words titling Boccioni's contemporary sculpture of 1913), yet also containing, in the secluded space of cerebral fiction, the deforming and disquieting loss of contours produced by the interpenetration of mechanical technology and human corporeity.

While reinventing the cycling iconography and latently exposing the inherent contradictions of the Futurist modernity through the bicycle,

Boccioni's *Dynamism of a Cyclist* also points to a broader philosophical context, opening up a space of reflection: first, on the re-evaluated meaning of the body and, second, on the relationship between culture and the new civilization of corporeity and movement brought forth by tourism and sports.

First, Boccioni's construction of the cyclist's body as a hybrid of vitalism and anxiety – as a "corps voyant-visible" (Buci-Glucksmann 71; seeing-visible body) reconfiguring perspectives in motion across the surrounding space, and a "corps pulsionnel" (Buci-Glucksmann 99; impulse-driven body) inflamed by an irrational drive towards metamorphosis – revealed an intellectual attitude of acceptance and disquiet with regard to the rediscovered cultural value of corporeity in the wake of Darwin's theories and Nietzsche's philosophy. On the one hand, Darwin's notion that "man is the modified descendant of some pre-existing form" (*Descent of Man* 5) and that "man is variable in body and mind" (53) had transformed the human body into a revealing site of human evolution (in the past and the future) and culturally constructed it as a metaphor of corruptibility and/or progress in its state of ongoing change. On the other hand, Nietzsche's critique of the Cartesian dualism of body and soul (which turned the body into a machine; Ulmann 14) and the Christian purification of Eros (which allegedly condemned corporeity to being inferior to the soul), had led Nietzsche (in the speech of his dancing prophet Zarathustra, "On the Despisers of the Body") to locate in the body the place of man's "great reason" (*Thus Spoke Zarathustra* 30), referring to it as the true self (the Dionysian), and the essence of modern man (Judovitz 170).[58]

Second, the excited dynamism and metamorphic motion of Boccioni's cyclist visually expressed a new concept of culture as transitory (along Darwin's idea of evolution) and corporeal (along Nietzsche's idea of embodiment). The cyclist, in fact, seemed to personify Nietzsche's association of the body with reason ("There is more reason in your body than in your finest wisdom"; *Thus Spoke Zarathustra* 30) and with a new culture in and of action ("It does not say I, but does I"; 30), revealed in Zarathustra's thinking while dancing, and further explained in the *Twilight of the Idols* ("'On ne peut penser et écrire qu'assis' [G. Flaubert]. Now I have you, nihilist! Assiduity is the sin against the Holy Spirit. Only ideas won by walking have any value"; *Twilight* 26).[59] Along with the new wandering culture of tourism and sport and in parallel with the diffusion of Nietzsche's thought in Italy, Boccioni's *Dinamismo* manifested the problematic shift from an old elitist conception of culture, immobile in its classicism, to a new idea (brought forth by the industrial revolution) of culture *in* and *as* movement, intimately embodied in the flow of contingent events. While offering the model of a new hybrid being in the cyclist, in his fusion with the machine and in his accelerated speed, Boccioni portrayed in him the experimental image of a new intellectual, able to "write while pedalling" (scrivere pedalando)[60] and to combine the eternity of art with the moving immediacy of modernity in a new, unexplored dimension of culture itself.

Boccioni's *Dinamismo* would have a lasting influence on the later iconography of bicycles. While foreshadowing the coincidence of energetic thrill (in the Futurist excitement for war, as many enrolled in the cyclist battalion) and reckless disquiet (as a mortal omen of Boccioni's own death in combat in 1916), bicycles would identify, in reference to the chaotic movement of modernity, a similar juxtaposition of elation and vertigo. On the one hand, as in Gerardo Dottori's *Ciclista* (*Cyclist,* 1914), Enrico Prampolini's *Dinamismo di un ciclista* (*Dynamism of a Cyclist,* 1921), or Fortunato Depero's *Ciclisti* (*Cyclists,* 1922), bicycles represented the whirl of modernity and the speed and lightness of a new world and symbolized collective regeneration after the darkness of the First World War. On the other hand, as in Mario Sironi's two paintings entitled *Il ciclista* (*The Cyclist*; 1916 and 1920), bicycles conveyed a gloomy representation of modernity, identified, in the 1916 canvas, with the image of a cyclist's uncontrolled descent (no longer a racer but a worker) against the dreadful background of an emptied city, and, in the 1920 canvas, with the image of a half-covered racer (seen from behind), painfully ascending (after the experience of the First World War) a steep hill.

With the advent of Fascism, bicycles definitively lost their cultural appeal and their aura of modernity. There are two main reasons for this diminishment. First, at the moment in which they spread to an ever-greater part of Italian society[61] and became familiar and popular, their capacity to signify a bourgeois distinction rapidly vanished. Second, with the ascent of Mussolini, bicycles came to evoke "an outmoded past, particularly in the eyes of the Fascists, who preferred to identify the regime with the new world of racecars, ships and aeroplanes" (Cardoza 369). Reluctant to support the Giro d'Italia since his anti-cycling campaign of 1912, the Duce discarded cycling because it displayed the country's underdeveloped areas and internal divisions (in the regionalism of its champions). He decided instead to invest in soccer, which, spreading in the same period as cycling (with the founding of Juventus in 1897, A.C. Milan in 1899, Lazio in 1900, and Inter in 1908 and the establishment of the national team in 1910), became the national sport par excellence only under Fascism, because of the sport's exaltation of teamwork and above all its capacity to bring victories (like the ones in the 1934 and 1938 World Cups) to a country searching for a positive affirmation of itself among other nations.

Chapter Five

Gramophones, Radio, and the New Languages of Sound

On 11 April 1902 in the improvised recording studio of a reserved room on the third floor of Milan's Grand Hotel, the young tenor Enrico Caruso, accompanied by the pianist Salvatore Cottone, sang ten arias in front of a gramophone.[1] A few weeks before, the American entrepreneur Fred Gaisberg, on his way to Rome to record the voice of Pope Leo XIII, had heard Caruso at La Scala during his second performance of Franchetti's opera *Germania* (which had premiered on 11 March under the baton of Arturo Toscanini). Enthused by his voice, the business agent and recording engineer for the British Gramophone & Typewriter Co. changed his plans and organized a recording session with the singer, at his own expense and without his company's permission. Despite the Neapolitan tenor's reluctance, shared by many of his colleagues, to sing "in the tube" (as in his reported words "nun me piace cantá dint' 'o tubo'"; quoted in Gargano 54), Caruso accepted the challenge and performed for an hour in front of the talking machine. The recording, which immortalized the unexpected encounter of Gaisberg's pioneering vision and Caruso's artistic audacity, marked an early point of contact between the world of operatic performance and the new gramophone technology and, at the same time, was an important milestone in music history. Sold at 12.50 lire and branded with the RCA Victor Red Seal label (first featuring the Nipper logo of "His Master's Voice"), Caruso's 1902 recording became a worldwide sensation (Steffen 169–71) and opened the way both to the birth of the flat-disc industry (in its legitimation of gramophones as instruments for private performance) and to the Italian tenor's American success (Caruso made his debut in Verdi's *Rigoletto* at the Metropolitan Opera House in New York on 23 November 1903).[2]

In a parallel event, on 19 January 1903, the Italian inventor Guglielmo Marconi completed the first two-way transatlantic wireless communication between America and Europe, successfully transmitting and receiving messages between the U.S. president Theodore Roosevelt and King Edward VII of England, from the Cape Cod radio station of South Wellfleet to the Cornwall station of Poldhu.[3] The episode marked a milestone in the development of the newly invented wireless telegraphy, launching it on a global scale. Following Herzt's discovery of invisible waves, Marconi had been experimenting with radio transmission since 1895 when he first sent a signal over a hill in his estate at Villa Grifone near Bologna. Ignored by the Italian government, his new application found immediate backing in England as well as the endorsement of the British Postal Service. In 1897, after patenting his technology, Marconi established the Wireless Telegraph and Signal Company Ltd. and opened the first wireless station on the Isle of Wright and the first wireless plant in Chelmsford the following year. Meanwhile, the Italian inventor had also invested in extending the range of radio communications by first sending a message across the British Channel (1899) and then wiring the letter *S* across the Atlantic from Cornwall to St John's, Canada (1901).[4] As British royal and merchant ships began to be equipped with radio sets operated by trained personnel (*marconisti* or "Marconi men"), and even armies adopted wireless telegraphy (starting with the Russian-Japanese conflict of 1904), this Italian-born technology grew into a profitable global business. By virtue of its successes – the first transatlantic transmission in 1903, and the later impact of radio in naval rescue operations – Marconi's Wireless Telegraph Company (the new name starting in 1900) gradually replaced cable telegraphy, inaugurating and monopolizing a new era of mass communications.[5]

As these facts illustrate, Italy played a leading role in the global growth of sound recording and transmitting technologies, as both an importing and an exporting nation. On the one hand, the nation heavily imported foreign-made recording devices (phonographs, gramophones, and even Marconi's British radio sets). Their import coincided with Italy's larger discernment, or negotiation, of its relationship to global modernity, hanging between cultural rejection (in the intellectual rebuttal of recorded music or in the missed recognition of Marconi's invention), and experimental reinvention in dialogue with tradition (as in Caruso's bold acceptance of the "tube" or in the later development of an aesthetic radio language). On the other hand, Italy offered a rich experimental soil for global export, in the successful advance of Marconi's technology and in the worldwide diffusion of the Italian operatic repertoire via the recording industry, symbolically marked by the premiere of Giacomo Puccini's *La fanciulla del West* in New York in 1910.

In this context, the parallel evolution of sound recording and transmitting devices in the 1890s and 1900s constitutes not just a chapter in the history of technology but also a cultural crossroads, mediating old and new languages at the border between Italy and the world. Against the back-drop of the ascending and descending popularity of opera, the Italian interaction with the global sound-recording and -transmitting industry, and the nation's social and intellectual reception of recorded sound, provide an unusual perspective on the industrialization of music, as well as on the evolution of new aesthetic idioms in composition, graphic design, and radio communications.

Writing, Reproducing, and Transmitting Sound

Edison's phonograph (1877) was an unexpected invention. Preceded in 1857 by Leo Scott's phonautograph (which lacked playback capability), this new sound-recording device was born out of the casual "attempt to optimize telephony and telegraphy" (Kittler 27) as Edison was trying to achieve the goal of obtaining both a high-speed telegraph transmitter and an improved version of Bell's telephone (1876). After the first recording (on 12 August 1877)[6] and the patenting of the invention (on 19 February 1878), Edison published an article for *The North American Review* during the summer of 1878, in which he described the phonograph as a device that "will perfect the telephone, and revolutionize present systems of telegraphy" (534). In the same article he also listed its applications, in this order: dictation without a stenographer, audio books, teaching of elocution, reproduction of music, family records, music boxes, toys, clocks, advertising, language preservation, and telephonic record-keeping (Edison 530). In Edison's intentions, the phonograph was indeed a scientific instrument for the recording of evidence (speeches, people's voices, business transactions), capable of providing the field of psychology with a suitable model for visualizing the brain or memory in the traces of noises and meaningless sounds; the field of anthropology with "a unique technology for the preservation of the past" (Kern 60) or of endangered languages; the field of criminology with a way of authenticating court hearings; and the field of human sciences (poetry, family history, archives) with an new, infallible mnemotechnology.

Edison's phonograph, with its tie to telegraphy and telephony, intrinsically related to the age's broader endeavour to capture, write, and transmit sound waves and human words.[7] The appearance of *phono*graphy was consistent with its century's ambition to record sensorial experience, documented in the achieved writing of *light* (*photo*graphy) since 1839 or of *movement* (*cinemato*graphy) in 1896. The phonograph's act of etching sound waves in reproducible patterns (*typoi*) also connected it to the contemporary endeavour to convert phonemes into transmissible wired or wireless signs (Morse code), as well as to the parallel development of new systems of automatic writing (from the invention of *type*writers in 1868 to the creation of the lino*type* process in 1886). Despite these connections, however, Edison's phonograph appeared to be a totally unexpected novelty, creating an unforeseen need for sound recording. As of 1877, in fact, "no open-ended market surveys of consumer needs would have listed a device with the capabilities of the phonograph" (R. Weber 150), and no mental model existed other than biological memory or the music box. The sudden appearance and rise of recorded sound thus lay at the intersection of science, business, and, later, aesthetics. Consequently, its evolution is at once "the history of an invention, an industry, and a musical instrument" (Gelatt 11).

After the foundation of the Edison Speaking Phonograph Company in 1878, phonographs met immediate success on the American market, mostly as curiosities offering audiences the attraction of an evening's entertainment. The invention's appeal, however, did not last long; as Edison began to work on the concomitant project of the incandescent lamp (after witnessing a solar eclipse in Wyoming in October 1878), the phonograph entered into a long phase of "torpid retirement" (Gelatt 33). The product re-emerged a decade later, in 1887, when, in response to Bell and Tainter's competing invention of the graphophone (patented as a new recording technique that adopted wax instead of tin-foil), Edison counter-launched on the market an improved version of the phonograph (without changing, however, its substance and purpose).[8]

The commercial contention between Edison's Speaking Phonograph Company and Bell's Graphophone Company ended in 1888 when the entrepreneur Jesse Lippincott acquired the rights and patents of both companies and founded the North American Phonograph Company (later Columbia Records), thus establishing control over the entire talking-machine industry (Welch et al. 26–7). By mistakenly following Edison's suggestion to sell phonographs exclusively as tools for stenographic dictation rather than as musical devices, Lippincott's company rapidly failed, however, and went bankrupt in 1894. Meanwhile, a parallel market of commercial musical recordings was slowly emerging, along with the success in California of coin-operated phonographs and records for advertising purposes. It was around 1894 (when Edison started to consider phonographs as sources of amusement) that Emile Berliner's gramophone (patented in 1887 and licensed from 1893 by the Berliner United States Gramophone Company) made its appearance in the American market. Its origins "can be traced directly to the failings of the phonograph" (Smil 237): poor reproduction, a two-minute recording limit, and the absence of any duplicating method, which obliged musicians to reproduce copies (an average of 305 cylinders per day) by constantly replaying the same piece. Thanks to the introduction of a horizontally acid-etched zinc disc that was engraved by a stylus, and of an original negative matrix that allowed easy duplication, Berliner's gramophone reduced production costs and reconfigured records as entertainment commodities, implicitly shifting the product's goal from the writing of sound (*phono*graphy) to the phonic enactment of its written sign (*gramo*phone). Berliner had always envisioned the gramophone as an amusement device, selling it to the German toy-maker Kämmerer & Reinhardt of Waltershausen in 1889. Over time, he increasingly invested in music (Osborne 34–5) by hiring Fred Gaisberg (accompanist, recruiter, and pioneer of coin-operated phonographs) in 1894 and pioneering the expansion of a European market in 1898, in the established partnership with William Barry Owen's London-based Gramophone Company (called Gramophone & Typewriter Co. from 1900).[9] Berliner's company was acquired by Eldridge Johnson (who had previously added to gramophones a clockwork motor and a new system for making master records) and renamed Victor Talking Machine Company in 1901 (Jones 89–91). Thanks to the popularity of its HMV brand and logo (Jacques Barraud's bull terrier Nipper gazing at the horn of a gramophone while listening to "His Master's Voice"),[10] and thanks to Gaisberg's role – in manufacturing the 1902 Caruso series (unanimously considered the first completely satisfactory gramophone records to be made) and in recruiting for the Victor Red Seal series talents like Adelina Patti, Francesco Tamagno, and Beniamino Gigli – gramophones and the flat record system rapidly conquered the music market, replacing Edison's cylinders as early as 1910.

In tandem with the spread of recorded music, Marconi's radio-telegraphy was gradually establishing a sound transmission technology, thereby engendering a new global market from 1895. Born out of the combination of electricity and the lack of wires, Marconi's radio-telegraphy did not intersect with the expanding record industry until the 1930s but rather focused (as in Edison's initial plans for the phonograph) on its scientific and practical applications. Radio-telegraphy mainly applied to navigation and had an important impact on the rescue efforts in a few naval accidents: in 1909 the Marconi Company saved 4,000 lives after the collision of the SS *Republic* with the SS *Florida* off the shores of Nantucket, and in 1912 the radio signals sent to the RMS *Carpathia* contributed to the rescue of 712 people from the sinking *Titanic*. The recognized utility of radio-telegraphy, confirmed by the Nobel Prize in Physics that was conferred upon Marconi in 1909, rapidly turned into a monopoly project (the so-called Imperial Wireless Scheme) that was launched in 1910 as a strategic plan to establish a worldwide chain of shore- and ship-based radio stations throughout the British Empire. The project, conceived as a means of global communication for the Royal Navy, found public support in 1910 after the capture of a murderer (notice of the arrest was wired to the London police from an ocean liner on which the criminal was escaping from England to Canada), and in 1912 after the shipwreck of the *Titanic*. Marred by the discovery of the company's bribes to government officials (the so-called Marconi scandal; Raboy 366–79), the Imperial Wireless Scheme failed in 1914, yet Marconi's monopoly of radio communications remained unaltered.

During these years and until the 1920s the Marconi Company applied global radio-telegraphy exclusively to maritime communication and navigation. In 1914 Marconi opposed the idea of turning radio into an instrument of entertainment and music diffusion, proposed by the company's vice-director of traffic David Sarnoff (who later became director of the Radio Corporation of America and a pioneer of television broadcasting), as confirmed in a letter written to a New York newspaper stating that his own invention was a scientific tool, not "a toy for lyrical music and popular songs" (un giocattolo per lirica e canzonette; quoted in Pezza). Despite Marconi's initial opposition, amateur experiments of radio diffusion of live (and later recorded) music began as early as 1910. In a singular coincidence, once again, the first ascertained radio transmission of music is considered to be that of 13 January 1910, when De Forest devices broadcast Enrico Caruso's performances of Pietro Mascagni's *Cavalleria Rusticana* and *Pagliacci* from the Metropolitan Opera House of New York, both to local stations and to the ocean liner *Avon* while it was reaching the city (Taylor 73).

The Social and Intellectual Reception of Recorded Sound in Italy

Phonographic activity developed in Italy in the late 1890s and exclusively relied on the import of foreign brands like Pathé or Columbia. By that time, the consumption of reproduced music was mainly confined "to the bar automatico – or coin-slot parlor – where popular *canzoni* and standard opera arias could be heard for ten centesimi each" (Gelatt 104). With the spread of phonograph cylinders and gramophone records in the early 1900s, national companies started their businesses: the Anglo-Italian Commerce Company (producing cylinders and early recordings of Caruso);[11] La Societá Fonografica Napoletana (Naples, 1901), promoting Neapolitan singers and folk music; the Societá Anonima Italiana di Fonotipia (SAIF, Milan, 1905), recording opera celebrieties and distributing La Voce del Padrone (HMV) records;[12] and La Societá Nazionale del Grammofono (SNG, 1912), distributing Columbia records from 1922.

Recorded music initially received an ambivalent response from the Italian market. At a social level, in the early 1900s phonographs and gramophones appeared as intriguing novelties, capturing the interest of elites and the curiosity of readers – as seen in the illustrated coverage of Pope Leo XIII's first recorded apostolic blessing in *La domenica del corriere* (29 March 1903) or in contemporary advertisements that presented the new devices as a magnetizing attraction (the "marvellous repeater of human voice") and a home alternative to theatre – as in the motto "Non andate più a teatro" (Never go to the theatre again!). At a cultural level, even though the new acoustic technology "moved beyond its pioneering phase in order to steadily enter the habits of a restricted part of society" (Capra 3), sound recording remained at the margins of intellectual reflection. After the initial curiosity, documented in the first tribute to the phonograph in *Illustrazione italiana*[13] and in its mention among the objects presented at the 1878 exposition in De Amicis's *Ricordi di Parigi* (44), Edison's device rarely found space in contemporary art, as opposed to the parallel appearance of a related invention like telegraphy – in Verga's *Malavoglia* (1881), in which telegraph wires are seen as a curse withholding rain in clouds,[14] or in De Amicis's *Costantinopoli* (1877), in which their "immense spider web over the roofs of the noisy city" (151) foresees an ominous future.[15]

In the early twentieth century the idea of reproduced sound met with strong resistance in cultural debates despite its wider acceptance in the market. Gramophones indeed represented hybrid instruments (as both pieces of furniture and automated music boxes)[16] and a radically new kind of aesthetic experience, which did not receive immediate legitimation either in musicological journals or in musical practice.

With the increasingly refined design of their horns and cabinets (since the Victrola model of 1906), gramophones subsumed both the Italian tradition in the aesthetic craft of instruments (dating back to the sixteenth-century artisanal production of strings in the Cremona workshops of Amati, Guarneri, and Stradivari) and the global manufacturing revolution of the nineteenth century (with the start of the mass production of pianos).[17] At the same time, performing with the infinitely repeatable precision of an automated music box, they achieved the century-long ideal of the musical automaton (from the sixteenth-century hydraulic organs to Antoine Favre's carillon with rotating cylinder of 1796 to the nineteenth-century self-playing pianos) and acquired, like the piano, a new function, both "as a universal accompanying and schooling instrument" (M. Weber 123) and as a prime status symbol of the bourgeois home.

As the sound quality of the gramophone improved and it became clear that the sound it reproduced best was that of the human singing voice, a new repertoire of vocal entertainment (ballads, folk-songs, and arias from famous operas) started to spread on a larger scale. Thanks to the easy transportability of records, gramophones also played a pivotal role in the worldwide diffusion of the Italian operatic repertoire, offering a linguistic and cultural support to Italian emigrants to the Americas (Mallach 8). The later expansion of the record industry into a complex production and consumption system led opera products to their broadest reach ever and launched the global stardom of Italian artists like Caruso, Puccini, and Toscanini (who first conducted in studio for the Victor Talking Machine Company during his U.S. tour of 1920). As a result, recorded music was gradually accepted and even started to impose its technical restraints on composers (like Puccini who "tailored many of his arias to fit the time-span of a ten-inch record"; Semeonoff 102) and on singers (as Caruso's performances became standard models for global imitation).[18]

Gramophones and radio made their way into the Italian literary scene around the early 1910s in the contemporary works of D'Annunzio and Marinetti. In the novel *Forse che sì forse che no* (1910), book 2, D'Annunzio alludes to an "unknown tool" producing "a dark hum" (701), later revealed to be "an old harmonic box, with a metal soul made up of a steel comb whose teeth vibrated while spinning around a bristly cylinder" (704).[19] In the rhymed exchange between the protagonist, Paolo, and his lover, Isabella, the object appears, through the man's sceptical rationalism, as "una stregheria" (witchcraft) and, in the woman's nostalgic evocation, as "una malinconia" (704; a melancholy). Alluding to reproduced sound yet avoiding any direct mention, the music box nonetheless mirrors the growing intellectual attitude towards gramophones, divided between the rejection a priori of mechanical reproduction and the attempt to apply a unique aura to it, as envisioned in its explicit relation to the personal space of Isabella's childhood: "Who can say why an old mechanical thing becomes dear to us, tied to some remote fibre of our interior life, so much so that we can no longer separate ourselves from it without dreading to let something in us die? [...] It was the *aura* of my childhood dreams. Even now, I can't hear it without my heart wavering. For no other thing have I had the feeling of ownership as I do for this little object" (705, emphasis mine).[20] Rather than simply pointing to the object's lost time, Isabella's music box pictures instead the struggle to reconquer poetry in a mechanical instrument. In light of the new recording devices, allegedly transforming music into an impersonal, standardized, and ultimately disowned product, the musical experience is embedded here with the attempt to reclaim a lost authenticity and unrepeatability. In the novel's allusion D'Annunzio initially explores sound *recording*, or a broader notion of mechanical music, with both curiosity and hesitancy. On the contrary, with regards to sound *transmission*, D'Annunzio later manifested an explicit interest, expressed by his pioneering will to expand the use of wireless telegraphy beyond its naval applications, by his visit to the radio station of Rome-Centocelle in 1915, and by his letter to Marconi, revealing that indeed "his [Marconi's] science and his [D'Annunzio's] poetry had become an instrument of war, a fighting force, and a promise of victory" (quoted in Cobisi).

During the 1910s, sound recording and transmitting devices also surfaced in the rhetoric of Marinetti's Futurism. In the manifesto "Destruction of Syntax, Words in Freedom, Wireless Imagination" ("Distruzione della sintassi Parole in libertá e immaginazione senza fili," 1913), gramophones were listed among the modern discoveries influencing human sensibility: "Anyone who today uses the telegraph, the telephone and the gramophone, the train, the bicycle, the motorcycle, the automobile, the ocean liner, the airship, the aeroplane, the film theatre, and the great daily newspaper (which synthesizes the daily events of the whole world) fails to recognize that these different forms of communication, of transport and information, have a far-reaching effect on their psyche" (Marinetti, *Critical Writings* 120).[21] Unlike the other single terms of the list, gramophones are oddly paired with telephones. The two-tiered syntagma certainly signals their implicit categorization as sound-transmission devices but also conveys a deeper unease. The rare adoption of gramophones both in Marinetti's writings (in spite of the central position of the phonic aspect in his program of aesthetic renewal) and in

Futurist performance (in spite of his praise of mechanical music in the "Technical Manifesto of Futurist Literature" that "nothing is more interesting to a Futurist poet than the dance of the keys of a mechanical piano"; *Critical Writings* 111)[22] reveals the same concomitant attitude of excitement and anxiety towards new sound technologies. In a way, paralleling Boccioni's dismissal of photography, Marinetti rejects recorded sound as a paralysis of life's movement. Against the idea of embalming sound in a fixed recording, a written score, or a standard form, he focuses instead on the unfiltered dynamism of noise (as in the onomatopoeic sounds of *Zang Tumb Tumb* or in supporting Russolo's noise attuners) and on the immediacy of performance (as in the Futurist *serate* and ever-different poetic declamations; Daly 2–4). In another way, embracing Marconi's wireless telegraphy as the model for a new wireless imagination, Marinetti espouses transmitted sound as the expression of a purely Futurist mindset, mirroring modern man's "desire to know what his contemporaries, in every part of the world, are up to," his "need to communicate with all the peoples of the world," and his "urgent need to determine, at every moment, our relations with the whole of mankind" (*Critical Writings* 122; "Destruction of Syntax").[23] Marinetti sees in Marconi operators (*marconisti*) a new model of dynamic writing, expressed as a reception and transcription of signals into new words or sounds. In their ability to become one with the apparatus Marinetti detects the image of a new poet, conceived as a "hybrid creature" (Campbell 13), removed from contingence, as his head detaches from the writing hand, and inhabiting a simultaneous "logosphere." In a similar way, the new art in transmission expressed by radio-telegraphy (and, only as sound-transmission devices, by gramophones) responds thus to the same goals, stated in the ideal of "infinite, omnipresent speed" (*Critical Writings* 14; "Manifesto of Futurism"; eterna velocità onnipresente), to overcome distances, to construct the world as a synchronized acoustic community, and to achieve ultimately dynamic ubiquity.

From Music to Graphic Design

By the time that flat records had replaced cylinders in the early 1910s, the Italian musical market was still under the overwhelming influence of opera productions and publishing houses. Opera was still a predominant business in Italy because of its important sponsoring partners, its inherent social value (theatre), and the recognized superiority of its aesthetic experience (performance). Casa Ricordi, founded in Milan in 1808, had acquired a leading position in the business[24] thanks to the vision and leadership of Giulio Ricordi (1840–1912) in pioneering lithography for score printing, in patronizing composers like Verdi and Puccini (Baia Curioni 165–7), and in developing a new visual language for music through poster designs.

The Milanese entrepreneur had invested in printing technologies since the early 1870s, when he was the first to bring back, from a business trip to Germany, some state-of-the-art machines and a team of specialized technicians (Fioravanti 263). Consequently, the house developed a new line of advertisements for opera performances in 1874, in addition to its series of printed scores. In 1884 the creation of Officine Grafiche Ricordi (a new factory just for the chromolithographic reproduction of posters and scores) eventually split the business, separating opera production from print manufacturing. After the hire in 1889 of former La Scala choreographer Adolfo Hohenstein as the atelier's first artistic director, the graphic laboratory rapidly turned into a "frontier company" (Mughini xi), attracting avant-garde domestic artists (e.g., Achille Beltrame, Aleardo Villa, Giovanni Mataloni, and Aleardo Terzi) and foreign talents (Leonetto Cappiello, educated in France; Franz Laskoff, from Poland; and Leopoldo Metlicovitz and his pupil Marcello Dudovich, from the then-Austrian city of Trieste). Hohenstein developed the collaboration with Ricordi's star composer Puccini by designing for his opera *Edgar* (1889) the first Italian poster to have true graphic qualities, albeit in a small format. In 1895 the German graphic designer worked on a poster for *La Bohème*, which was considered a milestone in Italian advertising for its accomplished transposition of the French models and for its acquired capacity to complement the opera with a matching visual language (Weill 84). By virtue of its success, Giulio Ricordi – who had previously (unsuccessfully) organized the first National Exhibit of Advertising in 1894 – resolved again to exhibit his graphic products at the first Biennale of Venice in 1895. On that occasion two enthusiastic visitors, Emiddio and Alfredo Mele, owners of a Neapolitan department store in search of innovative marketing solutions, noticed the quality of Hohenstein's work and commissioned Officine Grafiche Ricordi to work on its first posters for external distribution.[25] From 1895 to 1912 (the year of Giulio Ricordi's death) posters continued to be the main marketing tool for Casa Ricordi's opera productions – as in Hohenstein's advertisement for Franchetti's *Germania* (1902) and in Metlicovitz's promotion of Puccini's *Tosca* (1900) and *Madama Butterfly* (1904) – and a powerful source of income for Officine

Grafiche Ricordi, in addition to its postcard line (added in 1899) and catalogue of scores (including Verdi, Wagner, Chopin, Czerny, Liszt, Mendelssohn, Bach, and Beethoven; Degrada 9–23). In light of the commercial liaison with Mele (180 posters from 1895 to 1915), however, Ricordi launched its advertising business on a massive scale and set in motion a more ambitious cultural project. Drawing from the illustrations of the store's *novità* (clothing, perfumery, linen, etc.) by Cappiello, Terzi, Villa, and Dudovich among others, the Ricordi-Mele series assembled a systematic iconography of modern living, often identified with the images of joyful, elegant, and sensual women carrying purchases, enjoying outdoor relaxation, or crossing urban spaces. The quality and innovative designs of the Mele lithographic prints moved other businesses – electrical companies, expositions, newspapers, manufacturers of clothing and hygienic products, food industries, the mechanic sector (bicycles, cars, tyres), and cinema – to establish a commercial partnership with Officine Grafiche Ricordi.[26] Among them, Campari started a fruitful, long-lasting partnership with Ricordi in 1899, inaugurated by the early works of Villa (1900), Hohenstein (1901), and Dudovich (1901).[27] It is fair to say, then, that the Ricordi artists designed the look of Italian industrialism in these years, promoting its economic expansion (in the established association of wealth and well-being) and fashioning a visual imagery of its core values.

The Ricordi masters played an important role in transforming the repertoire of the French art nouveau[28] into an original graphic language in Italy. Hohenstein first filtered the imagery of Jules Chéret and Alphonse Mucha,[29] and, under his influence, other Italian artists (Villa, Bistolfi, Metlicovitz, Cappiello, Dudovich, Terzi, and Mazza) initially imitated the mix of *japonisme, byzantinisme*, symbolism, and allegorism typical of the French style of the 1890s. While importing the French culture of the affiche, they also contributed to the spread of the Liberty style in Italy (as their decorative solutions influenced the aesthetic design of objects and architectural spaces) and visually unified the Italian market by turning local shops into brands and promoting local products nationwide (Scudiero, "From Belle Époque to Balilla" xvi). In collaboration with its clients Officine Grafiche Ricordi also established an independent aesthetic language, which in turn acquired prestige in France (after the Italian Leonetto Cappiello transferred to Paris) and worldwide cultural recognition.[30] Although the vastness, diversity, and ephemerality of Ricordi posters impede a complete survey or classification of the company's production, a close investigation of three exemplary illustrations (by Hohenstein, Metlicovitz, and Dudovich) can provide valuable insights into the house's style, as well as into its "imagined" industrial modernity. Against the back-drop of Ricordi's operatic production and self-promotion, these posters – for *Il resto del carlino*, "Il varo della nave Roma–La Spezia," and Bitter Campari, respectively – manifest not just the house's intrinsic ties to Italy's early industrialism (in the representation of periodicals, current events, and brands) but also the characteristics of its mature aesthetics of the graphic image and its theatrical construction of modernity as a culture of rapid movement, technological advancement, and pleasurable enjoyments.

Adolfo Hohenstein's advertisement for the newspaper *Il resto del carlino* (1899; plate 10) pictures the intrinsic interactions of journalism and poster advertising. Against the contemporary rise of printed drawings in serial magazines and the creation of illustrated periodicals (like *La domenica del corriere*, established the same year), the poster documents the growing demand of the press to brand itself, by graphically rendering the reading experience. The numerous commissions of poster advertisements to Ricordi designers by local and national periodicals – like *Il corriere della sera* (Hohenstein), *La lettura* (Dudovich), *La sera* (Metlicovitz), and *La stampa*, *La tribuna*, and *Illustrazione italiana* (Giovanni Mataloni) – confirm the necessity of newspapers to promote and fashion themselves into an aesthetic form. Conversely, daily news provided poster advertising with a key metaphor, defining it as a visual genre of ephemeral and rapid consumption. Along with Metlicovitz's opera postcards of the same year, lively illustrating the key episodes of Puccini's *Tosca*,[31] Hohenstein's "moving" picture depicts an elegant woman, dressed in bright yellow, caught, with her partner, at the very moment of her reaction to a piece of information. Beyond its figurative debt to Alphonse Mucha and the Viennese *jugendstil* (in the frame's mixture of photographic realism and floral decorations), Hohenstein's image delineates an original vision of modernity, seen as a transient, instantaneous event, through the mental linking of the woman's fashion to the paper's ephemerality, or of the rapidity of news flowing on a ribbon ("finanza," "industria," "commercio," "politica," "arte," "scienza") to her cursory reading. In its dynamic rendering of a seemingly theatrical episode, the poster not only sets in dramatic motion the conventional imagery of contemporary opera advertising (as signalled by the rapid brushes on the journal's surface, the "quick" elements of cigarette and coffee) but also defines a metonymic representation of poster art itself by temporally positioning that art on the edge of a fleeting instant and spatially locating it on the street curb, where an urban subject in transit becomes its new spectator.

Leopoldo Metlicovitz's advertisement for "Il varo della nave Roma–La Spezia" (21 April 1907; plate 11) stages the impact of early advertising on the cultural depiction of contemporary facts. The "Launch of the Rome–La Spezia Line" provided the most versatile of Italian illustrators (Metlicovitz was a choreographer at La Scala, a designer for Puccini, Mele, and Fernet Branca, and a collaborator with cinema and the tourism industry) with the occasion to portray the relationship between Italy's modern present and its everlasting past. Metlicovitz's vision of a bare-chested mermaid leading a modern ocean liner shapes Italy's contemporaneousness as a hybrid of space and time. On the one hand, the designer analogically links the thrill of the present instant to the attractiveness of female nudity, and the ship's dazzling movement (powerfully parting waters and boldly conquering the marine space) to the charming fascination of a siren (wedging the poster's surface and ideally breaking into the street). On the other hand, by his reference to ancient Rome (in the acronym SPQR and in the founding date of 21 April) and the Italian flag, Metlicovitz explicitly links the image to an immutable past and an ever-present national spirit. By overlapping the ephemeral movement of waves and winds (in the woman's hair) on the stillness of Rome's eternal glory (in the laurel), and by transferring the mythical allure of ancient sirens to modern transportation, the Triestine designer stages Italy's modern present both as a renewed continuity of its unrejected past and as a forward-looking bid to join the process of global transformation. In its paradoxical mise-en-scène, advertising thus stages an experimental visual theatre in which new technologies interact with old myths, and progress and tradition merge, as iterated in a later poster by Aldo Mazza for the "Raid Parigi-Roma-Torino" (1911; Panzeri), similarly displaying a car racing on a dirt road, a boat sailing on the sea, and an aeroplane flying in the sky against the background of two monumental Roman columns.

Marcello Dudovich's advertisement for Bitter Campari (1901; plate 12) at last displays the ties between Ricordi's operatic style of brand advertising and the portrayal of modernity as a multi-sensory experience of pleasure. Alluding to the opera's repertoire, Dudovich relates the drink (minimalistically depicted in the glasses at the bottom right)[32] to the melodramatic scene of a kiss. Imitating the operatic mingling of arts (music, text, costumes, acting, etc.), he thoughtfully constructs the image as a synaesthetic experience, by emphasizing the tactile element (the man touching the woman, the velvety decorations on the pillow, the cloth's edge under the glasses), by associating the pleasure of taste on the lips to that of the kiss, as well as by transferring the red of the drink onto the overall picture (equally charging it with an atmosphere of sensuality and passion). In addition, the studied contrast between the dark red tonality of the image and the brightness of the *Campari* letters in white injects an "acoustic" element into the representation, along the lines of Davide Campari's explicit request to Ricordi artists to turn brilliant colours into sounds and advertising itself into a "music that requires hearing" (Cenzato, quoted in "Il Cordial Campari").[33] In parallel with his signature red, Dudovich's use of "singing" colours (which will be a key stylistic feature in the parallel series for Mele) will also appear in his later advertisement for Cordial Campari (1913; "Il Cordial Campari"), which features four fashionable ladies wearing colourful gowns and hats and drinking Campari in the company of two elegant gentlemen in black and white. By breaking the bidimensionality of poster advertising, Dudovich thus turns promotional images into a multi-sensory theatre, in which products are actors, and the allusions to different sensorial realms (or arts) fashion modernity as an all-embracing experience of *jouissance.*

As outlined in reference to periodicals, events (e.g., expositions, races, cruises), and companies in search of innovative branding strategies (like Campari and Mele), Ricordi's artists moved with an indelible operatic background, creatively fashioning a new three-dimensional visual language through their images. Beyond their promotional grammar and their debt to opera lithographs, the analysed advertisements stage a more complex tie to operatic aesthetics in their equal focus on the ephemerality of the performance, on the hybrid merging of present forms with tradition, and on the synaesthetic simultaneity of the arts. The technical and artistic experimentation of the Ricordi designers will have a profound impact on the verbal and visual languages of the Futurist avant-garde and on the later growth of the Italian advertising industry.

In the short term, advertising art provided a rich background for the Futurist typographic revolution, both in its technical aspects – the material construction of the page, the design of its architecture, the transformation of graphic signs (colours, font, spacing) into poetic signifiers, the letter's expansion into a "typographically designed image" as stated in Marinetti's manifesto "Geometrical and Mechanical Splendour" (*Critical Writings* 138) – and in its aesthetic language. While prefiguring the Futurist ideal of itinerant art – focused on the moment (Hohenstein) and on the rendering of a contingent atmosphere through the five senses

(Dudovich) – as well as sharing the same thrill for modern transportation (Metlicovitz), the visual language of poster advertising also informed the rhetorical strategies of Marinetti's *tavole parolibere* (words in freedom), such as the use of telegraphic language (with no adjectives and adverbs and with verbs in the infinitive); the synaesthetic clustering of simultaneous sensations; the graphic representation of the noise, weight, and smell of objects; and the creation of an intuitive knowledge through chains of visual analogies.

In the long term, Ricordi's pioneering investment in poster design also laid the foundations for the full-fledged development of Italian advertising in the 1920s and 1930s. Even though the house's graphic business declined in the mid-1910s after its involvement with Italy's war propaganda, its influence continued over the following decades as many of its former designers either schooled new artists or began independent collaborations with companies (e.g., Marcello Dudovich with Rinascente) and with the Fascist regime (as did Gino Boccasile, pupil of Aleardo Villa, who invented the graphic *stile del ventennio*). From the time of the early Ricordi-Mele partnership, which inspired Giuseppe Prezzolini's pioneering book *The Art of Persuasion* (*L'arte di persuadere*, 1907), to the first National Exhibit of Poster and Graphic Advertising (Mostra nazionale del cartellone e della grafica pubblicitaria) of 1936 in Rome, advertising evolved in Italian culture from a mere "discourse about objects" (Baudrillard 164), solely intended to transform a commercial relationship into a personal one, into "an object to be consumed" (164), capable of creating a "pattern of collective and mythological projection – in other words, a model" (184).

The Gramophone's Ascent and the New Consumption of Music

The Ricordi posters document the company's gradual shift from operatic productions to the advertising business. At the same time, in the paucity of other intellectual sources, they make visible the contrast and the juxtaposition of live opera production and the sound-recording industry, the progressive transformation of performance set in motion by gramophony, and the decade-long reconfiguration of the musical experience from the artisanal to the industrial dimension.

As phonographs and gramophones turned from a clever invention into "the machine that launched the record industry" (Morton 1), they gradually fashioned first a credible alternative to theatres and second a new form of mass production and consumption of music. By their unprecedented ability "to 'materialize' sound, capture it, and permanently store it" (Prato 6), they had realized a new kind of musical product, advancing new performing and composing styles and capitalizing on the individual's need to impress others, distinguish oneself, and make beauty permanent. As documented in the Ricordi advertisements for recorded sound of the early twentieth century, the ascent of gramophone technology over phonography thus ultimately developed a new notion of authenticity in music, deflating the aura of performance into an ephemeral event and transposing its lost uniqueness onto an infinitely replicable ideal model.

The earliest gramophone advertisement commissioned by Casa Ricordi in 1906 promoted the newly established record company Fonotipia (plate 13). The poster, designed by Marcello Dudovich, stages a pioneering vision of the rising music industry, in its attempt to move beyond a niche market, and alludes to some new elements introduced to the musical experience by sound reproduction. The lithographic print pictures a fashionable lady, seated in a relaxed pose on a bench outdoors, and a small fawn that is listening to a gramophone while holding discs and moving from one selection to another. As the woman winks at the advertised gramophone, her characteristics – a fascinating charm, an amused yet tempting expression, and the bright colours of her dress – transfer to the device, allowing the viewer to visually equate recorded music with an alluring fashion, a seductive entertainment, and a luminous (or, rather, "speaking") invention. As the fawn playfully listens, inviting the lady to enjoy his varied selection of discs, the gramophone transforms into a new Muse (as hinted by the laurel on its head) and creates a new kind of performance, faithfully reproducing the sounds of nature (in the alluded rustling of the foreground tree), yet acquiring an artistic value (as stressed in the motto "Dischi artistici" [Artistic records]), independent of theatre staging (in the depicted portability of discs) and ubiquitous (in the lady's outdoor location). Dudovich's image thus validates the gramophone's appeal to a larger market by expanding its reference from the domain of technology into the realm of art, as well as by transforming its inner features (portability, fashion, and ubiquity) into ideal concepts related to a new way of being modern.

Such a new model of listening slowly turned into a real alternative to theatre performance during the 1910s, as confirmed in the contemporary affirmation of gramophones on the market and in the increasing number of related advertisements by Ricordi. The singular paradox of a publishing house simultaneously patronizing opera and advertising recorded music documents the ongoing process of legitimizing such a new form of music consumption, long before the intellectual recognition of gramophones.

The poster for the newly founded Societá Nazionale del Grammofono (affiliated with the British HMV), designed by Aldo Mazza in 1913 (plate 14),[34] presents the hybrid figure of a gramophone embedded into a singer and transformed into a real person (his and its legs coincide). This living creature, acting both as a talking machine and a dynamic piece of furniture, embodies a transitional state between past and present, or human performance and mechanical reproduction. Mazza ideally suggests the trajectory of operatic tradition from Verdi's heroic melodrama to Puccini's *La fanciulla del West* (1910) in the image of a tenor simultaneously dressed as a knight (with cloak, armour, and helmet) and an American pioneer (with boots and spurs). At the same time, he implicitly ratifies the suggested equivalence of the actual opera singing to its domestic reproduction by overlapping the tenor's voice with the playback of the records in his left hand.

Despite contemporary criticism, this equation finds another confirmation in the 1915 advertisement by Leopoldo Metlicovitz "Il celebre tenore Comm. Cav. G. Anselmi canta esclusivamente per la Fonotipia" ("The Famous Tenor Anselmi Sings Exclusively for Fonotipia"; plate 15). The poster stages the achieved interchangeability of the singer and the gramophone's voice by metonymically substituting the missing advertised object with the image of a dressed tenor (presumably Rossini's *Barbiere di Siviglia*, as hinted by the scissors in his right hand). The absence of a stage for the tenor, replaced by a vague and neutral background, indicates that the advertised object is no longer the gramophone per se but rather recorded sound, faithfully emerging from silence (as suggested in the chromatic contrast of the background's dark blue and the actor's bright yellow) and indistinctly "happening" as a real performance in any undifferentiated setting.

The definitive shift from theatre to gramophone performance is a fait accompli in Aldo Natoli's contemporary poster for Fonotipia (mid-1910s; plate 16). In parallel with an advertisement for Grafofoni Columbia by the same illustrator, depicting a gramophone on the stage of an opera house, which is "directed" by a conductor and is replacing singers (Manzato 15), the Fonotipia poster pictures a similar image of a conducted performance, presumably within a domestic space, yet alluding to a fully fledged theatre experience. Both images recognize the gramophone as a new instrument, visualizing once again the juxtaposition of studio and theatre recordings in the perfect parity of singing and playback. At the same time they also introduce a reflection on the conductor as the innovative and decisive figure for the production of an authentic home performance, outlining the transformation, operated by recordings, of ephemeral executions into one standard, hence eternal, version and of a repeatable reproduction into an ideal event.

Recorded music turns into an idealized experience in Plinio Codognato's design for Grafofoni Columbia (1920; plate 17). His image of a naked mermaid standing in front of a gramophone – drawing from the previous iconography of recording fidelity (as in Barraud's painting *His Master's Voice*) or of juxtaposing modern technologies and mythical past (as in Metlicovitz's "Il varo della nave Roma–La Spezia") – ratifies the existence of an aesthetic third dimension in the industrial product. By breaking aural boundaries, gramophones offer, even in the context of a private room, not just a stimulation of all the senses – as hinted in the synaesthetic association of water and sound waves (moving the sea and the girl's hair, respectively) with sensual pleasure (as hinted in the labial expression of the woman) – but more importantly the new intense aura of a unique personal experience, paradoxically achieved through industrial reproduction.

As documented, these posters mirror (and often anticipate) the theoretical issues involved in the slow acceptance of recorded music. Gramophones were initially perceived as instruments that undermined the musical language, stripping it of the aura of "its unique existence in a particular place" (Benjamin, *Work of Art* 21) and devaluing its intimate relation to the ritual of the concert – that is, depriving the performance of its authenticity. In a persistent cultural resistance, intellectuals saw in recorded music an invasive and standardized product enacting the "ubiquity of commercialism" (Adorno 19), as well as a light entertainment allowing private people "to avoid straining themselves during the free time allotted to them for intellectual consumption" (10). Over the years, however, the gramophone industry not only opposed these objections with an ever-greater faithfulfulness in the reproduction of sound but also elaborated a new model of authenticity for the musical experience. According to this model, the aura of a live concert was slowly transferred to the "ideal event" of a studio execution (Eisenberg 89), and the emphasis of the performance gradually shifted from the singer to the conductor, now endowed with the new role of rescoring a piece for recording purposes and of crystallizing it into an eternal sound image.

The New Aesthetics of Music

In the essay "Mr. Bennett and Mrs. Brown" (1924) Virginia Woolf famously stated: "On or about December 1910, human character changed. I am not saying that one went out, as one might into a garden, and there saw that a rose had flowered, or that a hen had laid an egg. The change was not sudden and definite like that. But a change there was, nevertheless; and, since one must be arbitrary, let us date it about the year 1910" (Woolf 4). Along the same lines and with a similar degree of arbitrariness, it is possible to ideally locate the watershed of musical modernity around 1910. On the one hand, in conjunction with the success of Puccini's *La fanciulla del West* – featuring Toscanini as the conductor and Caruso as Dick Johnson at its New York premiere on 10 December 1910 – Italian melodrama started its slow decline (until Puccini's death in 1924 and his posthumous *Turandot* in 1926). On the other hand, in conjunction with the ascent of gramophones over cylinders, the record industry overlapped with the ongoing crisis of Western music, spurring the development of new aesthetic languages.

Gramophones undermined operatic language by devaluing live performance, as did the ready-made language and reduced costs of records. At the same time, gramophones instigated an unprecedented moment of rupture in contemporary Western music, leading, in one way, to the collapse of the tonal system, the symphonic grammar, and all the codified criteria for good or bad music (Adorno 7) and, in another way, to a new ethics of chaos, dissonance, and freedom (from tradition and from harmonic natural laws), envisaged in Schoenberg's dodecaphony and in Stravinsky's *Sacre du printemps* (*Rites of Spring*, 1913).

In Italy, in the early 1910s, the two Futurist composers Francesco Balilla Pratella and Luigi Russolo enacted a similar protest against academism and the harmonic tradition, vehemently directing it at opera and its production system. In the "Manifesto of Futurist Musicians" (1911) Pratella openly attacked the *misoneismo* (fear of the new) that was reducing "Italian music to a unique and almost unvarying form of vulgar melodrama" (32); the "fascinating mirage of opera under the protection of the big publishing houses" (33); and the dictatorship of bel canto, librettos, and Puccini. In agreement with him (and with Marinetti's scorn of the Winged Victory of Samothrace), Russolo discarded harmony and classical music in *L'arte dei rumori* (*The Art of Noises*, 1913) by stating: "We delight more in combining in our thoughts the noises of trams, of automobile engines, of carriages and brawling crowds, than in hearing again the Eroica or the Pastorale" (25). In reaction to the preformatted space of melodrama (with its burdensome tradition) and reproduced music (with its commercial homologation), Italian Futurist musicians experimented with the new language of noise sound, meant as a way to imitate the new acoustic background of the industrial age and to set musical scores in vital motion. By recreating mechanical vibrations in a quarter-tone scale and re-enacting different timbres (roars, swirls, explosions, whistles, etc.) through hand-crafted *intonarumori* (noise attuners), the two Futurist composers aimed to break the old musical ground with "combinations more dissonant, stranger and harsher for the ear" (Russolo 24) and, ultimately, to construct a new synaesthetic experience, intersecting acoustics with colour dynamics and the weight of objects. As a musician and painter himself, Russolo indeed put into practice (perhaps under the influence of poster advertising) Carlo Carrá's theory (and canvas) of *Sounds, Noises, and Smells* (1913), which outlines the necessity of colours, lines, and volumes to reproduce aural sensations. Furthermore, Futurist composers deliberately pursued ephemerality and inimitable performance as the core qualities of their pieces, as attested in the controversial premieres of Pratella's and Russolo's pieces in the *serate futuriste* of 1913 and 1914, which met with journalistic provocations and brawls.[35] Rejecting imitative recording through the ever-varying "fantastic association of the different timbres and rhythms" (Russolo 29), they conceived performance in fact as an unrepeatable event, created as a perishable spectacle (the portion of Russolo's *Il risveglio* that was published in *Lacerba* on 1 March 1914 still represents "today the only score of the piece"; Gangale 21) and as a unique encounter between the musicians and distinct, transitory audiences.

In synthesis, the new experimental phase, opened by the affirmation of recorded sound, implicitly yet deeply accelerated the crisis of opera and musical performance by introducing new repertoires on a massive scale, turning music into a standard product, and offering a portable alternative to theatre. At the same time, though encountering a longer and stronger resistance before their acceptance as cultural products, gramophones left a lasting imprint on the evolution of a new musical language, both in the theory and practice of composition (along the trajectory from the laboratory of Futurist music to the later, noise-based expressiveness of John Cage's *4'33"*) and in the manufacturing of music as a mass-produced form (leading from the early serial production to the post–Second World War form of consumption in rock 'n' roll).

In parallel with gramophony, the transformation of radio-telegraphy from science into entertainment had an impact on the post–First World War development of sound recording and transmitting technologies in the field of mass communication. The first application of Marconi's invention to public diffusion took place in 1920 during the Italian occupation of Fiume. On 23 September at 2:00 p.m., from the radio station set up on board the ship *Elettra,* D'Annunzio read the first public message, in the company of Guglielmo Marconi, and claimed the international recognition of Fiume's independence in opposition to the Versailles Treaty of 1919. The American broadcaster Franz Konrad had experimented with public radio transmissions a year earlier from Pittsburgh's KDKA (the first station of radio diffusion), yet D'Annunzio's communication from Fiume represents the first circular radio diffusion in the Italian language, taking place four years before Italy's first official radio diffusion ("Concerto d'inaugurazione") in Rome on 6 October 1924, after the foundation of the Unione Radiofonica Italiana (URI).[36] Marconi's involvement with D'Annunzio – motivated by the inventor's wish to end his marriage with Beatrice O'Brien by virtue of Fiume's liberal statute (promulgated by the Vate and allowing divorce, among others) – can be considered a significant turning point in the scientist's approach to his invention, and the start of a new form of mass communication, which the Fascist regime would later massively exploit (Cobisi). D'Annunzio's choice of radio diffusion visualizes its future connection to politics by his clear intent to use Marconi's figure as a way to metaphorically amplify the *questione fiumana* to the world. In light of the poet's blindness (due to a war wound, as reported in his diary *Il notturno*; *Nocturnal,* 1916), his communication also lay the theoretical ground for the construction of a new acoustic space, based on time and orality, and of a new deterritorialized language.[37]

Radio diffusion found legitimacy in the early 1930s, after the foundation of Ente Italiano Audizioni Radiofoniche (EIAR) and the merger (mirroring the British fusion of HMV and Columbia into EMI) of Marconi's company (Marconiphone) and Italy's two gramophone houses Voce del Padrone (formerly Societá Italiana di Fonotipia) and Columbia (formerly Societá Nazionale del Grammofono) into VCM (Alunni 296). By 1931, radio had acquired a broadcasting structure and introduced reproduced music as a part of its programming. As contents became more and more educational – radiophonic "reductions" of literary masterpieces, theatrical radio dramas, Futuristic *sintesi radiofoniche,* or reproduced concerts – and devices acquired a cultural reference (in their naming of radio-phonographs after Petrarch or Ariosto), some intellectuals started to investigate the possibilities offered by the new technology of developing a new aesthetic language. In 1931 the director of the journal *Il convegno,* Enzo Ferrieri, published an essay entitled "La radio come forza creativa" ("Radio as a Creative Force"), which launched the debate on the new language of spectacle and mass communication. Against the critics of radio diffusion (Guido Piovene, Emilio Cecchi, and Anton Giulio Bragaglia) Marinetti participated in the debate by co-authoring with Pino Masnata the manifesto "La radia," which was published in *La gazzetta di Torino* on 22 September 1933. In his reflection Marinetti envisioned radio as an experimental space of interaction between different arts and as a platform for the development of a new abstract language of entertainment and communication, not only abolishing any unity of action or a rootedness in tradition but also inventing a new ubiquitous poetics of "synthesis of infinite simultaneous actions," "spatial immensity," and "amplification and transfiguration of vibrations."[38] Inspired by "La radia" and his own broadcasting experience, the Futurist painter Fortunato Depero published in 1934 the poetic collection *Liriche radiofoniche* (*Radiophonic Lyrics*). In his vision, radio expressed not just a unique continuum of words and sounds (or "onomalingua") but also an experimental poetic form, representing the modern capacity of registration, transmission, and displacement of sound from its contingent location to a multitude of simultaneous spaces and atmospheres: "These radiophonic lyrics of mine are apt expressions for long-distance transmission. The listener is no longer only gathered in a silent and romantic parlour but rather is everywhere: on the streets, in the cafés, on an aeroplane, on a boat's deck, in a thousand different atmospheres. Therefore, the character of radiophonic poetry must be spatial, determined, sonorous, unexpected, and magical. In a word, this radiophonic poetry that I have invented must be the lyrical expression of a most pure state of mind" (Depero, *Liriche radiofoniche,* 8; preface).[39] For Depero, the recorded and transmitted language of radio sounds turned into a dynamic performance, embracing the most diverse subjects (as in the poems "Pioggia," "Acciaio," "La voce dell'antenna," "La febbre del telegrafo," and "New York nuova Babele") and conveying the ultimate experience of a "dinamondo" (*Liriche* 70; dyna-world), that is, of simultaneity in perennial movement.[40] As gramophone recordings had done, this new radio language abolished the distance between composer and consumer by becoming delocalized (in constant movement on a street, in a café, or on the new means of transportation), and therefore ubiquitous in the form of a pure state of mind.

Chapter Six

Cigarettes and Smoke: The Modern Lightness of Being

In an early photograph of the Futurist movement, taken in Paris in 1912 (fig. 6.1), the painters Carlo Carrà, Umberto Boccioni, Gino Severini, and Luigi Russolo proudly posed with the founder, Filippo Tommaso Marinetti.

Despite the seemingly casual shot, the configuration of the group enacts a well-studied visual grammar, purposefully locating its members on the street curb (on the threshold of an ephemeral instant), with confrontational poses (furrowed brows, menacing gazes, and lit cigarettes), handling their weapons as in the moment of battle: a book (Severini), a defiant glove (Marinetti), and a brandished cigarette (Carrà). In conjunction with the First Exposition of Futurist Art, at the Bernheim-Jeune Gallery in Paris from 5 to 24 February 1912 (Cohen 15), this singular display of cigarettes seemingly enforces the correlation of fashion (emphasized in the artists' trendy attire) with the ephemerality of art. Paralleling Marinetti's most recent writings – *La battaglia di Tripoli* (*The Battle of Tripoli*; published in the French journal *L'intransigeant* between 25 and 31 December 1911) and the "Technical Manifesto of Futurist Literature" (published in *Poesia* on 11 May 1912) – the exhibition of cigarettes suggests instead a more complex symbolism both of the avant-garde and of modernity itself.

In the war account *La battaglia di Tripoli* (translated into Italian by Decio Cinti in 1912 and reporting Italy's occupation of Libya) Marinetti portrayed a representation of the Futurist group through the depiction of a handful of Italian soldiers caught at the front line on the eve of the battle of Tripoli. In depicting the apparent normality of the night preceding the conquest of the city, Marinetti deliberately focuses on the comrades' "multiple eyes, long range, unloading their gazes like the bullets of our automatic rifle" (11),[1] as well as on the soldiers' war

Figure 6.1 Luigi Russolo, Carlo Carrà, Filippo Tommaso Marinetti, Umberto Boccioni, and Gino Severini in Paris (1912). Wikimedia Commons.

alertness as they smoke their last cigarettes. In the scene, cigarettes relate to the group's position (at the conflict's avant-garde) and brotherhood (in the imminence of the battle). They likewise represent a synecdoche of the war itself, as confirmed in the visual link between their burning ashes and smoke and the book's recurring imagery of inflamed shrapnels, bomb detonations, and cloud-covered skies.[2] As the writer flies over the conquered Tripoli in the book's concluding pages, these elements appear once again. While reporting history's first aerial bombardments, Marinetti adopts a smoking metaphor by comparing the city, seen from above, to an "enormous cigar, which the big lips of the sea are now smoking calmly" (56), and by dwelling on his aeroplane crossing an "immense chalky cloud" (52).[3] In this narration, smoke coincides with an act of conquest and elevation in the seizing of Tripoli and in the poet's unprecedented aerial vision of reality. At the same time, smoke also refers to an act of disintegration and recomposition, as the air filling the poet's lungs[4] mixes with the engine's discharge and the cloud's vapour in forming a new immaterial presence.

Modelled on a similar tale of ascent and reification, another flying scene opens Marinetti's contemporary "Technical Manifesto of Futurist Literature," as the poet rises "two hundred metres above the mighty chimney stacks [*fumaioli*] of Milan" (*Critical Writings* 107). In parallel with the flight's "material dematerialization of matter" (Schnapp, "Propeller Talk" 7), the clouds symbolize the poet's rarefaction into a bodiless spirit. In parallel with Milan's *fumaioli*, signifying the emerging force of Italian industrialization during the Giolitti era, the cloud of smoke visualizes not just the transformation of the poet into fluid matter but also the forging of a new Futurist self in a state of constant combustion. Throughout his flight, smoke, as an instance of fire, is a metaphor for the transforming energy of industrial labour and man's new form of transportation (already expressed in the steam-filled pages of Marinetti's "Manifesto of Futurism").[5] At the same time, smoke also indicates an incorporeal, light, and ever-moving being, evidencing Bergson's idea of the self's rarefaction into matter (*Matter and Memory*, 1896), the awareness of material objects (or the "intuitive psychology of matter"; *Matter and Memory* 112), and the poet's dispersion into matter. At last, smoke coincides with a new upward force – also reflected in Gazourmah's ascent to the heavens at the end of *Mafarka il futurista* (*Mafarka the Futurist*, 1910) – as well as with a new form of aerial writing, expressed as a moving chain of analogies and a detached composition, as Marinetti theorized in "Supplemento al Manifesto tecnico della letteratura futurista": "the hand that writes seems detached from the body and continues for a long time freed from the brain, which also, somehow detached from the body, having taken flight, looks down from on high, with an awesome clarity of vision, upon the unexpected expression coming from the pen" (*Critical Writings* 115).[6]

As smoke subsumed such complex imagery in the Futurist discourse, cigarettes also acquired aesthetic relevance. Coinciding with their industrial growth in the early 1900s, cigarettes become paradoxical symbols in Futurist art not just of a new modern being – burning and dissolving at the same time – but also of the modern present as a "self-consuming actuality" (Habermas 8). At the same time, paralleling the contemporary conversion of tobacco in Italy from an import-based good to a state-managed industry, smoking or smoke likewise became a core metaphor of Italian culture at the turn of the twentieth century, from Marinetti to Palazzeschi, from First World War narrations to Svevo. The image of the cloud of smoke also became a recurring theme, reconfiguring an old trope of industrialism into the experimental symbol of the new identity and psychology of the modern self.

Tobacco and Smoke in Italian Culture

The Futurist association of cigarette smoking with fashion, war, and a-corporeal lightness marks a radical break from the previous Italian imagery of tobacco. Marinetti not only asserted cigarettes as the new industrial way of smoking but also shifted smoke from a perfunctory quality of modernity to its very essence. Before Futurism, tobacco consumption had mainly related to chewing, taking snuff, and cigar or pipe smoking, and its symbolization, in either positive or negative terms, was merely instrumental, after Columbus's first voyages,[7] in the double-edged portrayal of the New World and the new age inaugurated by his discoveries.

Brought to Europe by royal ambassadors and unscrupulous sailors, tobacco represented an ambivalent synecdoche of sixteenth-century America, subsuming a "paradoxical identification with both empowerment and enslavement" (Pollard 43) and embodying either a miraculous remedy against all maladies (from plague to boredom) or an infectious poison damaging the body and inciting sinful inclinations.[8] First appearing in seventeenth-century painting as a novelty item from the New World, tobacco continued to reflect such ambiguous ties to modernity. In contemporary Dutch painting (Tempel 206–8) it referred to the pleasure, wealth, and well-being of modern life – as in Jan Miense Molenaer's *Smoker* (1635; Salerno 55) or in Adriaen Van Ostade's *Apothecary Smoking in an Interior* (1646; Tempel 208) – and to its vanity, ephemerality, and degeneration (as portrayed in Adriaen Brouwer's disquieting *Smoker* of 1636 [Salerno 31]).[9] Likewise, in contemporary Italian painting, tobacco appeared in conjunction with youth and pleasure (in the fluid form of smoke) and with the vanitas theme (in the solid form of ashes),[10] as pictured in the presumed *Portrait of Masaniello* (ca. 1647; Salerno 25), attributed to the Neapolitan painter Micco Spadaro, in which a young (in body) yet aged (in face) protagonist was significantly portrayed with a pipe in his right hand and an ash-tray in his left (reflecting the alternation of his relaxed posture and serious gaze).

During the seventeenth and eighteenth centuries tobacco products spread widely across Europe. Along with its increased production and consumption, tobacco started to be condemned (in the anti-smoking campaigns by King James I of England, Pope Urban VIII, and the Shah of Persia),[11] tolerated (e.g., by Cardinal Richelieu in France) as a means of raising the nation's income through taxation, and even exploited (e.g., by the American colonies in the trade of African slaves for European exports; Burns 87) as a way to fund wars. Against the background of this political resistance the cultural imagery of tobacco slowly veered from its previous association with vanity to the exclusive semantics of pleasure. In the early eighteenth century two Italian Arcadian poets, Girolamo Baruffaldi and Francesco Arisi, published two epic poems about the pleasure of snuffing or chewing tobacco – *La tabaccheide* (*Tobacco-eid,* 1714) and *Il tobacco masticato o fumato* (*Chewed or Smoked Tobacco,* 1725) – relating it to a mythical state of nature; to an idealized golden age of brotherly friendship, creativity, and a lack of boredom; and to the vision of the New World (in Baruffaldi's imaginary journey of a loaded galleon from America to Europe, and in Arisi's tribute to Cortes for bringing tobacco across the Atlantic). Such imagery of tobacco as a remedy for melancholy and as a poetic symbol of the spirit's primitive age continued until the early nineteenth century, as confirmed in the work of the poet Giacomo Leopardi, another consumer of snuffed tobacco (Veglia 170), who praised in the *Zibaldone* the pleasurable benefits of the plant in reawakening wit (19 August 1823) and empowering imagination and thought (29 September 1823).

Against the practice of snuffing or chewing, tobacco smoking prevailed in conjunction with the mid-nineteenth-century rise of industrialism. The growing success of pipes and cigars fashioned a social ritual (offering men an occasion to meet, converse, and bond) and shaped a new idea of leisure (in contrast to the fast pace of business), structured according to the format of separate clubs and distinct clothing (like the *smoking*-jacket; Hilton 126–8). In the Italian context, tobacco first appeared in relation to the practice of smoking and industrial smoke in Antonio Guadagnoli's *Il tabacco: Sestine* (*Tobacco: Sestets*) of 1834, in which the presented image of the poet smoking in a tavern significantly matched the representation of his own age as "the century of smoke / the century of steam machines" (verse 50).[12] The pairing of tobacco smoke with industrial clouds or industrialism in general found more consistent representations in Italy during the second half of the nineteenth century, as documented in anti-smoking publications and art. In the anti-smoking pamphlets of the age, the spreading of tobacco is often equated to the image of a marvellous yet poisonous cloud of smoke covering the world, as in Francesco Scalzi's *Il consumo del tabacco da fumo in Roma in attinenza alla salute pubblica* (*The Consumption of Smoking Tobacco in Rome with Reference to Public Health,* 1868) or in Vincenzo Palmesi's *Del tabacco: Specialmente del tabacco da fumo* (*On Tobacco: Especially on Smoking Tobacco,* 1876).[13] In contemporary art, recurring images of smoke clouds signify the excitement and anxiety about industrialization. In one way, against the back-drop of the construction of new train lines (like the

Trans-Siberian or the Berlin-Baghdad line) and the increasing speed of liners crossing the Atlantic, smoke denoted man's bold ambition to fill the geographical space, the oceans, and later the sky with his cloudy presence (of trains, ships, or aeroplanes). In the painting *Il treno* (*The Train*, 1880; plate 18), commissioned for the Turin exposition of 1884, Giuseppe De Nittis signified the thrill of Italian industrialism and man's self-confident audacity in conquering the virgin space of the earth through the deliberate image of a moving cloud of smoke covering the space of the countryside (and of the canvas), in contrast with the stillness of two peasants. In another way, smoke (of train stations and factory chimneys) also conveyed the desolate vision of modern living, filled with greyness and melancholy. In the poem "Alla stazione in una mattina d'autunno" ("At the Station on a Fall Morning," in *Odi barbare*, 1877) Giosuè Carducci accordingly related the cloudiness of a train and a fall sky to the "everlasting bore" at the departure of his beloved and to the vision of a colourless modernity, which "has lost the meaning of being" (verses 60, 57).[14]

In parallel with this imagery of tobacco and industrial smoke, which continued into the twentieth century,[15] the metaphor of the cloud connoted in D'Annunzio's work the simultaneous burning and vanishing of both pleasure and the author's whole age, suspended between its vitality (belle époque) and its perceived end (fin de siècle). In *Il piacere* a cloud, first of smoke and then of dust, visualizes the contrast between pleasure and vanity. In the tea scene the cloud of smoke from Sperelli's cigarette, illuminated by the sun, visualizes his thoughts of love, his renewed imagination, and an invisible force, binding together his soul, his friends, his house's interior design, and the aroma of tea: "the lifting smoke was coloured with the almost horizontal rays of the sun; tapestries *harmonized* in a warm and festive colour. The tea's aroma *blended* with the odour of tobacco" (236, emphasis mine).[16] In the novel's conclusion, instead, "a dense cloud of dust" (un polverío denso) in the dismantled house of his lover, Maria, pierced by the "inflamed splendour of the sunset" (357), mirrors Sperelli's vanishing desires and his inner realization of the end of pleasure.[17] Dusty clouds recur in D'Annunzio's journalistic writing of the same years as an essential metaphor for the ongoing destruction of Rome and its reconstruction into the new Italian state's capital city. As confirmed by D'Annunzio's piece in *La tribuna* on 12 May 1885, the clouds of demolition define an ambivalent site of nostalgia for the vanishing of the old and a thrill at the fashioning, or rather the springing up, of the new city: "Rome becomes the city of demolitions. The copious dust of ruins rises from every point in the city and spreads throughout the sweet May sunshine [...] but from the ruins a new Rome will rise up and shine forth, the clean, spacious, and healthy Rome" (*Scritti giornalistici* 311).[18]

The Emergence and Early Imagery of Cigarettes in Italy

In the late nineteenth century cigars and pipes dominated the global market of tobacco, whereas cigarettes were a very marginal product, exclusively related to feminine consumption. Until the early 1900s cigarettes evoked images of exoticism and lust in Italy. Originating from the Turkish siege of the city of Acre in 1832,[19] they suggested a picturesque vision of the Ottoman Empire and the charm of the Orient. Evolving from the *spagnolette*, produced in Spain since the seventeenth century, they also related to erotic desire and feminine independence – as seen in the title character of Bizet's opera *Carmen* (1875) who smokes with the women workers of a tobacco factory in Seville, while she powerfully seduces Don José (Hutcheon and Hutcheon 233).

A cigarette first appeared as a central object of investigation (and no longer as an accessory) in a short story by the young D'Annunzio (under the pseudonym *Il Duca minimo*) that was published in *La tribuna* on 20 July 1886 with the title "Autobiografia di una sigaretta" ("Autobiography of a Cigarette"). The text presents the journey of a cigarette (surprisingly turned into the protagonist of a literary fiction) from Constantinople (as a tobacco leaf) to the languid hands of a Roman noblewoman (Mariana). Capable of thought and self-awareness, the cigarette narrates in the first person the lady's flirtation with Gustavo during the tea ritual. As it is handed over from her to his mouth, the cigarette hangs between the vitality of the woman's seductive power (emphasized by its pleasure at her warm lips) and the threat of the man's imminent advance (foreshadowed by the horror at his cold lips). The cigarette being unable to see the end of the affair (later revealed by another burnt fellow), the narration deliberately remains incomplete during the brief interval of the cigarette's life, spanning its birth as burning fire ("My foot suddenly got wet and my head burned. I was born"; 593) and its rapid vanishing into cold ashes ("My strength faded. I could no longer see or hear anything [...] my poor limbs scattered among the cold ashes"; 596).[20] Although mainly relating cigarettes to nineteenth-century imagery (in their connection to femininity, the Orient, and the social ritual of tea), D'Annunzio's short story introduces to the literary imagination the unusual element of an "I cigarette" (bordering on sensuality and ruin, fluidity and solidity, smoke and ashes, warmth and coldness, life and death), prefiguring the notion of a suspended "cloudy" self and the equivalence of smoking and thinking in subsequent Modernist narrations (e.g., those of Marinetti, Palazzeschi, and Svevo).[21] At the same time, as a historical record, D'Annunzio's "Autobiografia" captures the contemporary emergence of cigarettes as an industrial product, foreshadowing (perhaps unconsciously) their later success over the handmade *spagnolette* and the more masculine cigar.

A few years before the "Autobiografia," in the early 1880s, the American inventor James Bonsack had patented a cigarette machine capable of manufacturing 120,000 pieces in ten hours (equal to forty rollers making five per minute by hand), revolutionizing production and dramatically transforming traditional tobacco culture (Brandt 30–1). Along with the evolution of chromolithography, U.S. cigarette companies started to experiment with text and other visual advertising as a way of charging the new product with an aura of desirability and matching the increased production with a significant revolution in consumption.[22] From the early marketing strategies of the tobacco entrepreneur James Buchanan Duke in the 1880s to the grandiose campaign for Camel, Reynolds' American blend, which was launched in Times Square in 1913 (Pampel 16–17), advertising slowly transformed cigarettes from a handmade to an industrial-made item, from a feminine to a masculine product. It did this by transferring the element of unicity coming from the individual craftsmanship of cigarettes onto the invented practice of inserting collectable cards (representing sports heroes, flags, or exotic half-dressed actresses) into packages, promoting trade and exclusiveness. At the same time, by representing alluring female images (but seldom women in the act of smoking), advertising separated cigarettes from a destabilizing femininity (associated with prostitution and emancipation)[23] and made them "a gendered artifact" (Schmitz 101), charged, for the male public, with attractiveness, beauty, and power. After Bonsack visited Rome to install his new machine in 1892, American-made cigarettes started to enter Italian society, removing their association with decadence and the Orient and enhancing instead (as a tobacco brought to Europe a second time) their new ideal connection to America and industrial modernity.

During the 1890s Bonsack's invention encouraged the Italian import and production of industrial-made cigarettes and the growth of Italy's national tobacco industry. As a way to lessen its heavy reliance on imports, the Italian state began to directly control the cultivation and manufacture of tobacco, by first entrusting its management to the Department of General Administration of Duties (1884) and then moving it to that of General Administration of Monopoly (1895).[24] The Italian government also promoted the consumption of smoking products by quickly settling the number of social disputes regarding the health conditions of workers in the cigar factories and by establishing the journal *Il tabacco* in 1897 as the official guide of national tobacco selections and production.[25] At the beginning of the twentieth century the state-owned industry implemented production with the introduction of new tobacco varieties throughout the national territory and launched, in 1901, the popular cigarette Macedonia, which would become the cornerstone of its monopoly. As cigarettes (which comprised 3.1 per cent of tobacco manufacturing in 1900) turned from foreign-imported fashionable items into national products and emerged from a niche artisanal market (with a production of 1,000–1,200 handmade pieces per day) into a revenue-generating industry of their own, they started to appeal to a much broader market.

As of the early twentieth century, cigarettes continued to represent feminine symbols of emancipation, fashion, and seduction in the Italian imagination. In response to their common association with prostitution, contemporary manuals of behaviour by female writers allowed cigarettes solely within the self-contained realm of bourgeois etiquette by either tolerating women smoking – as in Marchesa Colombi's *La gente per bene* (*Well-Mannered People,* 1877) – or domesticating the idea through a balanced code for female smokers – as in Matilde Serao's essay "Le donne possono fumare?" ("Can Women Smoke?"), published in her 1900 book *Saper vivere* (*Knowing How to Live*).[26] In parallel with the suffragette movement, cigarettes continued to identify a feminine space of emancipation and freedom, either in connection with reading and imagination, as in the visual imagery of Federico Faruffini's painting of a smoking *Lettrice* (*The Female Reader,* 1865), or in connection with erotic desire, as later documented in the poem by Sibilla Aleramo, "Fumo di sigarette," which staged cigarette smoke as the virtual trait d'union of two lovers seeking each other at night across two facing windows (1921; *Momenti,* 111–12). In connection with the advertising imagery by Jules Chéret and the art nouveau master Alphonse Mucha (promoted and divulged all over Europe by the French manufacturer of Job flavoured rolling papers), cigarettes more commonly displayed the association of femininity with fashion, seduction, and style. Chéret's 1889 poster of La

Cherette related a cigarette to a brightly dressed woman flaunting her sexiness. Mucha's posters of the "blond smoking lady" in 1896 (Mitchell 298) and of the smoking "brunette" in 1897, inspired by the Sibyls of Michelangelo's Sistine Chapel and turned into visual icons, were even sold as independent lithographs. The Job iconography would influence the smoking advertising of the Italian painter and illustrator Aleardo Villa (mediator of the French style and collaborator of Hohenstein at Officine Grafiche Ricordi from the early 1890s).

In his first advertisement for Los Cigarrillos Paris in 1901 (plate 19), depicting a sensual woman smoking a cigarette and lying on a bed of flowers, Villa cites the key elements of Mucha's style in the studied contrast of colours and the emphasis on decorative patterns (flowing from the cigarette's smoke into the woman's hair or the background space), yet also charges the image with an operatic allusion in the link established between the Spanish text – "Los Cigarrillos Paris son los mejores" (Paris cigarettes are the best) – and the iconography of lust and exoticism related to Bizet's smoking diva from *Carmen*.[27] In his later poster for Papiers Job in 1906 (plate 20), which added his name to the French company's prestigious collection, Villa also related cigarette smoking to the image of a seductive and fashionable lady. The advertisement stages the hand and the face of a woman brightly emerging from a blurry haze, as well as the image of a cigarette releasing a white trail of smoke that traces the characters of the word *Job* in the air. The depicted smoke waves extend their moving lines from the cigarette onto the woman's hair, establishing the connection between her action of smoking and her experience of enjoyment. At the same time the grey colour of smoke merges with the turquoise of the lady's moving veil, serving as a vision and as a concealment (of the woman's nudity), enacting a chromatic synaesthesia (in the implicit evocation of the woman and tobacco's fragrance to the beholder), and dispersing the aroma of smoke and her sensuality onto the surrounding space (thus extending the constructed atmosphere of pleasure to the overall image).

In parallel with this feminine association, a new cultural imagery of smoke developed, as the cigarette industry gradually expanded its reach to men during the 1910s and 1920s. In a way, with their ever-growing share of the market, cigarettes acquired a more direct masculine connotation, as seen in the feature of an elegant man smoking a cigarette while riding a motorized bicycle, on the February 1908 cover of *La rivista del T.C.I.* (Breda); in the smoking sheriff of Giacomo Puccini's *La fanciulla del West*;[28] in Emilio Salgari's adventurous figure of Yanez (always represented with his "eternal cigarette" at Sandokan's side); or in the Italian illustrator Leonetto Cappiello's poster of a smoking pasha, all dressed in white, designed for Papiers Job in 1914 (plate 21). In another way, along with the diffusion of a very popular social habit, smoke assumed a new meaning in contemporary literature through the symbolic figure of a cloud, turned from a marvellous yet frightful element into a metaphorical site for negotiating and forging a new modern self. In Futurist theatre and narratives the image of a cloud of smoke symbolized a daring plastic complex of the experimental fusion of an a-corporeal and ubiquitous man. In Aldo Palazzeschi's novel *Il codice di Perelà* (*Man of Smoke*, 1911) the cloud of smoke becomes an emblem of lightness and the enigma of a bodiless and torn being. In Italo Svevo's *La coscienza di Zeno* (*Zeno's Consciousness*, 1923) smoke mirrors the annihilation caused by the First World War and offers a powerful metaphor for a new way of thinking.

Smoke, Cigarettes, and the Futurist "I" Cloud

Cigarettes constitute an indispensable tool in Marinetti's self-portrait and a polyvalent Futurist metaphor. On a personal level, the cigarette acquires in Marinetti a new masculine implication, as attested in his numerous "smoking" photographs (in which he is often in uniform), and a clear link with his intellectual activity, as drawn in Carlo Carrà's 1911 *Portrait of Marinetti*, which depicts him writing and smoking. On a metaphoric level, the cigarette symbolizes the vital and rapid burning of advertising, as hinted in the quoted réclame for Manoli in the manifesto of "The Variety Theatre" ("Smoke Smoke Manoli Smoke Manoli Cigarettes"; *Critical Writings* 191), and of Futurist art itself, as desired since the "Manifesto of Futurism" ("when we reach forty, other, younger, and more courageous men will very likely toss us into the trash can, like useless manuscripts. And that's what we want!"; *Critical Writings* 15–16).[29]

Cigarettes constituted a recurring theme in other Futurist works. In Anton Giulio Bragaglia's photodynamism *The Smoker, the Match, and the Cigarette* (1911; Greene 99), a cigarette visualizes the concurrent combustion and dissolution of a contingent action in connection with the character's gesture of smoking and lighting a match. In Fortunato Depero's photographic *Self-Portrait While Smoking* (1915; Minghelli, "Eternal Speed" 110) the cigarette reflects the painter's will to deride or break the bidimensional space of the image, in connection with his irreverent posture and his drawings above the photograph. In Giacomo

Balla's design of a "smoking stand," *Mobiletto per il fumo*, ca. 1916 ("Italian Futurism, 1909–1944" 15), cigarettes relate to the designing of objects and the fashioning of space, as their lines of smoke transfer onto the furniture structure and define the association with a new colourful and creative lifestyle.

Cigarette smoking assumed a particular role in Futurist theatre, turning into an essential scenic device of the *serate futuriste*. As Marinetti theorizes in "The Variety Theatre" (1913), "cigar and cigarette smoke" (*Critical Writings* 187) represent a fundamental wit of the *serate* because of their ability to express symbolically the Futurist ideal of "pure action" (*Critical Writings* 185; "attualità veloce," *Teoria* 81), to embody an unrepeatable dynamism, to reflect the contortionism of perfomers (on the model of the French *funambules*, Leopoldo Fregoli's transformism, and Ettore Petrolini's clowneries), and to re-enact the lightness of their improvisation and illogical laughter. In this sense, the *fumisterie* of the *serate* relates smoke to the actors' "body-madness" (*Critical Writings* 189; "fisicofollia," *Teoria* 87) and to the laughter and acrobatics (*funambolisme*) of the variety theatre and the commedia dell'arte. The dematerialized "body" of smoke indeed visualizes Marinetti's aesthetic ideal of a new self that is in a constant state of agitation, dispersion, and recreation, as well as the electricity of a new theatre, intended as a pure, unscripted, unrecorded performance, and bound by the absolute impossibility "to fall into stagnation or repeat" (184). At the same time, thanks to its capacity "to fuse together the atmosphere in the auditorium with that on stage" (187), smoke shapes a new kind of audience, which "does not sit there unmoving, like some stupid voyeur, but noisily participates in the action" (187). Through its visual blankness, smoke binds the participants in theatre in a multisensory experience of touch, taste, smell, and hearing, involving them in a common and individual work of "mental synthesis" (*Critical Writings* 187; "sintesi cerebrale," *Teoria* 84) or, in other words, in the creative reconstruction of a new universe.[30]

In parallel with the endless transformism, comic lightness, and creative improvisation of Futurist theatre, cigarette smoking ultimately defined an "allegoric equivalent of the poetic act" (Starobinski 33), as manifested by Francesco Cangiullo in two *tavole parolibere* (words in freedom) of 1914. In the first, *Ho molte parole in libertà in riserva di molti nuovi paroliberi* (*I Have Lots of Words in Freedom in Store for Many More Collections*, 1914; Sansone 181), the histrionic image of Marinetti smoking words (or uttering smoke) in front of his admirers portrays the evanescence and rapidity of his poetic imagination. In the second, *Fumatori IIa classe* (*Smokers, Second Class*; *Lacerba*, 1 January 1914), the fading or evaporating accent of the word *velocitá* and the word-act FUMARE reflect, in their association with the speed of a train, the Futurist ideal of writing as a self-vanishing performance and endless "actionism."[31]

Beyond its direct connection to tobacco, smoke becomes a crucial symbol in Futurist aesthetics with reference to war and flight. In the experimental report from the front line of the First Balkan War (1912–13), published in 1914 with the title *Zang Tumb Tumb*, Marinetti accompanies the sound-filled description of the Bulgarian siege of the city of Adrianople with recurring images of smoky clouds, methodically enacting a rhetoric of disgregation, elevation, and fusion of elements. A cloud of smoke appears, for example, in the crumbling rubble after a howitzer's explosion – "whirrrrrrl revolution of rubble smoke acrid pungent sulfide gas ammonia smell of something burnt" – rendering the impact of war through the five senses (*Zang Tumb Tumb* 621).[32] Clouds identify generally in the book the concomitance of burning, disintegrating, and dissolving. As implied in the "vampe" (flames) displaced along the Turkish line in a visual and verbal reproduction of the battlefield (695), smoke first refers to burning as a "power of deodorization" (Bachelard 103), purifying the world of its passéist institutions, like the Ottoman Empire, or, in other works by Marinetti, libraries, bourgeois morals, and Italy's monumental past. By virtue of the fire's capacity to separate substances, the smoky cloud then signifies an active force of *disgregazione* (disintegration), liberating the vital energy of a new moving body, as documented in the "crumbling whiteness" and "dispersion of 40 million billion molecules" following the explosion of a bridge (632).[33] In conjunction with the dispersion of matter into atoms (analogically mirrored in the graphic crumbling of the letters on the page; 627), smoke signifies at last the achieved *sparpagliamento* (scattering) of the self into a universal vibration, as described by Marinetti in his own comment on the scene in the manifesto "Geometrical and Mechanical Splendour and Sensitivity towards Numbers": "we are systematically destroying the literary 'I' so that it is dispersed throughout the universal flux, and we are coming to the expression of the infinitesimally small and molecular agitation. E.g., the meteoric shifting of molecules in the hole produced by a howitzer (see the final part of "Fort Cheittam-Tépé" in my novel *Zang Tumb Tumb*)" (*Critical Writings* 136).[34] In the words of the "Technical Manifesto of Futurist Literature," the whirling of smoke ultimately embodies the conscious "movements of matter" (*Critical Writings* 111) and the "illogical sequence" (115) of a disaggregated "I": "its compressive and its expansive forces, what binds it, what breaks it down, its mass of swarming molecules or its swirling electrons" (111).[35] Its white and illogical arabesque hence reflects not just the Futurist ideal of a dynamic plastic complex but also the formless, goalless vaulting movement of a new light "I," slowly elevating to the heavens.

In connection with the experience of flight, smoke imaginatively realizes this utopic being by allowing the author to rise above reality (acquiring its divine vision) and rarefy his corporeal "I" into an ethereal, ubiquitous self. Along the lines of his previous flights above Tripoli or Milan, Marinetti offers the most explicit representation of this new being in the scene of his flight over Adrianople in *Zang Tumb Tumb*, identifying it as follows in the enigmatic substance of an "I" cloud:

> to be the only lord of the SUN to have one's own blue castor oil agility monopoly of the heavens to have under one's feet prairies prairies prairies prairies of clouds clouds clouds clouds VRRRRRRRR softness of gaseous fur coats
>
> NAVIGATION OF MALLEABLE MOUNTAINS
> Deformation
> incorporation
> layers layers layers of clouds
> decompo-
> sition reincarnation
> analysis of cumulus clouds contortions
> shredding scraps condensation
> SYNTHESIS pain and regret of
> UNITY prodigality of forms
> competition of masses
> fusion anatomy of clouds
> Crumbling fraying at top
> Speed
> FLUID SCULPTURES
> Steam
> Ice dust
> In the ears
> 1300 m.
> Thermometer 0.2
> Identity of 3 billion small waves of wind
> Concentric = 3 billion wadding pieces
> Geometric pavement
> (*Zang Tumb Tumb* 602–3)[36]

In the compenetration in flight of air, gaseous emissions, and the poet's dematerialized spirit, the smoky cloud expresses the image of the modern writer and man as a light metamorphic being. In the blending of fires – of the engine and the arsonist poet (*poeta incendiario*) – smoke depicts a destructive power, as reflected in the page's semantics of *penetration* (the aeroplane's wedging of "softness of gaseous fur coats" and "malleable mountains") and *sbrandellamento* (the crumbling of matter in "3 billion small waves of wind"). In the blending of air and gas – the poet's dematerialized spirit, steam, and the cloud – smoke manifests an invisible process of dissolution and creative interpenetration of mind and matter, as mirrored in the page's rhetoric of fluidity (the clouds shaped as "FLUID SCULPTURES") and reconstruction (echoed in the capitalized words *synthesis* and *unity*). Through the image of this "I" cloud, Marinetti thus identifies lightness with a new incorporeal subjectivity and a new divine condition, allowing modern man to conquer heavens. By virtue of speed, lightness coincides with a new knowledge in movement, of the earth, no longer tied to logical models (e.g., systematic philosophies, figurative paintings, omniscient novels) but rather focused on the illogical and analogical succession of fragmentary instants, alternating moments of vision and "layers of clouds," words and graphic blanks, consciousness and intuition.[37] In his lightness in flight, man ultimately acquires a new power over reality by achieving the godlike inebriation of simultaneity and ubiquity.

Man of Smoke, Smoking World

During the same period, smoke became an aesthetic symbol in Aldo Palazzeschi's *Il codice di Perelà*. Published as a "Futurist novel" in 1911 and later re-edited as *Perelà: L'uomo di fumo* in 1954, the book revolves around a mysterious man of smoke, Perelà, born out of the ashes of a chimney, thirty-three years after a fire, and living in the fluid form of a cloud. The "most singular oxymoron" (Perella and Stefanini 106) of Perelà's light "body" connects him to the vitalism of his burning and his exhilarated laughter ("Ah! Ah! Ah!"), along the lines of Palazzeschi's own poem "E lasciatemi divertire" ("And Let Me Have My Fun," 1910; *Tutte le poesie* 236–8). At the same time, his smoky nature also relates to the gloom of a past combustion (as revealed by his boots, the only pieces left of his former corporeity) and to the hidden pain inscribed in his name, made from the initial syllable of each of his stepmothers: Pena (Pain), Rete (Net), and Lama (Blade). Perelà's ambiguous state is made manifest right at the novel's beginning when two policemen notice a "cloud of dust" (1) on a dirt road, a blend of the character's smoke and of the specks of sand raised by their riding horses: "did you see how we covered him with dust? You could no longer tell what he was —When we got close to him, I thought I saw him disappear —It's true, so did I. —But that wasn't a man at all, you know! —What was it then? You tell us. —It looked like a cloud. —Of course, we covered him with dust. We're the ones who look like a cloud, on this damn road! —No, no, I saw him before the road became thick with dust, he's a man of smoke!" (Palazzeschi, *Man of Smoke* 2).[38] In the policemen's report, Perelà's cloud, made of fluid spirit (smoke) and incinerated matter (dust), defines him as a present-absent being, excitingly self-transforming and painfully hiding.

After this episode Perelà offers his own definition of himself during a conversation with an old woman. In response to her question "Just what are you?" (*Man of smoke* 1), he purposely omits his name, relating his being instead to his quality of lightness: "I am light ... a light man ... very, very light ..." (*Man of Smoke* 1; io sono ... io sono ... molto leggero. Io sono un uomo molto leggero). Perelà's "I am ..." – in reply to the woman's question and, implicitly, to the "Chi sono?" of Palazzeschi's own poem ("Who Am I?," 1910) – reveals, yet conceals, his identity in smoke. In one sense, Perelà relates his lightness to movement (an analogy with the transformism, mockery, and acrobatic dances of the *serate futuriste*), to humour, to a process of elevation to a higher view of reality (as in his own ascent to Mount Calleio), or even to a new divine ubiquity (as revealed in his final ascension to the heavens). In a second sense, Perelà's "I am ...," allegedly mocking God's self-identification as claimed in the Bible, defines an elliptical self, suspended between an evaporated presence and a declared absence. The man of smoke enacts not just the Futurist ideal of pure self-perception, configured a few years later in the image of Marinetti's "I" cloud, but also the ultimate Modernist utopia of eternal consumption without exhaustion. Unlike the biblical burning bush, his divine condition as *incendiario* (one who burns, arsonist) ends up in his tragic self-erasure, which reveals Perelà's nature as a burned being (*incendiato*). On the one hand, as in Palazzeschi's own poem "L'incendiario" ("The Arsonist," 1910; *Tutte le poesie* 181–8), dedicated to Marinetti, Perelà appears as a divine outcast, welcomed in triumph at the court of King Torlindao and entrusted with the task of writing the kingdom's new code. Thanks to his burning and provocatory laughter, he gains omnipresence in the forefront of society by being separated from the masses. On the other hand, like the Marinetti of the poem, hung in a cage and exposed to public scorn, Perelà also constitutes a threat to the tranquillity of Torlindao's kingdom. After the suicide, by self-incineration, of the servant Alloro, justified by his will to "become light" (*Man of Smoke* 153) like Perelà,[39] the man of smoke is indeed put on trial and condemned. Unlike the protagonist of "L'incendiario," Perelà acquires the traits of a man in pain, as foreseen in the tea episode in which the women of the court confide to him their sorrows (melancholy, homicide, violence, deaths, betrayals). In anticipation of Palazzeschi's later manifesto of "Controdolore" ("Counter-Pain," 1914), Perelà's smoke, as a by-product of a fire, refers to an a posteriori process of comprehending human pain through laughter. Accordingly, lightness is the privileged way to decipher the depths of heaviness, as envisioned in the character's dancing walk, in contrast to the gravity of the final procession towards his cell ("in the centre the condemned man walks briskly, almost rising from the ground with each step he takes, while the procession forms and moves ahead gravely, slowly, at a funereal pace"; *Man of Smoke* 225).[40]

Palazzeschi's metaphor of a smoking cloud or a man of smoke represents the forging of a new metamorphic, modern being, whose identification and elaboration find expression at many levels throughout *Il codice di Perelà*. First, in his tragic laughter and rarefied or erased body Perelà reveals or hides Palazzeschi himself or, rather, the *monstrum* (as both portent and monster) of his difference as a *saltimbanco*, a homosexual, and a neutralist. The similarities between Perelà and the poet appear in the poems "Chi sono?" and "L'incendiario." In "Chi sono?" Palazzeschi identified with an acrobat clown (*saltimbanco*), "leaping" on stage with a deforming lens over his heart, provokingly distorting his codified role as poet – "am I perhaps a poet?" (verse 1; sono forse un poeta?) – yet tragically hiding in his performance the "follía," "malinconía," and "nostalgía" (*Tutte le poesie* 367; folly, melancholy, and nostalgia) of his soul. In "L'incendiario" he referred Marinetti's condition as an arsonist to his burning clownery – "he burns for fun" (*Tutte le poesie* 183, verse 63; brucia per divertimento) – presenting his attempt to destroy the old codes and inflame the world with new life, as a divine prodigy worthy of marvelling and as a threatening diversity worthy of public scorn.[41]

Second, in his cloud Perelà reveals Palazzeschi's ongoing negotiation of the poet's role in industrial societies, expressed in his Modernist portrayal as a "sacrificed saviour" (Starobinski 117). The combination of his clownish "code of lightness" (Saccone 67) and his Christ-like attributes (thirty-three years of age, an unfair trial, the final ascension) depicts him as a mystical and tragic circus performer. The identification of the modern poet with a sad *saltimbanco* draws its origins from the French contemporary reappreciation of the commedia dell'arte (Green and Swan 7–8), and, most likely, from Baudelaire's short prose "Le vieux saltimbanque" ("The Old Acrobat Clown," in *Petits poems en prose*, 1869), which presents the gloomy figure of a clown after the "uninhibited overflow of the vital energy" (80) of a circus show.[42] The association of the modern artist with circus, ballet, and *funambolisme* (as an island of marvel and childish spontaneity in opposition to the standardization produced by industrialization; Starobinski 8) will also appear in the works of Edgar Degas, Georges Seurat, and Henri Toulouse-Lautrec, in Pablo Picasso's reinterpretation of popular figures like Pierrot and Harlequin into innocent victims (in *Family of Saltimbanques*, and *The Death of Harlequin*, 1905), and in Georges Rouault's Christ-like clowns.

Third, in his mix of *fumismo, funambolismo*,[43] and tragedy, Perelà "embodies" a modern Italian archetype, in his original synthesis of Pulcinella and Hamlet. His smoky "costume" (similar to the white robe of Pierrot or Pulcinella) matches the insatiable urge for action and the illogical and analogical movement of deformation and transformation typical of the commedia dell'arte. At the same time, as the character is suspended in its endless vault over the abyss of its absence and the inexorability of fixity, his body represents the Hamletic dilemma of a presence and an absence.[44] Unlike Marinetti's "cloudifications," Perelà represents a paradoxical hero, on the edge of fluidity and solidity, ubiquity and nowhereness, self-exhibition and self-erasure. His body – lightly elevated in the heavens, flying in the abyss of emptiness, and rarefied into a laughing cloud in the book's conclusion – represents a new experimental being, analogically conceived as an open-ended space of chaotic interaction and a creative synthesis of endless possibilities.

Palazzeschi's own association with a laughing and self-erasing cloud of smoke prefigures his construction of himself as a comic victim of the later war memoir *Due imperi mancati* (*Two Failed Empires*, 1920), which fictionally evokes a posteriori the "dirty fire" (7; fuoco immondo) of the First World War with a mix of comedy and a tragic sense of its apocalypse. Palazzeschi's creative characterization of Perelà in the form of a man of smoke also precedes a common topos of war narratives, which similarly equated mankind and the world with a vibrating or destructive smoking cloud. The image of a cloud pierced by an aeroplane represents the Futurist excitement for war in Ardengo Soffici's poem of 1915,

"L'aeroplano" ("The Aeroplane"), which was dedicated to the flying ace Francesco Baracca and reflected (on the model of *Zang Tumb Tumb*) the poet's elevation in combat, his dissolution into matter (as he "swallows" air and "penetrates" steam, in an explicit sexual metaphor), and his transformation into a ubiquituous being.[45] Conversely, clouds of smoke also suggest the dread of war in the photographic and verbal narratives of the conflict by their association with fog (during the battle), the greyness of trench warfare, or the soldiers' clouded perception of an infernal life in death. The metaphor of the cloud of smoke thus presents the war both as the "apocalypse of modernity" (apocalisse della modernità, in the words titling Emilio Gentile's book) and as a collective experience of blindess – as indirectly symbolized in contemporary fiction in the pathological sightlessness of Pietro, the protagonist of Federigo Tozzi's 1919 novel *Con gli occhi chiusi* (*With Closed Eyes*), or in the moral paralysis of the title character in *Rubè*, Borgese's novel of 1921.

Cigarettes and the Language of Psychoanalysis

The outbreak of the conflict coincided with the boom of the cigarette industry. The sudden halt of provisions from Turkey, Bulgaria, Macedonia, and Russia, caused by war, forced Italy to reduce its reliance on imports and to increase its own tobacco production. As the United States entered First World War in 1917, cigarettes, which Americans widely distributed to Italy's population in the aftermath of the Caporetto defeat, became associated with a renewed spirit of freedom, a reawakened national cohesion, and a strong sense of empathy towards America's democracy. As they became more popular on the front, cigarettes turned from being "a new commodity of morale" (Brandt 53),[46] or a marketing tool of America's capitalistic modernity, into a mass-consumed industrial product. For Italian soldiers, cigarettes became emblematic elements of war experiences, enacting pleasurable breaks from the conflict's drama, a spirit of comradeship (even among enemies), a space of creativity (as for Giuseppe Ungaretti, who wrote the war poems of *L'allegria* on cigarette packages), and free laughter, as seen in the image of Soffici's 1917 war diary *Kobilek* (*Kobilek*) of soldiers mocking the imminence of death while smoking their last cigarettes.[47]

Against the back-drop of the First World War and the industrial affirmation of cigarettes, smoke became a defining symbol of modernity in Svevo's novel *La coscienza di Zeno* (1923), identifying not just its intrinsic capacity for self-destruction and reconstruction but also a new modern self. The symbol of an atomic cloud, or *nebulosa*, destroying the earth with its fatal explosion, significantly ends the novel: "there will be an enormous explosion that no one will hear, and the earth, returning to its gaseous state, will roam in the heavens bereft of parasites and diseases" (474).[48] The image represents the total self-destruction of mankind in war and, at the same time, defines the paradox of modernity itself as a "maelstrom of perpetual disintegration and renewal, of struggle and contradiction, of ambiguity and anguish" (Berman 15). While condemning the hecatomb of the First World War as the tragic outcome of the uncontrolled growth of Western capitalistic societies, Svevo portrays with bitter irony the end of Zeno's story as his narration, once again, "concludes rhetorically in order not to conclude" (Minghelli, *In the Shadow* 194) and as the character's own smoking cloud extends to the whole universe. The final cloud is thus an outward expression of Zeno's own "cloudification," manifested – in his passion for and obsession with cigarettes – as a process of both pulverization and metamorphosis.

Zeno's cloud of smoke becomes a metaphor of psychoanalysis as a new way of being. From an ideological perspective, the figure of Zeno is the subject of Svevo's attempt to filter Freud's theories into Italian culture through the voice of Svevo's own Triestine doctor Edoardo Weiss.[49] Despite the novel's later success, psychoanalysis did not develop in Italy until the late 1940s, for scientific, philosophical, ecclesiastical, and political reasons related to the Italian dominance of Lombroso's positivist approach,[50] the ideological opposition of Croce and Gentile (who, as minister, suppressed the teaching of psychology in secondary schools), the prolonged scepticism of the Catholic Church,[51] and the Fascist resistance to Freud's thought, which was discarded – in light of the "healthier" ideology of *latinità* – as Austrian and Jewish.[52] At the same time, from an aesthetic perspective, the figure of Zeno also embodies Svevo's attempt to depict the forging of a new, malleable self, whose traits coincide with the secularized conscience and self-reflexive thought brought forth by psychoanalysis. In this sense, along the lines of Palazzeschi, Svevo's "man of smoke" constitutes a creative reinvention of Perelà. Like Perelà, Zeno is light and fluid in his (self-) irony and his ever-shaping narrative movement. He is also an outcast, swinging between his bizarre role as *saltimbanco* and his nature as *inetto* (unfit), his mask as a buffoon and his latent anguish (foregrounded by the previous suicidal characters of Svevo's *Una vita* and *Senilità*). As a *fumiste* and a *funamboliste* he is a "man of infinite possibilities" (Minghelli, *In the Shadow* 165): in his endless transformation, in his risking to laugh at himself, and in his refusal to be imprisoned in a stable form. As *in-aptus* (or unassimilated), he personifies the poet's diversity in contrast to the homologation of industrial society or bourgeois institutions, by his stressed deviation from normality and his taste for deformation (of others' facial traits or behavioural codes). Unlike Perelà, however, Zeno's smoky mask does not instigate an exploration of pain but rather prolongs the hypocrisy of bourgeois tranquillity, safely suspending the trauma of his father's death as a way to keep moving without falling into stillness or heaviness. The lightness of smoke therefore coincides with Zeno's attitude of derisive irony, manifesting through humour the aggressive tensions (against the father, Guido, Ada) that would not be allowed explicitly because of moral and social restrictions.

Against the back-drop of this general imagination of smoke, Svevo refers smoking exclusively to cigarettes. From a historical viewpoint, his choice of cigarettes mirrors their post-war transformation into one of Italy's main industries, as confirmed in their growing share of the Italian tobacco market (from 0.05 per cent in 1880 to 40.5 per cent in 1925; Diana 99) and in the increased national production of tobacco plants (from 6,247 tons in 1900 to 46,844 tons in 1924; Diana 92), shifting Italy's reliance on export (from 81.8 per cent of total consumption in 1919 to only 24.6 per cent in 1928; Diana 97), and later constituting one of the financial engines of the Fascist state. From a literary viewpoint, Svevo's choice of cigarettes deliberately outlines a metaphor for writing itself. For Zeno, they become a pause of virtuality, a space of recovery from hypocrisy, and the ideal locus for testing and exploring different moralities. At the same time, they re-enact the dialectic of self-affirmation and self-deceit of Zeno's *coscienza*, revealing both its loss (in the Christian sense of the word) and its fictional recreation. As the smoking theme develops throughout the book – from the chapter "Il fumo" ("Smoke") to "Morte di mio padre" ("Death of My Father"), until the final "Psicanalisi" ("Psychoanalysis") – cigarettes become external appendages of Zeno's "soul" and constitute the existential and narrative engine of his own self-told story. On an existential level, they stage the Freudian tension between the excitement of desire (in overcoming a prohibition or in affirming one's absolute freedom) and the character's malaise, expressed in his disgust and physical suffering. At the same time they also reveal Zeno's *empasse* between the excitement of his metamorphoses and the remorse at the loss of his father. The succession of final cigarettes indeed manifests Zeno's continuous effort to cover his misdeeds (a loveless marriage, the adultery with Carla, the responsibility for Guido's death), to suppress pain in the exposed anti-heroic mask of sickness, to annul the perception of his own guilt, and to deceitfully create "an artificial peace of conscience, lacking a real one" (Spagnoletti 124). On a narrative level, cigarettes reflect the complex mechanism of Zeno's memory. In the chapter "Il fumo," cigarette smoking sparks his work of remembrance and writing, in between present discovery and past recovery, enacting his fluid and continuous idea of *scribacchiare* (scribbling). In a similar way, Zeno's choice of constantly postponing his last cigarette to an ever-changing major date reflects an ambivalent relationship between history and narration, where any attempt to order the events according to a definitive meaning or a coherent plot is constantly deferred to a goalless and open-ended fictionalization (Savelli 102).

While propelling Zeno's existential and narrative self-investigation, cigarettes ultimately refer his disease to his way of thinking. Staging a *querelle* between the father's cigars and the son's cigarettes, Svevo applies the contrast between old and new ways of smoking to the difference of their philosophies. A cigarette marks the split between their two ways of perceiving the world, as ironically underlined at the father's death: "'15 April 1890, 4:30 p.m. My father dies. U.S.' for those of you who don't know, those last two letters don't mean *United States*, but *ultima sigaretta* [last cigarette]" (91).[53] As hinted in Zeno's transfer from legal to chemical studies,[54] cigarettes reflect the son's distinctive (and repeated) rite of passage from his old self-definition in relation to authority (law) to his new self-ascertainment in relation to contamination (chemistry). This transition illustrates the shift of mentality from the father's idea of *uomo finito* to Zeno's new essence as an unfinished man. In relation to the father's authority, which is imbued with seriousness and religiosity, cigar smoking represents the idea of a *vita naturalis*, adequate to being and grounded on its motionless foundation. Within his horizon of "sincere acceptance of virtue" and "no movement" (94), cigars symbolize a positive remedy and an ineffable pleasure, which is impossible to grasp or define (like the word for which the father is searching), reflecting the unknowable mystery of Being. In relation to Zeno's discredit of religion and his "habit of laughing at everything" (101; abitudine a ridere di tutto), cigarette smoking visualizes instead the new idea of a light and transitory being, ironically redefinable through words and constantly adjustable by way of the character's "impetuous impulse towards betterment" (93; impetuoso conato al meglio). In their association with Zeno's poisoning vice, leading him to psychoanalysis, cigarettes define the son's disease in the father's eyes as a "disease of the word" (134; malattia della parola), that is, as a philosophical disease, reducing Being to the *loquabilis,* and the "I" to self-reflexivity. In the contrast between the father's attitude "to look above" (115) and Zeno's ironic and distracted self-centredness, Svevo portrays the epistemological difference between the two as a gap between the old idea of a created self (in relationship to the otherness of Being) and the new idea of a self-created or -creating "I" (as both subject and object of being). After the father's last smack, non-verbally attempting to wake him from his poisonous philosophical dream, Zeno's identity will be defined, after the trauma over a lost Being, only by the lightness of his thinking and being thought, watching and being watched, writing and being written. His *coscienza* will coincide hence with a presence or an absence and with the endless movement of his verbal acrobatics (*loquor ergo sum*) – that is, with the very act of smoking.

As documented in their various associations with fashion, war, lightness, self-annihilation, pleasure, and thought, cigarettes ultimately envisaged the malleability, ubiquity, and a-corporeality of a new modern being. In parallel with these qualities, the metaphor of the cloud of smoke similarly reveals, along with the Italian fascination for and resistance to industrialism, an "unfinished" space of trial, action, fusion of opposites, and self-reflexivity, which constitutes the hidden and creative laboratory of a new experimental modernity.

Chapter Seven

Toys, Clothes, Furniture, and the Aesthetic Power of Play

In the spring of 1919 the Futurist painter Fortunato Depero opened an art house (Casa d'Arte) in his native town of Rovereto, in the Northern region of Trentino, which had been incorporated into Italy after the end of the First World War. The nation was in tumult. In March of the same year Benito Mussolini launched his Fasci di combattimento (Combat League) with his speech in Milan's Piazza Sansepolcro; in April Prime Minister Vittorio Emanuele Orlando left the negotiation table in Versailles (before resigning in June); and in September D'Annunzio occupied the Istrian city of Fiume, in opposition to U.S. President Wilson (who had assigned it to Yugoslavia), to redeem Italy after the nation's "mutilated victory." In addition, an unprecedented wave of strikes and social unrest was agitating Italian industrial cities, fomenting a widespread fear of an incoming Soviet revolution in Italy – as reflected in the labelling of 1919 and 1920 as the "two red years" (biennio rosso). Within this context Depero's move from Rome to the former Italian warfront marked a symbolical counter-itinerary in the spirit of post-war reconstruction, vis-à-vis the collective paths from the North to the capital that had been represented over the same years by the national caravan of the unknown soldier from Aquileia to Rome (1921)[1] and by the Blackshirts' March on Rome (1922). At the same time, Depero's post-war attempt to "reconstruct" in his art house an aesthetic space for modern living (following Giacomo Balla's model in Rome) aspired to give visible form to the ideals stated in his manifesto "The Futurist Reconstruction of the Universe," co-authored with Giacomo Balla and published on 11 March 1915. Over the following years Depero's Futurist house, born as a workshop for the production of art tapestries and later turned into a laboratory of decorative art (and a permanent gallery since the 1940s), would represent not only a unique platform for the evolution of new aesthetic forms but also the sole experiment of its kind "to have any real success with the public and to last over time" (De Guttry and Maino, "From Artist-Artisans" 49).

In the house in Rovereto, Depero first stored his handmade tapestries and the puppets of his theatrical show *Balli plastici* (*Plastic Ballets*), designed with the Swiss writer Gilbert Clavel and performed eleven times in Rome in the spring of 1918.[2] Both ideas – of the plastic theatre and of the "quadri in stoffa" (paintings on cloth) – had come to him as unexpected evolutions of a failed project for Igor Stravinsky's ballet *Le chant du rossignol* (*The Song of the Nightingale*), commissioned by the Russian producer Sergei Diaghilev in 1916 and later assigned to Henri Matisse. Diaghilev's rejection of Depero's costumes had left the artist with a large quantity of unused sketches, materials, and fabrics (Scudiero, *Depero* 15).

After his encounter with Clavel in Rome and his stay at the writer's house in Anacapri in 1917, Depero converted some of the ideas for Stravinsky's ballet into the new project of *Balli plastici*, conceived as an anti-naturalist children's theatre, "idealistically directed toward a global reconstruction of the world through a reassertion of values associated with primitivism, childhood, dreams, magic, fantasy, and play" (Belli, "Gilbert Clavel" 193). In the design of its playful space Depero experimented with the free combination of colours, materials (tin, fabrics), and geometrical forms (cones, cubes, cylinders). In the design of its plastic characters – a mixture of enchanted puppets and mechanical marionettes – he conveyed the abstract image of a childish, robotic universe, mirroring his aesthetic ideal of infinite invention, endless self-multiplication, and unlimited automatic motion. By recreating the same imagery and transferring the show's repertoire of themes onto a bidimensional support, Depero drew *Balli plastici* simultaneously as a poster advertisement and as an equally titled canvas (based on the play's fifth scene; Scudiero, *Depero* 19) in 1918.

While working on his plastic theatre, Depero also adapted the preparatory work for *Le chant du rossignol* and the leftover fabrics from the project as a starting point to develop a new technique of cloth painting. In 1920 he designed and realized the tapestry *Il corteo della gran bambola* (*The Great Doll's Parade*; plate 22), displayed, with the puppets of the ballets, in the house's *sala delle tarsie* (room of marquetry works). The tapestry stages the image of a doll, carried on a podium by three tin puppets and honoured by three-unit ranks of marching heralds, fairies, and knights. As in the rhythmic procession of fairy-tale toys and bellicose figures (alternating fluidity and rigorous movements), the "body" of the doll (in between a puppet and a marionette) subsumes a similar tension of magic and mechanics, turning into an abstract site for fusing and testing new decorative solutions (as in the design motifs of its costume applied to the tapestry's frame).

As wished in "The Futurist Reconstruction of the Universe," Depero aspired, with his puppets, marionettes, and toys (whether actual or represented), to achieve what Marinetti defined in the manifesto as "the new Object, the new reality created with the abstract elements of the universe" (Balla and Depero 198). By virtue of his play and childlike imagination, Depero refashioned toys as living objects out of "infinite systematic invention" (199) and the free combination of their materials, forms, and "abstract equivalents" (197). As they migrated across different supports (children's theatre, poster advertising, painting on canvas, painting on cloth), object toys turned from choreographic elements into active figures, capable of "overflowing" their energy onto the surrounding environment and of merging different art forms into a common ambiance. Blurring the divide between high and low art, objects and artworks alike were thus both actors and miniatures of the larger aesthetic space of the house, as indistinct plastic "bodies" and as concurring elements of a new aestheticized universe. As the house turned into a museum in the 1940s, Depero's experimental playthings would become not just the malleable and incomplete testimonies of his own ever-evolving creativity but also autonomous pieces of a new modern art within a new expositional space.

Against the back-drop of the early years of the Fascist regime Depero's work would have an impact on the development of the so-called Second Futurism and the slow transformation of Italian artisanship into industry. Depero and Balla's theory of play, their childish technological imagery, and their aesthetic research into toys, clothes, and furniture would also lay the groundwork for the evolution of Italian plastic arts and the start of national production in these sectors.

Modern Play and "Artidesign"

By virtue of Balla and Depero's deliberate application of play to artwork, daily objects like playthings, clothes, and furniture became the epicentres of a new investigational curiosity and the engines of a new creative attitude in the planning and designing of forms. Against the widespread notion of playthings as tools of the containment, education, and discipline of children, and in line with the renewed Modernist interest in games, toys, and puppetry, Balla and Depero elaborated a radically new idea of play as a chaotic act of creative deformation and an adult anti-method of aesthetic experimentation.

By the early twentieth century the notion of play mostly referred to children's pedagogy. Following ancient roots from Plato – who first related child (*páis*), play (*paizéin*), and playthings (*páigma*) to education (*paidéia*) – and from Rousseau's illuminist rediscovery of the child, play represented an essential form in the aesthetic and moral formation of the modern citizen.[3] After centuries of rejection, in which it was identified as a distraction from prayer life (during the Middle Ages), as a useless frivolity (in the Humanist era), or as a non-serious activity (in Protestant ethics), play powerfully re-emerged in nineteenth-century industrial societies as the vehicle of a new *paidéia*, offering children miniatures of bourgeois codes and a space of apprenticeship in adult life (Sutton-Smith 152). In parallel with the age's technological advancements, toy factories increasingly produced ludic replicas of modern machines, in the fields of mechanics (with early reproductions of trains and carriages, testing new solutions for the design of actual locomotives and railways) and optics (with the production of kaleidoscopes and magic lanterns, mirroring the actual evolution of photography and cinematography). After the adoption of metal and tin in the 1850s turned the manufacture of playthings from a limited-scale artisanal craft into a mass industry, mechanical toys, automata, and dolls (like the French *poupées* or the German *sonneberg*) invaded the European market, offering a more democratic experience of play and modelling a standard form of leisure (Denisoff 6–10). Over the same period, playthings became primary tools of discipline and sites to vent childish instincts in relation to schooling, as reflected by the foundation in 1907 of Maria Montessori's first Casa dei bambini (Children's House) in Rome. In her pedagogical method, play promoted a rational discipline based on social consciousness and scientific judgment (by way of its empirical training of "the tactile, thermal, baric, stereognostic, visual, auditory, chromatic, and olfactory senses of the child's body"; Stewart-Steinberg 345), and also represented a way to build a cohesive national bond (by virtue of its capacity to foster disciplined movements in a disciplined body).

Meanwhile, detached from its association with childhood, play had also acquired an important cultural function in the Italian literary imagery of the age as a core metaphor of the new experience of *jouissance* (pleasure, in D'Annunzio's terms) brought forth by modernity, and as an adult space of fictional investigation.[4] The thrill of adult play related to competition (*ágon*), as documented in the growing success of sports culture or in the practice of sword duelling, evolved into a literary topos in Serao's *La conquista di Roma* (1885; 4, book 2), D'Annunzio's *Il piacere* (1889; 5, book 1), and Pirandello's *L'esclusa* (*The Outcast*, 1901; 2, part 1).[5] As an extension of this dimension, play took the form of an excited vertigo (*ílinx*) for speed and extreme danger, as emblematized in Marinetti's daunting car race on the verge of Death in the "Manifesto of Futurism" (1909) or in D'Annunzio's flying contest of *Forse che sì forse che no* (1910), opposing the pilots Paolo Tarsis and Giulio Cambiaso in a fatal game for the bold conquest of a record and their woman's heart.[6] Simultaneously, play represented a separate space of self-contained invention, as reflected in the bourgeois practice and literary trope of the *seduta spiritica* (séance) in Fogazzaro's *Piccolo mondo antico* (*Little World of the Past*, 1895; 1, book 3), Pirandello's *Il fu Mattia Pascal* (chapters 13–14), or Svevo's *La coscienza di Zeno* (1923; chapter 3).[7] Finally, in connection with a status of detachment and mental rapture, play referred to gambling, as envisaged in the addiction and ruin of Cesarino in De Marchi's *Demetrio Pianelli* (1890) or in the "fever of gaming" made manifest in the roulette scenes of Pirandello's *Il fu Mattia Pascal* (chapter 6) and Antonio Borgese's *Rubè* (1921; chapter 15).[8]

Concomitant with these developments, play was also being reconfigured in the late nineteenth century as a chaotic force, expressing a productive "discharge of superabundant vital energy" (Huizinga 2), connected not only to contest or illusion (from the Latin *in-ludere*, "at play") but also to artistic experimentation, Bacchic irrationality, and foolish laughter. In his famous article "La morale du joujou" ("The Morale of the Plaything"), published on 17 April 1853 in *Le monde littéraire*, Baudelaire associated child's play with a non-conformist aesthetic attitude by identifying in it the aspiration to reproduce a "life in miniature," a metaphysical "desire" to overcome strict codes (e.g., the bourgeoisie's mute adoration in front of objects, or their rigid denial in Protestant ethics), and the "first initiation to art," by virtue of a child's daydreaming capacity to re-elaborate mere objects into "living toys" or "actors in the great drama of life" (681).[9] A few years later, by retrieving its original link to ancient sacred rituals, in the *Birth of Tragedy* (1873), Nietzsche envisioned play as a Dionysian force of creative destruction and as an irrational energy unveiling the mendacious harmony of static norms.[10] In continuity with the carnival's "logic of the 'inside out'" (Bakhtin 11) in medieval buffoonery and humanistic court masquerades,[11] play also appeared as a destabilizing energy of subversion in late-nineteenth-century circus and children's theatre. Although confined to a "segregated society with its own costumes, pride, and laws" (Caillois 136), circus play, clowneries, and rodeo spectacles (like Buffalo Bill's immensely popular "Wild West" show)[12] captured the imagination of crowds, writers, and painters (like Baudelaire, Degas, Toulouse-Lautrec, and Seurat, to name a few) for their capacity to express a powerful combination of primordial energy, inventiveness, metamorphosis, and movement. Likewise, children's theatrical plays inspired the fantasy of Modernist artists by mimicking a subconscious state of active imagination (later theorized in Freud's 1907 essay "Creative Writers and Daydreaming") and opening an autonomous space of genuineness, creativity, and non-conformism.

Against this background, Balla and Depero elaborated play as an adult form of rediscovered childhood,[13] as an anti-bourgeois force of rupture, and as an inventive fiction, germinating new art forms out of its bold fusion of elements. By applying it to daily items, they dismantled their previous codifications (deemed immobile or ruled by the unquestionable harmony of symmetry and good taste) and recreated the items as unrepeatable art objects. Play subsumed the function in order to overcome the perceived homologation and anonymity of industrial objects, by fashioning them in a distinctive style. In parallel with the contemporary growth of the Italian toy, fashion, and furniture industries, Balla and Depero's new planning attitude in the making of playthings, clothes, and furnishing items foresaw a fundamental shift in these sectors – namely, the passage from excellence in execution to originality in style making – which would have a profound impact on their extraordinary success in the post–Second World War years. Along with this gradual transition from artisanship to authorship, their "artidesign" (De Fusco, *Made in Italy* 13) also anticipated a new awareness of the designer as a hybrid artist at the crossroads of traditional crafts and industrial planning. By relating themselves to the archetypical figure of Geppetto in Carlo Collodi's *Le avventure di Pinocchio* (*The Adventures of Pinocchio*, 1883), Balla and Depero envisioned the artist and designer as a creative puppet master, able to recognize in raw material a "thin little voice" (Collodi, *Adventures* 83), to see in a shapeless piece of wood "a wonderful puppet who can dance, and fence, and make daredevil leaps" (89), and to fashion its body with clothes made out of the available resources ("Geppetto, who was poor and didn't even have any penny in his pocket, then made him a modest little suit out of flowered paper, a pair of shoes of tree bark, and a cap out of bread crumbs"; 133).[14] At the same time, Balla and Depero embodied in their research a new form of artistry in designing ordinary objects, conceived as an "antimethodical method" (Schnapp, *Modernitalia* 59) of chaotic deformation (by way of their subversive play) and inventive recreation (by way of their free interplay of elements, materials, and arts). In one way, their crafting of unique pieces expressed an element of resistance to the logic of industrialization, perceived as foreign (e.g., German toys, French clothes) and standardized (as opposed to the Italian artistry in furniture making). In another way, however, their ludic combination of singled-out materials[15] functionally created new stylistic variations and constructed the object as a unique but replicable *type* (as a remedy to the limited capacity of artwork to circulate). At the border of art and the market, the proto-design of Balla and Depero would gradually become a rich laboratory for the experimental evolution of a new industrial culture.

Futurist Toys and the Miniature of Industrial Modernity

In the manifesto "The Futurist Reconstruction of the Universe" Balla and Depero not only dedicated an important section to the Futurist toy but also pointed to it as the new creative model for making art. In opposition to traditional playthings, reduced to "immobile objects, stupid caricatures of domestic objects [...] fit only to cretinize and degrade a child" (199), the Futurist toy enacted the aspired "fusion of art and science [...] into a new creature" (200). Thanks to its capacity to generate abstract associations and its implicit tension with materiality, the Futurist toy was embedded with the potential to subvert realism in laughter, stretch creativity, and stir new "imaginative impulses" (199).

Along the lines of the manifesto and in continuity with his work on *Le chant du Rossignol* and *Balli plastici*, Depero started to design his own toys during the war years, producing a repertoire of wooden animals, dancing figures, Pinocchio models, and tin marionettes. Over the same period Balla also created similarly conceived playthings as experimental objects of décor and sites of artistic investigation (as their shapes and motives migrated onto his platters, vases, and chairs), rather than as children's toys for actual use.

The pioneering activity of the two artists coincided with the start of Italy's toy industry, which grew after the establishment of Metalgraf (Milan, mid-1910s; tin dolls), Lenci (Turin, 1919; washable dolls), Industria Nazionale Giocattoli Automatici Padua (INGAP, 1919; spring-motioned trains and puppets), and Cardini (Omegna, 1922; tin trains and cars). Despite the extraordinary growth of the European toy industry since the mid-nineteenth century, Italy did not develop serial manufacturing of playthings until the First World War (with the exception of Furga, founded in Mantua in 1872, producing papier-mâché masks and dolls). Several concurrent factors related to the belated development of the Italian toy industry, such as the preference for the artisanal production of playthings, the long tradition of crafting puppets and marionettes (in local folklore and in popular theatre),[16] the nation's economic backwardness, the perpetuation of an unequal distribution of wealth,[17] and a school system that heavily privileged ruled discipline over personal creativity. At its deepest level, however, the slow start of an Italian toy market and industry ultimately related to post-unification Italy's rooted cultural resistance to the perceived homologation or "impersonalization" of play (and, by reflex, of society) brought forth by industrial modernity.

In this context, industrial-made toys had made their way into the larger Italian market in the early twentieth century by way of the magazine *Il corriere dei piccoli* (*The Courier for Little Ones*), founded in 1908 as a children's supplement to *Corriere della sera.* In its advertisements, stories, and vignettes the illustrated periodical not only shaped an imagined community of young readers around playthings but also manufactured a cultural demand for them. While showcasing modern marvels in its fantasizing articles (on electricity, telegraphy, radio, or aeroplanes),[18] the magazine cultivated children's receptive minds as fertile grounds to overcome grown-ups' *misoneismo* (fear of the new). In presenting toy versions of new technologies, it also introduced new products for adults in a less constrained space, as seen in its advertisements for gramophones and cameras (or even unchildlike items like guns and cocaine). Over the years its writers and designers created *Il corriere dei piccoli* as an experimental adult platform of play, fiction, satire, and improvisation around infancy, which influenced contemporary Futurist art and the evolution of Italian comics.

Depero's research into toys was certainly influenced by the work of Antonio Rubino, an illustrator, a writer, and a painter for *Il corriere dei piccoli,* who had developed over the years a recognizable visual style, malleably combining childish clowneries, nonsense poetry, grotesque caricatures, and humorous vignettes. In parallel with him, Depero's interest in playthings also found roots in the fertile ground of the avant-gardes: Italian Futurism, which identified childlike fantasy and clownish improvisation with a pure state of artistic creation; and European Dadaism, which saw in play an overriding non-homologated force, disrupting the bourgeois "means-ends rationality" (Laxton 19).[19] Dialoguing with these sources, Depero's design of plastic toy figures (actual or represented) thus reflected the ideals of wonderment and rupture. His playthings ultimately staged this (unsolved) coexistence of marvel and disquiet, or rather this tension of creative freedom and chaotic overturning of rules, through the tentative elaboration of a synthesis between the opposing symbolisms and cultural imageries of puppet- and marionette-making.

Puppets related to popular culture and fictionalized the caricature of an imperfect world. Since the time of Burattino (a stock character in the commedia dell'arte and a puppet in nineteenth-century humorous theatre) they had expressed a carnivalesque tension to overturn the world's hierarchy through free improvisation and satire. The Italian archetype of Burattino was, of course, Pinocchio, whose creation in the opening of Collodi's book is deliberately equated with his hysterical laughter ("the mouth wasn't even done when it quickly began to laugh and mock him [Geppetto]" (*Adventures* 99)[20] and his acrobatic running and leaping from place to place. Pinocchio's association with the world of puppetry and the acrobatics of the commedia dell'arte is made even more explicit in the episode of Mangiafuoco's theatre (chapter 10), when Arlecchino and Pulcinella recognize him in the middle of a comic show and invite him to join their spectacle. Leaping on stage and being welcomed by the actors, Pinocchio enters a separate life, playfully overturning the strict ranks, hierarchies, and codes of behaviour of bourgeois society, until the menacing arrival of Mangiafuoco, who re-establishes order and wakes them from their dream. On this stage of free improvisation and childish wonder Collodi thus enacted not just Pinocchio's transformation from a puppet into a living character (anticipating his final metamorphosis into a child) but also the metonymy of a parallel space of adult play, which found similar expression, in contemporary humorous and satirical journals, as a fictional deformation or desecration of ordinary life.[21] *Marionettes,* by contrast, related to high culture and defined an abstract space of mechanical perfection (in dancing), a religious ideal of self-overcoming (as they were perfectly manoeuvred from a distance), and a mesmerizing technological attraction. In Futurist aesthetics, mechanical dancing represented an idealized model of dehumanized perfection, achieved through its accomplished fusion of man and machine. In parallel with the frantic movement of actors in the *serate,* the mechanical "bodily madness" of marionettes envisaged the physical electricity, lightness, plasticity, energy, and thrill of modern life. Along the model of Georges Seurat's painting *Chahut* (1889), associating the forceful rhythm of the female dancers with the excited voyeurism of the spectators, the ludic space of marionettes also related to explicit sexual attraction, as envisioned in Marinetti's controversial *Poupées électriques* – published in French in May 1909 and translated into Italian as *Elettricità* (*Electricity*) in 1913 (*Elettricità sessuale* since 1920).[22] While realizing the ideal vision of perfect humanoid automatons (anticipating the later creation of "robots" by the Czech writer Karel Čapek in his 1920s *R.U.R.*),[23] the play's electric dolls also visualized, in their correlation with childish enchantment and erotic excitement, the artist's wonder and thrill at the new industrial civilization.

In this context, Depero's paradoxical "geometrical puppets," often composed of disjointed pieces (cones and cylinders, etc.) of mechanical marionettes, envisioned the attempt to reconstruct a new industrial universe as an ideal synthesis of automated and human beauty. In the contrasting space of his toys, which ended up fashioning a less than celebratory image of industrialism as an ambivalent force of endless creativity and subtle dehumanization, Depero ultimately found the source of Italy's first "neo-plastic code" (Branzi 101). His imagery and style would have an impact on contemporary children's narrative (as seen in the continuous dialogue with Antonio Rubino) and on the early developments of the Italian toy industry (in the work by Attilio Mussino and Giuseppe Riccobaldi del Bava).

In Rubino's popular children's book *Tic e Tac, ovverossia l'orologio di Pampalona* (*Tic and Tac, or Rather Pampalona's Clock*), written and illustrated in 1922, the former illustrator of *Il corriere dei piccoli* proposed an equally ambivalent vision of industrial modernity in the "toy city" (9) of Pampalona, populated by both men and playthings. The book narrates the story of the toys' rebellion against their creator (Master Odilone) and men. After forcing the toy gears Tic and Tac out of the mechanism of the tower clock and imposing their backwards "toy hour" (72), playthings proclaim themselves equal to men and transform Pampalona into a carnivalesque mechanical kingdom. After losing control of the situation, Odilone finally rescues Tic and Tac and re-establishes normality in the city. Despite its happy ending, Rubino's nightmarish vision – of automatic toys revolting against their inventors, replacing human intelligence, and losing grasp of their acquired omnipotence – highlights the similar paradox of industrial modernity, seemingly reducing overpowered men to automated toys or gears in a larger mechanism.[24] In their fantastic power and their obscure, anarchic force Rubino's toys also refer to the dual nature of play as a magic and chaotic energy, ultimately revealing the coexistence in hyper-technological societies of order and symmetry (as in Tic and Tac's mechanism) and a latent state of uncontrollable destructiveness.

In parallel with Rubino, two other designers of *Il corriere dei piccoli* formed new plastic languages around playthings, in partnership with the contemporary expansion of the toy industry. Attilio Mussino, illustrator of the acclaimed 1911 Bemporad edition of *Le avventure di Pinocchio,* collaborated with Cardini, a company based in Omegna, which started in 1916 as a manufacturer of metal sheets and then converted, during 1922–9, into a factory of tin toys. Leading the way into Cardini's strategy to turn playthings into aesthetic pieces, Mussino advertised the factory's toys on the pages of *Il corriere dei piccoli,* mastered tin lithography, and designed packaging as functional and aesthetic scenarios for play (e.g., in the reconfiguration of boxes into train tunnels or car garages). Like Mussino, Giuseppe Riccobaldi del Bava (illustrator for *Il corriere dei piccoli* from 1916 to 1927) also collaborated with the toy industry to organize and promote the first Exhibit of the Italian Toy in 1929, an event that would be replicated in 1935 at the Trajan's Market in Rome under Mussolini's patronage. His work on children's illustrations and his association with the journal *Valori plastici* would then constitute the basis for his advertising career and for his cinematographic collaborations – with Ambrosio Films in the early 1920s and later with First National Film (one of the leading American production houses in Italy).

From Futurist Clothing to National Fashion

By virtue of the new centrality of play in the production of objects, and of the renewed attention to materials outlined in "The Futurist Reconstruction of the Universe," Balla and Depero's manifesto also set in motion a new aesthetic investigation on clothing, which anticipated and promoted the early developments of Italian fashion in the 1920s. Even though painters (such as Giovanni Boldini), writers (for example, Gabriele D'Annunzio), illustrators (like Marcello Dudovich for the Mele series), and journals (such as *Margherita,* named after the queen in 1878) had designed or imagined a space for a national style, it was still "impossible to speak of an Italian fashion independent of the French" (Gnoli 9) between the nineteenth and twentieth centuries.

Balla first attempted to construct a new aesthetics of fashion and an Italian clothing style in the 1910s. In his 1913 "Futurist Manifesto of

Men's Clothing" he advocated for "brilliant colors and dynamic lines" (132) in the design of men's suits and experimented with tailoring in accordance with his stated ideal of a new Futurist fashion defined as "dynamic," "aggressive," "shocking," "energetic," "violent," "flying," "peppy," "joyful," "illuminating," and "phosphorescent" (133). In 1914, in parallel with the interventionist debate preceding Italy's entrance into the First World War, Balla released the manifesto "The Anti-neutral Clothing," similarly arguing against the neutrality of bourgeois menswear (imprisoned in the codes of good taste, symmetry, pale colours, and balanced mediocrity) and advancing the need to redefine clothes – according to the Futurist ideal – as short-lived works of art. In a similar way, Depero envisioned clothing as a creative space of free imagination and as an ideal ground for breaking pre-established bourgeois codes. From his initial intuition of turning fabric into a new support for art, elaborated while working on the costumes for *Le chant du rossignol*, he developed the idea to overlap cloth painting (his tapestries) with clothing design (in the creation of decorative Futurist waistcoats for Azari and Marinetti; Greene 214–15).[25]

Based on Balla and Depero's "Futurist Reconstruction of the Universe," other Futurist artists took interest in clothing design in the immediate aftermath of the First World War. In 1919 the painter and sculptor Thayath (Ernesto Michahelles) designed a Futurist mono-size garment for everyday life (the TuTa), which brought him critical acclaim and led him to begin collaborating with the Parisian couturière Madeleine Vionnet. In 1920 the Futurist writer Vincenzo Fani Ciotti, known as Volt, published the "Futurist Manifesto of Women's Fashion" (featuring Futurist painter Tullio Crali's illustrations of gowns), in which he first designated fashion as an art, equal to architecture and music,[26] and woman as "a walking *synthesis* of the universe" (quoted in Braun 40, emphasis mine). In the manifesto Volt also laid the ground for a new aesthetics of women's fashion by indicating the qualities of the Futurist dress (identified in ingenuity, daring, and economy) and by formulating the Futurist ideal to transform the elegant lady "into a real, living three-dimensional complex" (40). Like Volt, who further re-elaborated Balla and Depero's theory, Marinetti also developed an indirect reflection on Italian modern fashion in the early 1920s. In his "Contro il lusso femminile" ("Against Feminine Luxury," 11 March 1920) he attacked the corrupted morals of the post-war period (female prostitution and male sexual folly), also targeting, albeit indirectly, the extravagance, chic, and lavishness of French fashion that still dominated the Italian market.[27] Once again without being explicit, Marinetti alluded to fashion as a new aesthetic science of materials, in the manifesto on tactility ("Manifesto sul tattilismo," 11 January 1921), by classifying four types of tactile sensations in relation to a detailed taxonomy of fabrics and their endless combinations.

Paralleling Futurism, a new interest in original Italian fashion developed after Lydia De Liguoro founded the magazine *Lidel* in 1919. The magazine, whose acronym indicated both the initials of its founder and the list of its sections – "Letture," "Illustrazioni," "Disegni," "Eleganze," "Lavoro" ("Readings," "Illustrations," "Drawings," "Elegance," "Work") – was an instrument to educate women in good taste, promote the Italian model of a sober elegance (vis-à-vis the French luxury), and launch Italy's fashion industry (separating it from the reliance on British and French imports). The magazine, whose covers were illustrated by prominent designers like Marcello Dudovich, Brunetta Mateldi, and Bruno Munari, hosted important contributions by well-known intellectual figures like the later Nobel laureates Grazia Deledda and Luigi Pirandello; the founders of the Novecento movement Margherita Sarfatti and Massimo Bontempelli; the Futurist painter Carlo Carrà; the fashion pioneer Rosa Genoni; and the feminist writers Ada Negri, Matilde Serao, and Sibilla Aleramo. In the context of the post-war transition and early Fascism, *Lidel* was a fundamental tool not just in supporting national production (as it was directed to a female readership with spending power) but also in constructing a critical discourse and a relatively uncensored space of creative imagination around women's fashion (as opposed to the controlled homogeneity of men's fashion, dominated by the standard clothing canons of business attire or military uniform).[28]

After the first National Congress of the Clothing Industry organized at Rome's Piazza del Campidoglio in 1919, the Exposition des Arts Décoratifs et Industriels Modèrnes (Exposition of Modern Decorative and Industrial Arts) in Paris in 1925 was a crucial moment for the development of Italian fashion. The event launched the international career of Maria Monaci Gallenga, who was awarded the Grand Prix for her gowns and fabrics. During the 1910s the Italian designer had worked to perfect a new technique for printing fabrics and to create an original set of decorative moulds for textiles that elaborated on the repertoire of Italy's medieval and Renaissance paintings.[29] After her initial recognition at the Panama-Pacific Exposition of San Francisco in 1915, and the success of her itinerant exposition of modern art in the Netherlands and Belgium (organized in 1921 in collaboration with the glass artist

Vittorio Zecchin), Monaci Gallenga's definitive acknowledgment of 1925 led to the opening of her *boutique italienne* in Paris in 1928. In the same Exposition of Modern Decorative and Industrial Arts of Paris in 1925, Depero presented his advertising posters, tapestries, and toys, and the waistcoats designed for Marinetti, Prampolini, and Balla (which were worn in the celebratory photographs of the group on top of the Eiffel Tower). The success of his gallery granted him international appeal and the first artistic commissions by American buyers (which would take him to New York in 1928). His works also captured the attention of the Italian noblewoman Elsa Schiaparelli (a friend of Marcel Duchamp and Jean Cocteau), who also visited, in the same complex of the Paris exposition, the first Surrealist exhibit (featuring the works of, among others, Giorgio De Chirico, Paul Klee, Joan Miró, and Pablo Picasso). Inspired by Futurism, Dada, Surrealism, and the radical style of the Parisian couturier Paul Poiret, Schiaparelli established her maison in Paris in 1927 and also developed, along with her visionary idea of fashion as art, her revolutionary style in designing clothing: in applying the Futurist theory of "colori parlanti" (speaking colours) to fabrics and inventing the famous "shocking pink"; in transposing the pictorial artwork of Joan Miró onto gowns; and in collaborating with Salvador Dalí in the realization of her trademark Lobster Gown for Duchess Wallis Simpson (Riello 99–102).[30] Concomitant with the transformation of Italian fashion into a style-making enterprise, operated around the artistic mastery of Monaci Gallenga, Depero, and Schiaparelli, Italian accessory fashion also started to emerge in those years, along with the reconfiguration of specialized local workshops – of leather goods (Gucci, 1921), handbags (Fendi, 1925), and shoes (Ferragamo, 1927) – into recognized brands for a larger market (Steele 91).

Against the back-drop of these developments, Fascism openly promoted the creation of an Italian clothing industry in the 1920s as a way to establish a distinctive national style. The evolution of Italian fashion during the Fascist years represents "a particularly illuminating window" on the dictatorship, enabling the observation of "mechanisms both of social control and resistance to that control" (Paulicelli, *Fashion* 148). Mussolini encouraged the project of Italian Fascist fashion (both industry and style) as a consensus- and nation-making tool. He identified the military uniform as a visual system of unification for both children – the *Balilla* boys and the *Piccole Italiane* (little Italian girls) – and party officers and invested in new materials like rayon (with the foundation of SNIA-Viscosa, which became, under Senatore Borletti's direction, the world leader in the production of artificial silk). He also instituted the Ente Nazionale della Moda (National Fashion Agency, Turin, 1932) for the Italianization of fashion and promoted the construction of a cultural discourse around the new industry by publishing Cesare Meano's *Commentario dizionario italiano della moda* (*Italian Commentary-Dictionary of Fashion*) in 1936, translating all French terminology into Italian and relating all Italian products to their literary antecedents (Paulicelli, *Fashion* 75–8). Despite the Duce's influence, however, fashion, and in particular women's fashion, silently acquired over time a creative space of freedom, less controlled by censorship, and sometimes even offering opposing ideological views to the those proposed by the regime, as documented in the advertising work by Marcello Dudovich and Gino Boccasile. In his series for La Rinascente, Dudovich designed images of emancipated women – joyfully walking on the street, putting on their make-up, or playing on the beach in their bathing suits – which were shockingly in contrast to the Fascist rhetoric of the woman-mother educating children at home. In his famous advertisement for the popular Fiat Balilla in 1934 (Cogeval and Avanzi 27), he similarly portrayed a new space of unsubordinated femininity through the image of a woman, dressed in elegant clothes and having a slim silhouette, confidently walking towards a car as if to drive it. In the same way, Gino Boccasile, official illustrator for Mussolini and inventor of the so-called *ventennio* graphic style, represented women's fashion with a less controlled space of unbridled fantasy. In his renowned character of the Signorina Grandi Firme (Miss Famous Brands), replicated in endless variations during the 1930s, Boccasile designed the sensual and hedonist image of a modern Italian woman, represented with long exposed legs or in a bathing suit (or even bare chested, defying the regime's ban of nudity). Captured in all but maternal poses – while weighing herself, running errands, reading newspapers, travelling, singing, or playing tennis – the Signorina Grandi Firme was not the angelic woman idealized by the regime but rather a diva, charged with eroticism and desire, envisioning in her ever-changing clothes a revolutionary idea of modern femininity.

The Art of Furniture

In 1914, while working on the designs for Futurist menswear and the anti-neutral suit, Balla started to execute playful, geometric children's furniture for his daughter Elica's room. Over the following years he designed and realized magazine racks, chairs, decorative objects, patterned-wood furniture, painted tiles, and tapestries (Bosoni, *Modo italiano* 156; Cogeval and Avanzi 98–9), which later appeared in his Roman art house and playfully reconfigured the space of his own residence in line with the theory of the Futurist reconstruction of the universe. Depero also worked during the war years on aesthetic forms of fabric, tin, and wood, producing toys (and their costumes) that had been conceived as the furniture items, or rather the theatrical apparatus, of his Casa d'Arte.

In parallel with Balla's and Depero's work, the idea of Futurist furniture making evolved after the war years. On 16 December 1916, Arnaldo Ginna, co-author with Balla and Marinetti of the avant-garde film *Vita futurista* (*Futurist Life*, 1916), published "Il mobilio futurista" ("Futurist Furniture") in *L'Italia futurista.* The text is a generic tribute to Marinetti focused on the clichéd praise of the age of machines. The first substantial contribution on the subject came only in 1920, when the Futurist intellectual, poet, and cabinet-maker Francesco Cangiullo published the manifesto "Il mobilio futurista: I mobili a sorpresa parlanti e paroliberi" ("Futurist Furnishing: Surprise Speaking Furniture in Freedom") in *Roma futurista* on 22 February. In contrast with the deprecated immobility of traditional furniture, Cangiullo elaborated "mobile a sorpresa" (a surprising piece of furniture) as a new category, offering as its examples the Zang cabinet (polyvalent and asymmetric pieces of furniture paying homage to Marinetti's *Zang Tumb Tumb*) and the so-called *sedia a sbalzi o sedia nevrastenica* (bumpy or neurasthenic chair), envisaged as "a chair made of twitches and bumps, which will disrupt any attempt to sit down" (quoted in Albanese, xix). In continuity with the aspired "bodily madness" of the Futurist variety theatre (aimed at agitating spectators in their chairs by way of a powder that caused itching and sneezing), as well as with the theorized asymmetry of the Futurist clothing (sketched in both Balla's and Volt's manifestos), Cangiullo located the experimental site of a new dynamic art in furniture, simultaneously breaking old representational codes and energizing a new aestheticized universe.

Inspired by Cangiullo, Depero actually realized the project of the *sedia a sbalzi* in 1926 and transposed it onto a bidimensional representation in the tapestry *La festa della sedia* (*The Chair's Feast,* 1927; plate 23). *The Chair's Feast,* which was positioned next to the *Great Doll's Parade* in the *sala delle tarsie* of the Casa d'Arte, shared some common elements with it. Both tapestries stage a parade, constructed using multiple perspectives and moving planes, which centres on the toy object (a chair or a doll) as the gravitating point between two converging lines (of worshipping figures and dancing or marching puppets). Both representations deliberately juxtapose opposing tensions: of magic and war, serialization and singularity, play and fear. Lastly, both scenes visualize in the "body" of the main object or actor a space of ludic creativity and experimental decoration, extending its energy onto the surrounding choreography – of the tapestry (reconfigured as an improvised childish theatre) as well as of the art house (objects, furniture, and architecture). In the case of *The Chair's Feast,* the chair's intrinsic tension between triangles and curves is reflected in the equivalent contrasts of the composition: of mechanical and fluid movements in the puppets and marionettes, of round and square shapes in the trees, of arches and stairs in the architecture of the letters (and buildings) *B* and *T.* As confirmed in a 1936 letter, in which Depero equated its composition to the micro-theatre of *Balli plastici,* the chair represents a plastic toy "around which dance automata resembling the mechanical marionettes of my *Plastic Ballets*" (quoted in Belli, *Depero* 178), as well as a new living object, turned by force of play into an animated doll. By virtue of this "ontological" change, parallel to the contrary reconfiguration of the doll (or any other plaything) into a furnishing item, all the artwork of the house (toys, decorative items, pieces of furniture, tapestries, posters, paintings) transformed into concurring elements of a reconstructed universe and into real presences co-inhabiting the house (as in the ancient idea of the Lares ever-residing with their descendants in the Latin *Domus,* or household).[31]

In the work of Balla, play constituted a similar force, transforming daily objects into pieces of furniture, and pieces of furniture into artwork. The artist playfully deformed the standard form of common furnishing objects (chairs, tables, or desks) and imbued them with an autonomous life by way of stark colour contrasts, asymmetrical designs, and pictorial decorative models. In his aesthetic planning of objects and furniture Balla engendered a new tension in the space itself (of the house and of modern living) as a form of art – as documented in his Casa d'Arte or in the bar Tic Tac that he decorated in Rome in 1921 (inspiring, the year after, Depero's Cabaret del diavolo). Balla's identification of furniture items as art pieces encouraged the research of contemporary artists, like Enrico Prampolini (who designed the Table for the Artist in 1925), Felice Casorati (who realized a sculpture chair in 1925), Thayath (who presented his furniture art pieces at the Biennale of Monza in 1923 and 1927), and Julius Evola (who produced Dada tables in the mid-1920s; Cogeval and Avanzi 118). Likewise, Balla's idea of the house as an art space inspired the creation of other Case d'Arte – by

Prampolini, Bragaglia, and Depero – and competed with the parallel model of D'Annunzio's Vittoriale, similarly designed as an expositional architecture of the Vate's works and feats.

Against the back-drop of Balla's and Depero's research, three contemporary processes revealed the ongoing configuration of an aware aesthetics of furniture: the artistic recreation of household goods, the symbolization of furnishing items in the space of contemporary painting, and the evolution from furniture to architectural design. First, Balla's and Depero's artistic reconversion of household objects found contemporary artisanal equivalents in the ceramic artwork of Tullio d'Albissola, who opened his workshop of Futurist ceramics in 1927 under the inspiration of Gió Ponti's collections for Richard-Ginori (Barisone 287–94); in painting, in Giorgio Morandi's abstract reconstruction of glasses, bottles, and vases that had been painted and rearranged in ever-different combinations; and, in industry, in the laboratory of Giovanni Alessi, opened in 1921 in Omegna, which produced aesthetic silverware in brass and metal. Second, Balla's and Depero's attention to furniture as art coincided then with its explicit appearance in contemporary painting, as seen in Felice Casorati's *Eggs on a Chest of Drawers* (1920), in which the solid immobility of a chest of drawers counterbalances the perceived instability of modern living (denoted in the precariously balanced eggs placed on it), or in Giorgio De Chirico's 1926–7 series, *Mobili nella valle* (*Furniture in the Valley*), in which chairs and armoires become metaphysical symbols of the artist's displacement (as he continuously moved from one place to another). Lastly, the Second Futurism's planning attitude towards objects and house interiors preceded and informed the evolution of furniture design both as a key component in architectural design (as attested in the work on chairs and tables by the contemporary architects Marcello Piacentini, Ivo Pannaggi, Giovanni Muzio, and Guglielmo Ulrich) and as an independent industrial sector, as emblematized by the launching of Emilio Lancia and Gió Ponti's furniture line Domus Nova for La Rinascente in 1927.[32]

Arts and Industry at Play

The Futurist manifestos on toys, clothes, and furniture focused their attention on daily items as the starting point of a new art and as the concurring actors of a new aesthetic universe. As they playfully deconstructed, reconstructed, and rearranged elements in ever-different plastic combinations, the artists of the so-called Second Futurism gradually shifted their experimentation from their artisanal workshops to the larger space of industry. Conversely, as toys, fashion, and interior design evolved from small-size businesses into larger industrial structures of production and consumption, entrepreneurs increasingly engaged artists, endowing them with the task to author serial products and to test innovative models of industrial art. Among the increasing number of collaborations between arts and industry during the 1920s, the partnership of Depero and the Milanese entrepreneur Davide Campari constituted a singular case of cross-fertilization of languages, as the artist turned his play into an industrial style, and the product turned into a polyvalent aesthetic site.

In the footsteps of his father, Gaspare (inventor of Campari Soda and founder of the Campari company in 1860), Davide Campari deliberately related his product to fashion and art. Along with the trendy Caffé Campari (opened in 1867 in Milan's Galleria Vittorio Emanuele and regularly visited by Giuseppe Verdi, Arturo Toscanini, Carlo Carrà, and Marcello Dudovich), he opened the Camparino in 1915, an aperitif bar (later decorated by the Liberty artists Eugenio Quarti and Alessandro Mazzucotelli) fitted with a special system to continuously flow soda water and guarantee consistent freshness in the drinks. Building on the successful partnership with Ricordi (which had started in 1899 and developed in the early 1900s), in the 1920s Davide Campari commissioned the design of advertising posters from first-rate illustrators like Leonetto Cappiello and Marcello Nizzoli as a way to expand the appeal of his products beyond Milan and launch the brand on an international scale. In 1921 Leonetto Cappiello designed a renowned poster for Campari that associated the brand with the image of a joker (*spiritello*) emerging brightly from a dark background and playfully handling a bottle of the aperitif while being surrounded by a peeled orange skin (in a synaesthesia of tactile sensations). In 1926 Marcello Nizzoli designed two parallel posters for Campari Cordial and Campari Bitter, displaying, in the one, a "Cubist" vision of the product, seen from the observer's perspective above and, in the other, the soda seltzer as the company's quality trademark (from the Camparino bar to the rest of the world).

While poster advertising emblematized the commercial strategy of tying the brand to one colour (red, along the lines of Dudovich's 1901 poster), to a logo (joker), and to quality (soda),[33] the collaboration with

Depero, starting after the Paris exposition of 1925, led the company over the following years to explore new possibilities, transforming the product into an experimental platform of plastic research. In the same way, although the artist had already been working for the industry (in his advertisements for Richard-Ginori, San Pellegrino, Liquore Strega, Bianchi, Linoleum, Vido Mandorlato, and Rimmel), the partnership with Campari from 1925 to 1932 led to the unpredictable evolution of his multidimensional language as he broke the barriers between painting and advertising in his aggressive "play" of fonts, images, plastic forms, and graphic architectures.

Following Cappiello's jolly rhetoric, Depero first realized the plastic figure of a playful monkey drinking Campari (1925), based on a wooden toy puppet from *Balli plastici*, which would later reappear as the subject of his covers for *Vanity Fair* in 1929. In 1926 Depero elaborated another Campari puppet or marionette in the drawing *Pupazzo che beve Camparisoda* (*Puppet Drinking Campari Soda*; Massoni 32), cheerfully composed of conic or cylindrical pieces and surrounded by explosive chromatic combinations. The conic shape of the puppet's drinking vessel would later "overflow" into a plastic complex, inspiring the iconic design of the Campari Soda bottle as a reversed chalice, commissioned in 1932 to launch the product globally. At the same time, the use of aggressive colours and the geometrical composition of figures also constituted a major stylistic element in *Squisito al selz Campari* (*Delicious Campari Seltzer*; plate 24), which was presented in 1926 at the XV Biennale of Art of Venice. In this experimental painting Depero boldly played with perspective, synthesizing concurrent actions (the move from the seltzer to the finalized drink) on multiple planes, and elaborated a new genre of "advertising painting" ("quadro pubblicitario," as he defined it; Scudiero, *Depero* 35) as an independent synthesis of poster advertising and traditional pictorial art. In continuity with his plastic theatre, he constructed the theatrical architecture of the space, objects, and graphic signs of *Squisito al selz Campari* as a new tridimensional language, obtaining the illusions of depth through character doubling, coloured shades, and contrasting asymmetrical geometries. Depero's study on the architecture of objects went along with his investigation into the graphic architecture of letters and words, as attested in 1927 by the parallel design of the letters *T* and *B* in the background of *The Chair's Feast*, and the typographic construction for the pavilion of the publishing houses Treves, Bestetti, and Tumminelli at the Biennale of Monza. The study on the graphic architecture of words would later lead the artist to create new typographical fonts for *Vogue*, *Emporium*, and *Il popolo d'Italia* and to design in 1931 a draft for the never-realized Campari pavilion at Milan's Triennale of 1933, converting the word *Campari* into a three-dimensional architectural complex.

In conjunction with these projects Depero expanded his visual experimentation into publishing and wrote and designed two unique volumes for Campari. In 1927 he presented a limited edition "libro imbullonato" (bolted book) entitled *Depero futurista 1913–1927*, which was conceived as an object of art and constructed through the montage of several different "fabrics" of paper (white, coloured, thin, thick, etc.) as a crafty combination of images and theoretical writings (including Depero's "Manifesto to the Industrialists"). In 1931, after his first experience in New York (where he presented a personal collection and collaborated with the magazines *Vogue*, *Vanity Fair*, *The New Yorker*, and *Atlantica*), Depero released *Numero unico futurista Campari* as the sole issue of a Futurist Campari magazine, featuring, among other texts, the artist's manifesto "Futurism and the Advertising Art" ("Il futurismo e l'arte pubblicitaria"). Influenced by the city of New York, defined by the painter as an immense poster advertisement, the text openly theorized advertising as the new aesthetic language of modern times, distinct from the pictorial traditional language and conceived instead as "a living art, multiplied, and not isolated or buried in museums – an art free from academic restraints – carefree – smug – exhilarating – optimistic – an art of difficult synthesis where the artist is dealing with authentic creation" (quoted in Belli, *Depero* 150).[34]

In Depero's view, advertising represented an intermediate art, destined to serial production and mass consumption, yet also characterized by a clear authorship, an independent aesthetic intention, and a capacity to connect different languages in a creative synthesis. Although his manifesto would propel the organization of the first (and only) Exposition of Advertising Art in 1936, his theory and the event itself received lukewarm critical reaction. After the establishment of the Ufficio Stampa (Press Secretariat) in 1934 and the invasion of Ethiopia in 1935, the regime had tightened its restrictions on advertising, imposing set standards (the abolition of foreign words and any form of nudity) and limiting the space of creativity to the sole autarchic call to buy exclusively Italian products. Despite its failed take-off in the 1930s, advertising continued to develop in Italy as an a-systematic project carried on by the isolated visual quest of artists like Depero, Boccasile, and

Dudovich. While Depero's work fell into obscurity after the Second World War, his experimental approach to art and play would leave a profound mark on the evolution of Italian design, influencing the creativity of important designers like Bruno Munari (in his early Futurist phase), Marcello Nizzoli (in his transition from Campari to Olivetti), Ettore Sottsass (who succeeded Nizzoli at Olivetti after 1956), and Enzo Mari (who started to work on toys in the mid-1950s).

Chapter Eight

The Industrial Laboratory of Italian Modernity

On 15 January 1928 the architect and designer Gió Ponti released the first issue of the journal *Domus: Architettura e arredamento dell'abitazione moderna in città e in campagna* (*Domus: Architecture and Furniture of Modern Living in the City and in the Country*). The publication (which also included texts in English, French, Spanish, and German) proposed a methodological reflection on modern Italian style in the planning of objects and spaces, giving mature expression to the new culture of industrial design.

In its graphic space *Domus* combined theoretical essays, blueprints of new architectural projects (for dining rooms, gardens, kitchens, and villas), photographs of furniture and house interiors, presentations on decorative items (ceramics, glass-ware, silverware, lighting, flower vases, book binding, etc.), and visual advertisements for new products and materials (e.g., refrigeration systems and linoleum floors). In its narrative the journal explicitly linked the reflection on objects and spaces to the elaboration of a corresponding way of living, such as associating cookware items with recipe suggestions, shelving units for records (*discoteca*) with listening recommendations, and floral compositions with ideas to live "the style of today." In Gió Ponti's view, the design of objects and space specifically related to the construction of an Italian lifestyle, as outlined in his editorial "La casa all'italiana" ("The Italian-Style House") in the first issue of *Domus.* In this early quasi-manifesto of the design movement Ponti advanced the idea of a modern Italian style of architecture and outlined its original characteristics in the notions of the "indoor and outdoor" house – reflecting the Italian balance of inner (spiritual and cultural) and outer (natural) life – and of *conforto,* identified with practical comfort and the additional Italian ideal of recreation (offering "measure for our thoughts [...] a healthy balance for our habits [...] and the sense of a confident and prolific life").[1]

As the journal acquired national relevance during the 1930s, Ponti continued to elaborate in its pages an a-systematic philosophy of Italian industrial design. His reflection revolved around two ideas: the one identifying the designer as a new, self-conscious, modern intellectual; and the other associating his or her planning attitude with the design of a new modern cosmos. First, as outlined in the June 1932 essay "Art and Industry" ("Arte e industria"; *Domus*, no. 54), Ponti called artists to involve themselves directly in industrial production and reconfigure their roles subsequently. While pointing to industry as the new modern style maker or as "the manner of our time," Ponti also underlined the impact of aesthetics on its expansion, assigning to modern artists the task of making "authentic and vital a production that otherwise would just be, in the best of cases, a perfected imitation."[2] He similarly advanced the idea of the industrial designer as a new authorial figure, turning serial items into typical products, and unique prototypes into infinitely replicable goods. Second, Ponti tied the designer's aesthetic attitude in decorating objects and planning spaces to the broader elaboration of a new Italian way of being modern, moving from a lifestyle to a cultural order. Ponti's tension in connecting material and intellectual life is reflected in his August 1928 article "The Fashionable House" ("La casa di moda"; *Domus*, no.8). Presenting the Italian home as a mix of material and ethical functions, of contingent needs and the aesthetic urge for a long-lasting "testimony,"[3] he also delineated the cultural ideal of Italy's modernity as a similar synthesis of usefulness and beauty, practicality and humanity, and tradition and innovation. As indicated in the March 1932 article "Death and Life of Tradition" ("Morte e vita della tradizione"; *Domus*, no. 51), Ponti located the creative source for this new Italian way (of design and of being) in the fusion of opposing forces (from the past and the present), in the open-ended search for a unique form (or a variation from the standard), and in the pursuit of an authentic life. By discarding both the assumption that "modernity equals a mortification of tradition," and the a-critical veneration of the past (mortifying instead "those living energies that gradually built up the elements of great tradition"), Ponti thus reassessed design as a new cultural force, simultaneously enlivening the latent forces of tradition (operating in those who are "most lively") and authenticating industrial objects with its genuineness.[4]

Rather than elaborating an organic theory of design, Ponti delineated in *Domus* the designer's core traits, emerging from his own experience as artistic director for the ceramics company Richard-Ginori during the 1920s.[5] In his work on ceramics Ponti had shaped a "neo-archeological fashion" (De Guttry and Maino, "Le straordinarie avventure" 35), imagined as a dynamic style (reinventing past elements in the present) and as a language of mixedness that playfully combined different types, materials, realities, and aesthetic statutes in new fluid forms. After participating at the expositions of decorative arts in Monza (launching his ceramic line for Ginori in 1923[6] and his furniture line Domus Nova for La Rinascente in 1927) and in Paris (where he presented his Perspective vase in 1925; Prospettica, La Pietra 23), Ponti helped to organize similar events in 1930 (when he coordinated the Triennale Exposition of Monza with the painter Mario Sironi) and in 1933 (when he launched the first Triennale of Milan with the architect Giovanni Muzio). In his multifaceted work as an architect, intellectual, and organizer, Ponti thus embodied the new creative role of the designer ("at least fifteen or twenty years before this word even appeared in the professional language"; Bosoni, "Per una 'profezia'" 199), and his own practice reflected the contemporary shift, across different productive sectors, from traditional craftsmanship to industrial planning.[7] At the same time, Ponti also became a recognized spokesman for the new Italian design movement, as his journal and expositions provided a communications platform for artists, an experimental site for the "cross-fertilisation" of their ideas (Hockemeyer 130), and a window for displaying the new products in the industry of applied arts.

In parallel with Ponti's towering contribution, the Italian design project found fertile ground for its evolution in the relationship with Fascism (before the crisis of 1929) and, most importantly, in connection with the multifaceted industrial culture developing around the objects examined throughout this book.

Fascism and Design

After the end of the First World War in 1918 Italy faced a period of social unrest and industrial reconversion, mainly focused on the dual urge to transition from a war to a peace economy and to settle the war debt (a result achieved in 1925). After Mussolini's ascent to power in 1922, first the National Fascist Party (until the first elections under the Acerbo law in 1924) and then the Fascist state (after the passing of the *leggi fascistissime* in 1925 and 1926) openly partnered with the Italian bourgeoisie to launch the project of a national industry. Mussolini eliminated any form of dissent by dissolving trade unions (1922) and abolishing strikes

(1926), and supported the domestic market, indirectly through the regime's inherent ideology of self-sufficiency (made official after the proclamation of autarchy in 1936, in response to the embargo imposed upon Italy after its invasion of Ethiopia), and directly through the foundation in 1933 of the Istituto per la Ricostruzione Industriale (IRI, Institute for Industrial Reconstruction), a public holding company that aimed to refinance struggling Italian businesses during the Great Depression.

The Fascist suppression of press freedom and the nationalization of media contributed to the establishment of uniform consent (as symbolically epitomized by the "plebiscite" of Mussolini's first "re-election" in 1929) and, consequently, a uniform market. Mussolini controlled the news after making *Il popolo d'Italia* (the newspaper he founded in 1914) the official voice of the regime and establishing L'Unione Cinematografica Educativa (LUCE) in 1923 as the official source of *cinegiornali* (news diaries). He imposed direct control on publishing houses (e.g., through the ties of Mondadori with the Fascist senator Borletti) and universities after the publication of the "Manifesto of Fascist Intellectuals" in 1925 and the establishment of the Accademia d'Italia in 1929 (which assigned grants and research incentives through the Mussolini Prize). Despite his initial disinterest, he invested heavily in radio and cinema in the 1930s. Mussolini launched regular radio broadcasting in 1931 (in partnership with Marconi), and on the occasion of Italy's invasion of Ethiopia in 1935 he invented the *adunate radiofoniche* (radio gatherings), ideally extending his physical presence and voice from Rome's Piazza Venezia to the national territory thanks to loudspeakers located in the piazzas of all Italian cities. After the bankruptcies of the 1920s Mussolini also empowered the cinematographic industry by passing the first protectionist law in 1931, funding Italian dubbing of foreign movies (after the transition to sound), and establishing the studios of Cinecittà in Rome in 1937 under the motto "Cinema is the strongest weapon" (Tacchi 99). While enacting the regime's consensus strategy, these mediating channels also defined a homogeneous marketplace for Italian products and indirectly supported Italian industry. In addition to backing specific companies (Olivetti for typewriters, and SNIA-Viscosa for military uniforms) with public orders, the regime encouraged industrial growth by advertising Italian-made products in the press or by exhibiting its industrial infrastructures through cinematography: as in Mario Camerini's *Gli uomini che mascalzoni...* (*What Scoundrels Men Are...*, 1932), displaying the world's first highway (Milano-Laghi, 1923–5); or in Walter Ruttmann's *Acciaio* (*Steel*, 1933), based on a text by Luigi Pirandello and showcasing the state-owned steel manufacture in Terni. In particular, Fascism strongly endorsed the Italian automobile and aeronautic industries through its journalistic elaboration of the myth of racing. By the mid-1920s, Italian car makers, mostly operating since the turn of the century – Bianchi (1899), Fiat (1899), Isotta Fraschini (1900), Itala (1904), Lancia (1906), Alfa Romeo (1910) – were starting to reach a broader market, thanks to races and decreased prices.[8] The Lancia Lambda, launched in 1922, created a new demand for cars by virtue of its innovations (unitary body, suspensions, and elevated speed; Sessa 178–9). Over the same years, Isotta Fraschini and Alfa Romeo invested in racing by way of their affiliated mechanic workshops, respectively run by two emerging pilots (then entrepreneurs): Alfieri Maserati, who opened his business in 1914 and presented its first model for Isotta in 1926, and Enzo Ferrari, who opened his officina for Alfa in 1929. Meanwhile, Fiat also tied its production to racing, as seen in the creation of its pioneering Lingotto plant in 1923 (designed by the architect Giacomo Mattè Trucco), featuring a racetrack on the roof of the factory. At the same time, however, this Turin manufacturer continued to reduce production costs and in 1932 launched the Fiat Balilla as the first Italian popular car (following the 1912 Fiat Zero, the first car to be mass produced for an affordable price; Gregotti 34). In parallel with the automobile industry, the creation of the Italian Royal Air Force (Regia Aeronautica Italiana) in 1923, the Futurist celebration of flight (in aerial poetry and painting), and the first transatlantic flights by Italo Balbo in the early 1930s also led to the rapid expansion of the aeronautic industry. Caproni started its production of airliners in 1924 (in addition to its military fleet), and Aeronautica Macchi emerged in the sector as a new influential business (before turning into the major provider of Italian aeroplanes during the Second World War).[9]

The key expression of Fascist support for the national industry – in line with post-unification Italy – consisted in the organization of expositions (Rocco 179): of decorative arts, for example, the Biennali of Monza starting in 1923 and the Triennale of Milan starting in 1930; of cinema, like the cinematographic festival of Venice, inaugurated in 1932; and of self-celebration, as in the case of the Mostra del Decennale della Rivoluzione Fascista (Exhibit of the Tenth Anniversary of the Fascist Revolution) of 1932, and the planned (yet never realized) Universal Exposition of Rome (EUR) of 1942. In light of the Fascist "disregard for international trade" (Hockemeyer 130), Mussolini conceived expositions as a way to promote a strong visual identity for the nation, to aestheticize the regime, to showcase Italy's goods, and to pragmatically direct artists in a network of exchange and commissions. At the same time Mussolini also envisaged them as "impermanent sites of volatile memory" (Schnapp, *Modernitalia* 146) and as "concealing spectacles," offering both a counter-cultural alternative to the durable memorial of museums and a distraction from the political reality of dictatorship (Rocco 180). In the apparently monolithic culture of the 1920s, Italian expositions nonetheless represented a singular and eclectic space of artistic development and free experimentation – at least until 1933, when Mussolini started to abandon his initial policy of laissez-faire towards the arts and veered towards Germany's model of state-organized propaganda (with the establishment of the Ministry of Popular Culture in 1935).

Unlike the heavy censorship it imposed on writers, journalists, and intellectuals in its first decade, Fascism allowed the decorative arts a significant degree of autonomy.[10] The Italian design movement found fertile soil in which to grow, in light of the regime's flexible approach (which favoured mergers of traditional craft practices with industrial methods of production; Schnapp, *Modernitalia* 148) and its hybrid nature, constantly overlapping ancient and modern codes. Given its promise to "forestall the spread of standardization and degeneration while bringing to Italians the benefits of contemporary life" (Ben Ghiat 3), it is no accident that Fascism represented a suitable context for the birth of the Novecento movement and rationalism in the years preceding the Lateran Pacts and the financial crisis of 1929.

The Novecento ("1900s") movement grew out of the initiative of Margherita Sarfatti, a Jewish writer, an art critic for *Il popolo d'Italia*, director of the magazine *Gerarchia* (*Chain of Command*), a key architect of Fascist cultural policies, and, last but not least, Mussolini's mistress and biographer.[11] In Milan on 26 March 1923 she gathered seven painters (Mario Sironi, Achille Funi, Anselmo Bucci, Guido Marussig, Leonardo Dudreville, Ubaldo Oppi, and Gian Emilio Malerba) in Mussolini's presence for the inauguration of their first exhibit at the Galleria Pesaro. Rejecting the experience of the avant-gardes, the group promoted not just a "return to order" but also a renewal of traditional art, reinterpreted in a "classically modern" style (Bonito Oliva 93). Although emerging within the cultural horizon of Fascism, Novecento did not represent art of the regime, because Mussolini in his inaugural speech had distanced himself from "encouraging anything like a state art" and had indicated that providing conditions for artists to work and flourish was the state's only duty (Paulicelli, "Art in Modern Italy" 253). Despite later disagreements with the movement, Mussolini allowed the group to influence contemporary architecture (in the work of Gió Ponti and Giovanni Muzio), painting, and literature. In 1926 Novecento organized an exhibit in Milan featuring, in addition to its members, all the leading figures of Italian painting, including Giorgio De Chirico, Giorgio Morandi, Carlo Carrà, Giacomo Balla, Fortunato Depero, and Gino Severini. In the same year, the writers Massimo Bontempelli and Curzio Malaparte established the international journal *900: Cahiers d'Italie et d'Europe,* published in French until 1927, addressing all the cosmopolitan intellectuals of "stracittà" and launching the poetics of magic realism as a way to explore and uncover the human unconscious against the constraints of mass society. Owing to its international dimension, the regime closed the journal in 1929.

In parallel with Novecento, a group of seven architects (Carlo Enrico Rava, Luigi Figini, Guido Frette, Sebastiano Larco, Gino Pollini, Giuseppe Terragni, and Ubaldo Castagnoli, who was replaced later by Adalberto Libera) launched a new form of rationalist and functional style of Italian modern architecture (modelled on the influence of Gropius and Le Corbusier) in 1926. Rationalism (promoted as Gruppo 26) competed with other currents for representing the new "modern imperial style" (Sparke, *Italian Design* 63) to which Fascism aspired. While embodying the regime's continuity with the glorious past of Rome (e.g., by the recovery of the architectural elements of arches and columns), rationalism also expressed its discontinuity in the present (as an Italian reinvention of the Modernist movement). After the first Italian exposition of rationalist architecture, held in Rome in 1928 and leading to the foundation of Movimento Italiano per l'Architettura Razionale (MIAR), the members of the group (which formally dissolved in 1931 under the pressure of the Italian old academia) contributed to refashioning or designing from scratch the urban space of Fascist Italy, as documented in the rationalist design of the Torre Littoria as the tallest residential building of Turin; the Olivetti factory in Ivrea (by Luigi Figini and Gino Pollini); the city of Sabaudia, founded in 1933 after the reclaiming of the Pontine marshes; and the colonial city of Portolago (Lakki), named after the Italian governor of the Greek island of Leros. Rationalist architects also shaped the new space of private and public houses in a modern key, as emblematized in the Casa Elettrica (1930), designed in Milan by Gruppo 7 and Piero Bottoni; in the famous Casa del Fascio (1932–6) and the Casa Rustici (1935), designed in Como and Milan respectively by Giuseppe Terragni; and in the Casa Malaparte (1938), designed in

Capri by Adalberto Libera (who was also the architect of Littoria and, later, the Palazzo dei Congressi in the Universal Exposition of Rome).

Rationalism and Novecento visualized the experimental space of Italy's classical modernity (in opposition to the imitative repetition of Roman architecture as exemplified by Enrico Del Debbio's Foro Mussolini) and paralleled the development of Marcello Piacentini's simplified neoclassicism as a signature Fascist architecture. Defining his eclectic architectural language as a synthesis of the two styles, Piacentini used it to design the public façade of Fascism, as documented in his public buildings (e.g., the University of Rome and the Palace of Justice in Milan), in his work on urban planning (e.g., the reconfiguration of Piazza della Vittoria in Brescia and Genoa, and the design of Via della Conciliazione in Rome, staging the symbolic reconciliation of the two sides of the Tiber), and, lastly, in his coordination of the site of the universal exposition of 1942.

The Laboratory of Italian Design

Scholars conventionally trace the origins of Italian design to the development of architectural rationalism and the Novecento movement in the 1920s (Ambasz, Zevi, Sparke), or to the evolution of Second Futurism (Bosoni, Lees-Maffei, Fallan, Merjian) after the publication of Balla and Depero's "Futurist Reconstruction of the Universe" in 1915.[12] Some others (Avanzi, Cogeval) find a deeper source in the late-nineteenth-century recovery of the Italian Renaissance heritage of crafts and the early-twentieth-century practice of decorative arts in the Liberty style.[13]

In this context, *The Art of Objects* locates the ultimate foundation of Italian design in Italy in the early industrial age, or rather in its hybrid culture, generated around the irregular encounter of modern production systems with the philosophical and aesthetic legacies of the humanistic tradition. Rather than being a period of decadent, incomplete modernization, the early industrial age (1878–1928) constitutes a seething laboratory of failed attempts, experimental ideas, and open-ended exploration, as well as a complex time of incubation for the later success of Italy's productive model. In the context of Italy's discontinuous industrialization, the culture surrounding the objects that have been taken into consideration in this book precedes the self-conscious evolution of design, and documents the "behavioral" origins of "conceiving and making 'things' in Italy" (Bosoni, "Of the *Modo Italiano*" 22).

From the historical point of view, two conditions of the early industrial age favoured the development of an independent industrial culture in

Italy: the nation's prolonged political division and its delay in production. Italy's state of permanent instability and warfare had generated over the centuries a unique diversification of material cultures and a decentralized system of art patronage. In the absence of a main revolution the productive structure of the newborn Italian nation had maintained this old configuration, organized around an irregular galaxy of self-differentiated manufacturers (rather than a common project). While the inherited Italian tradition of excellence stalled the expansion of a national industry (in its quest for first-rate materials, in its obsession with quality, and in its artisanal crafts),[14] it would, however, help to define its core elements. In the context of Italy's industrial delay (compared to England, France, Germany, and the United States) the quality and variety of Italian-made products, drawn from the rediscovered repertoire of the Renaissance decorative arts, not only created a clear element of differentiation but also went along with the experimental quest for new solutions in dialogue with tradition.[15] In the soil of Italy's imperfect modernization, early Italian design grew equally as an intrinsically malleable culture and as a self-adjusting mediating practice,[16] drawing energy for innovation from its unrejected past and compensating for its lack of competitiveness by the experimental addition of art to production.

From a behavioural point of view, the slow formation of an independent aesthetics around certain industrial products provides a traceable background (and an experimental poetics) for the later development of design. The objects analysed throughout the book reflect the image of early Italian industrialism as a hybrid culture, on the edge of tradition and modernization, constantly dealing with opposing tensions: eternity versus ephemerality (watches), movement versus immobility (photographs), art versus mechanics (bicycles), and materiality versus immateriality (cigarettes). In their pursuit of a recognizable style (already hinted at since De Amicis' vision of the Italian pavilion at the 1878 Exposition of Paris), the objects mirrored the tension of Italian industrialism's efforts to construct an aestheticized modernity, bypassing homologation or ephemerality through the unique quality of materials, custom-made handicraft, and artistic intention, and achieving an inimitable lifestyle out of the balance of luxury and comfort, elegance and lightness, and taste and fitness. Lastly, in their dynamic urge to interact with each other, to animate space, and to merge different artistic realms (gramophones, cigarettes, toys, clothes, furniture), the objects revealed not just the anxiety of overcoming Italy's backwardness or immobility but also the enduring sense of restlessness within Italian industrialism, expressed as a perpetual state of creative incompleteness.

While the decorative arts reinvented a previous figurative heritage, the cultural discourse or additional aesthetic element constructed upon ordinary objects fashioned instead a new, original attitude in the planning of forms. The development of this "design vision" was prefigured in the fictionalization of objects (as in photographic wondering), in their creative state of suspension (in the metaphor of the cloud of smoke), in their emotional charge (as in play), and in their transfer of old cultural iconography onto new commercial languages (as in watches or cigarettes). The formation of an aesthetic gaze upon objects would find a parallel expression, in the contemporary figurative arts, through the language of Giorgio De Chirico's and Giorgio Morandi's still-life paintings. De Chirico configured objects as sites of a non-conformed *vita consapevole* (conscious life) and as interrogative or disquieting riddles, overlapping different spaces and temporalities. In contrast, Morandi staged in his bottles and vases an abstract space of open-ended investigation, deliberately erecting a serial construction as a way to explore and detect in it an unforeseen moment of instability, difference, and artistic innovation.[17] Against this background, design evolved over the years as an eclectic new culture, absorbing, interrogating, and remixing heterogeneous materials into new, unpredictable solutions, and as a mental attitude, in which creative suspension and intellectual exploration preceded execution. Within the aestheticized and controlled space of Fascism, the culture of design matured at the same time as an "inventive reaction" (Branzi 131) to limited political freedom and the lack of financial resources and materials for supplies, as ascertained in the creative utilization of materials (rayon, aluminium, linoleum) during the years following the embargo.[18]

In conclusion, Italy's diversified traditions, imperfect modernization, and aesthetic lifestyle, as well as its designers' eclecticism, mentalism, and creative sense of limit, constitute the key ingredients in the realization of the Italian "graceful ease" – or *sprezzatura*, as defined in Baldassar Castiglione's *Libro del cortegiano* (*Book of the Courtier*, 1528). The success of Italian fashion and design (related to their unique style), the positioning of Italy's industrial products in the high-quality niche of the market, and the ongoing appeal of Italian glamour are the most noticeable outcomes of the nation's so-called incomplete modernity.

Plate 1 Giovanni Segantini, *Mezzogiorno sulle Alpi*, 1891.
© FotoFlury. Segantini Museum St Moritz.

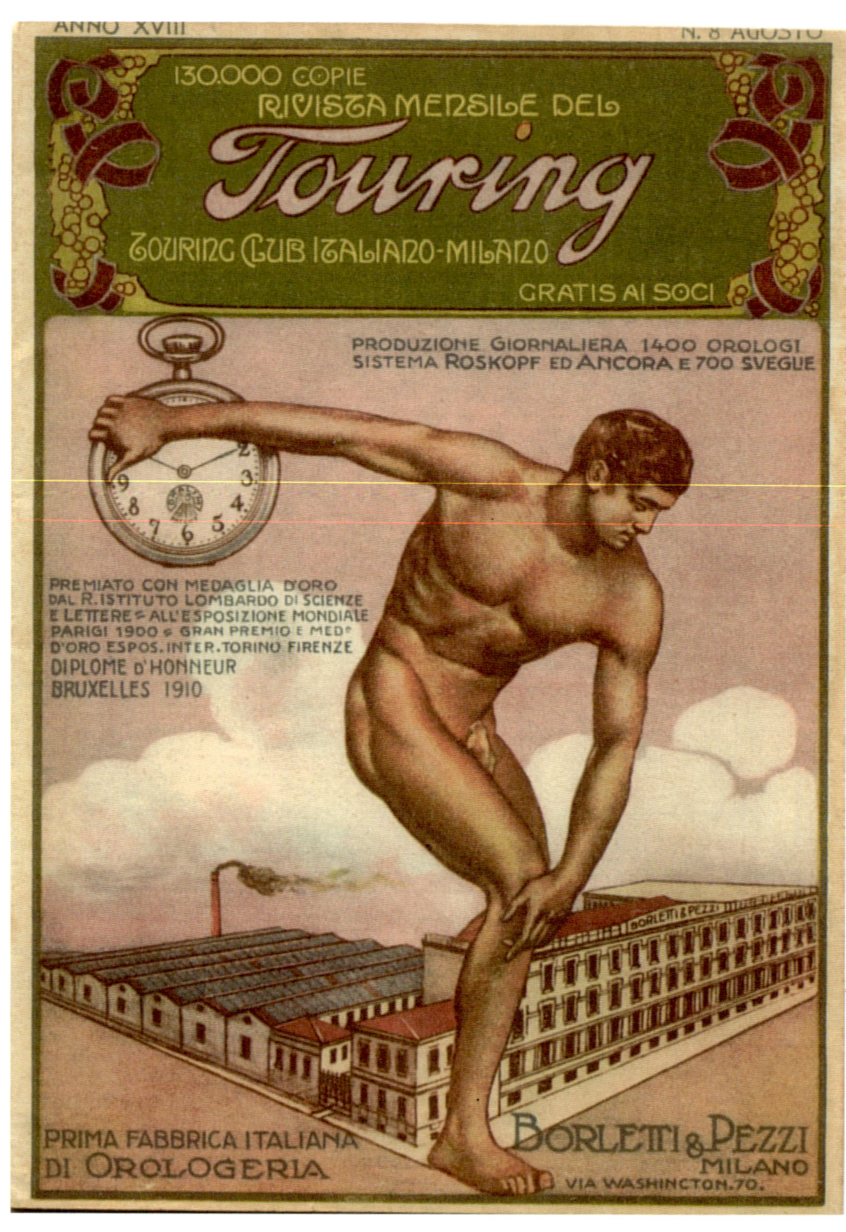

Plate 2 Anonymous, cover advertisement for Borletti & Pezzi, *La rivista mensile del T.C.I.*, August 1912. Archivio Touring Club Italiano.

Plate 3 Giorgio De Chirico, *Enigma dell'ora*, 1911. Milan, private collection. © 2017 Photo Scala, Florence. © 2017 Artists Rights Society (ARS), New York / SIAE, Rome.

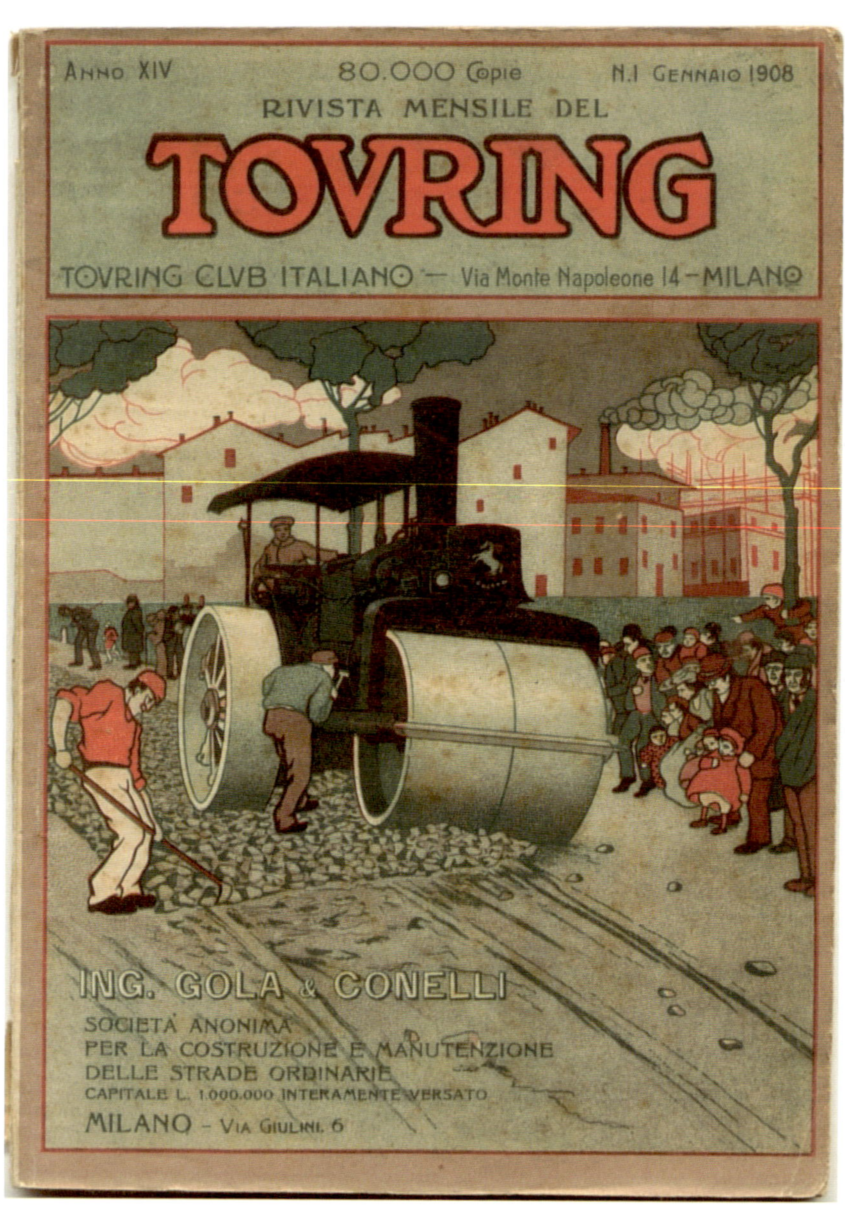

Plate 4 Umberto Boccioni, cover advertisement for Gola e Canelli, *La rivista mensile del T.C.I.*, January 1908. Archivio Touring Club Italiano.

Plate 5 Gian Emilio Malerba, cover advertisement for Pirelli, *La rivista mensile del T.C.I.*, December 1908. Archivio Touring Club Italiano.

Plate 6 Umberto Boccioni, cover advertisement for Bianchi, *La rivista mensile del T.C.I.*, March 1908. Archivio Touring Club Italiano.

Plate 7 Umberto Boccioni, cover advertisement for Frera bicycles, *La rivista mensile del T.C.I.*, April 1908. Archivio Touring Club Italiano.

Plate 8 Plinio Codognato, poster advertisement for Cicli Fiat, 1910. Museo Nazionale Collezione Salce – Treviso, Polo Museale del Veneto (Inv. 11913). Image used by permission of the Italian Ministry of Cultural Heritage and Activities and Tourism.

Plate 9 Umberto Boccioni, *Dinamismo di un ciclista*, 1913. Milan, private collection.

Plate 10 Adolfo Hohenstein, poster advertisement for *Il resto del carlino*, 1899. Museo Nazionale Collezione Salce – Treviso, Polo Museale del Veneto (Inv. 09384). Image used by permission of the Italian Ministry of Cultural Heritage and Activities and Tourism.

Plate 11 Leopoldo Metlicovitz, poster advertisement for "Il varo della nave Roma–La Spezia," 1907. Museo Nazionale Collezione Salce – Treviso, Polo Museale del Veneto (Inv. 02650). Image used by permission of the Italian Ministry of Cultural Heritage and Activities and Tourism.

Plate 12 Marcello Dudovich, poster advertisement for Bitter Campari, 1901. Museo Nazionale Collezione Salce – Treviso, Polo Museale del Veneto (Inv. 03976). Image used by permission of the Italian Ministry of Cultural Heritage and Activities and Tourism.

Plate 13 Marcello Dudovich, poster advertisement for Fonotipia, 1906.
Museo Nazionale Collezione Salce – Treviso, Polo Museale del Veneto (Inv. 04044).
Image used by permission of the Italian Ministry of Cultural Heritage
and Activities and Tourism.

Plate 14 Aldo Mazza, poster advertisement for Società Nazionale del Grammofono, 1913. Museo Nazionale Collezione Salce – Treviso, Polo Museale del Veneto (Inv. 03471). Image used by permission of the Italian Ministry of Cultural Heritage and Activities and Tourism.

Plate 15 Leopoldo Metlicovitz, poster advertisement for Fonotipia, "Il celebre tenore Comm. Cav. G. Anselmi canta esclusivamente per la Fonotipia," 1915. Museo Nazionale Collezione Salce – Treviso, Polo Museale del Veneto (Inv. 02601). Image used by permission of the Italian Ministry of Cultural Heritage and Activities and Tourism.

Plate 16 Aldo Natoli, poster advertisement for Fonotipia, mid-1910s. Museo Nazionale Collezione Salce – Treviso, Polo Museale del Veneto (Inv. 02308). Image used by permission of the Italian Ministry of Cultural Heritage and Activities and Tourism.

Plate 17 Plinio Codognato, poster advertisement for Grafofoni Columbia, 1920. Museo Nazionale Collezione Salce – Treviso, Polo Museale del Veneto (Inv. 11931). Image used by permission of the Italian Ministry of Cultural Heritage and Activities and Tourism.

Plate 18 Giuseppe De Nittis, *Il treno*, 1880. Museo Civico Barletta.

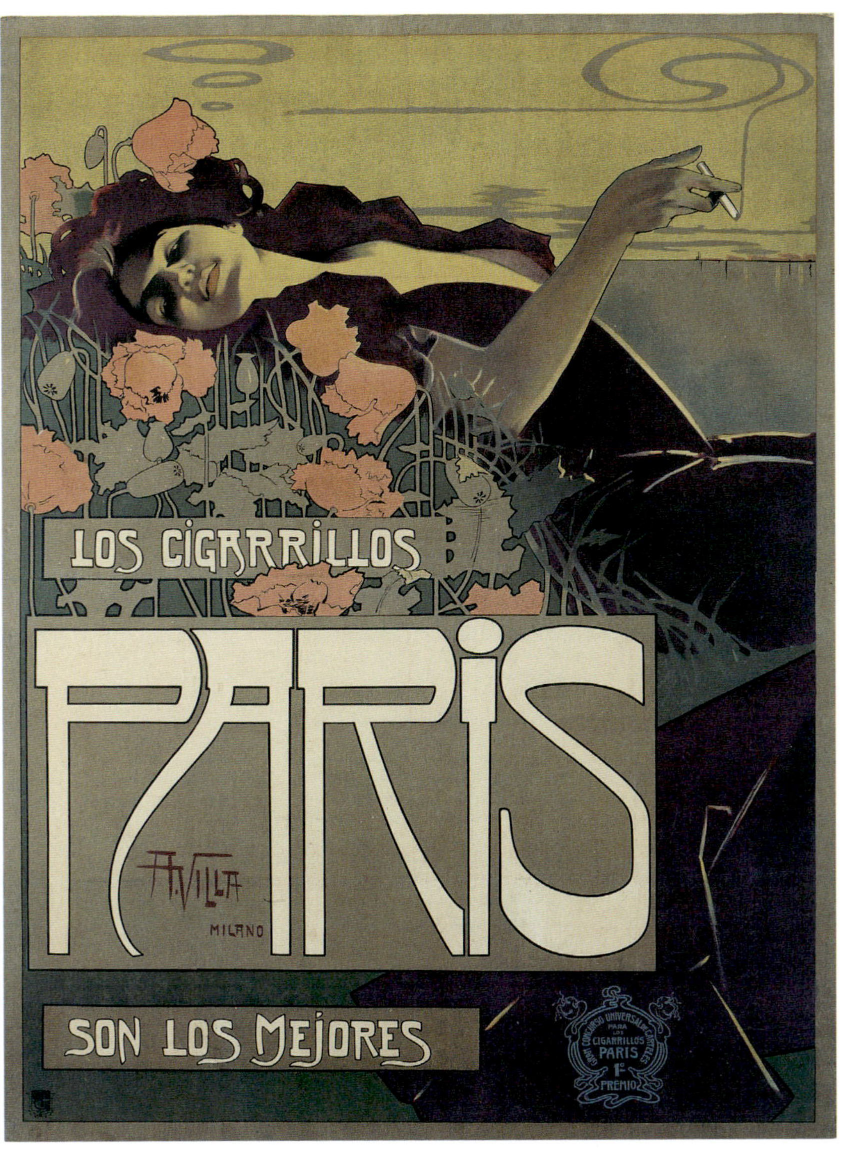

Plate 19 Aleardo Villa, poster advertisement for Los Cigarrillos Paris, 1901. © 2017 Museu Nacional d'Art de Catalunya, Barcelona. Photograph: Calveras/Mérida/Sagristà.

Plate 20 Aleardo Villa, poster advertisement for Papiers Job, 1906. Museo Nazionale Collezione Salce – Treviso, Polo Museale del Veneto (Inv. 01476). Image used by permission of the Italian Ministry of Cultural Heritage and Activities and Tourism.

Plate 21 Leonetto Cappiello, poster advertisement for Papiers Job, 1914. Museo Nazionale Collezione Salce – Treviso, Polo Museale del Veneto (Inv. 12203). Image used by permission of the Italian Ministry of Cultural Heritage and Activities and Tourism.

Plate 22 Fortunato Depero, *Il corteo della gran bambola*, 1920. MART, Archivio Fotografico e Mediateca.

Plate 23 Fortunato Depero, *La festa della sedia*, 1927.
MART, Archivio Fotografico e Mediateca. © 2017 Artists Rights Society (ARS), New York / SIAE, Rome.

Plate 24 Fortunato Depero, *Squisito al selz Campari*, 1926. Milan, Arte Centro.

Notes

Introduction

1 In "The Metropolis and Mental Life" (1903), Georg Simmel defines from within some of the changes brought forth by the industrial revolution, identifying them in the affirmation of the monetary economy (or the reduction of "all quality and individuality to a purely quantitative level"; 326) and in the levelling of distinctions among things (leading to a blasé attitude and ultimately "to the strangest eccentricities, to specifically metropolitan extravagances of self-distanciation, of caprice, of fastidiousness, the meaning of which is no longer to be found in the content of such activity itself but rather in its being a form of 'being different' – of making oneself noticeable"; 336).

2 The rapid industrialization of this age – moved by the new power of electricity, and bringing forth new types of materials (steel, aluminium, linoleum), new ways of transportation (with the improvement of internal combustion engines), new urban architectures (after the invention of elevators and the verticalization of buildings), and new accessibility of information (through the linotype, cameras, films, and reproduced sound) – had a profound impact on the contemporary formation of new cognitive systems (after the crisis of Euclidean geometries), new art languages (from Impressionism to the avant-garde), and new philosophies (e.g., Marx's materialism, Nietzsche's perspectivism, Freud's psychoanalysis, Husserl's phenomenology, or Bergson's intuitionism).

3 "Croce's influence over the whole area of the humanities in early twentieth-century Italy was exceptional, wholly without parallel in Britain or the United States" (Forgacs 5).

4 Bosoni's exhibit aims at correcting "a distorted cultural vision that has for a long time led people to see Italian design as a mere appendage of the architectural culture of the Modern Movement" (*Modo italiano* 18).

5 Annicchiarico associates industrial design with its Latin, Christian, and Renaissance roots, which are seen as parts of a common development, explaining the complexity of contemporary reality. This common story "is not recognizable from an individual style (but rather from many styles) or industrial strategy (which continuously changes over time)" but is characterized "by the permanence of a few, often very ancient, intellectual or political 'obsessions,' which render the evolution of Italian design completely unique with respect to that of other countries" (Annicchiarico and Branzi 43).

6 Crepas notes that, from the point of view of production, "the Italian delay contributed to configuring different models of industrial growth" (96). For a comparative approach to different forms of industrial development in European countries, see Tom Kemp's volume *Industrialization in Nineteenth-Century Europe.*

7 "Modernity has developed in Italy in what could be described as an uncertain, imperfect and incomplete way, with an eternal sword of Damocles hanging over its head, always caught between the desire to shrug off the burden of history and the difficult but intense attempt to hold a dialogue with it" (Bosoni, *Modo italiano* 22).

8 As suggested by Grace Lees-Maffei, the emphasis on mediation offers a third current in design history, "which brings together issues of production and consumption, not through the examination of designers' intentions or actual consumption practices, but rather through the analysis of the cultural and social significance of designed objects, spaces and processes to reveal shared ideas and ideals" (Lees-Maffei, "Production" 366).

9 Cars and aeroplanes reached a larger audience only in the 1930s after the release of the Fiat Balilla.

10 Despite their importance, objects like elevators or refrigerators do not generate a significant cultural reflection. Telephony and cinema are included in a reflection on telegraphy, phonography, and photography as elements of a broader investigation into simultaneity and vision.

11 "We charge objects intellectually and affectively, we attribute meaning and sentimental qualities to them, we enclose them in treasure chests of desire or shells of repulsion, we frame them in systems of relations, we insert them in stories which we can reconstruct and which have to do with ourselves and others" (Bodei 23).

12 Dorfles writes: "The mechanization of the modern world has impacted not only the social and economic component, but also the aesthetic component of human life" (153).

13 Objects interact with the nation-building project by creating collective markets (in the nationwide consumption of advertisements, postcards,

and records), behaviours (the time-money equivalence, a renewed attention to house décor), hobbies (smoking, photographing, listening to music at home), or practices (photo-portraiture, tourism, and sport).

14 While generating disputes at conferences (on timekeeping, photographic vision, or cycling) or in specialized clubs and journals (for cyclists and photographers), objects also catalyse a broader intellectual investigation into modernity, against the back-drop of contemporary European debates on modern time (Einstein versus Bergson), space (the acceleration in communications), body (Darwin's evolutionism), and the nature of the self (Freud's theories).

1 At the Origin of Italian Industrialism

1 The 1878 exposition featured among its attractions the Statue of Liberty's head (on display in the garden of the Trocadéro Palace), Alexander Graham Bell's telephone, and Thomas Edison's phonograph. The Statue of Liberty, meant to celebrate the centennial of the American Declaration of Independence in 1776, was completed and transported to New York in 1886.

2 The original reads: "uno sterminato teatro aperto" (De Amicis, *Ricordi* 52); "una varietà infinita di tesori, di ghiottonerie, di giocattoli, di opere d'arte, di bagatelle rovinose, di tentazioni di ogni specie" (39).

3 "Un museo enorme, dove gli ori, le gemme, le trine, i fiori, i cristalli, i bronzi, i quadri, tutti i capolavori delle industrie, tutte le seduzioni delle arti, tutte le gale della ricchezza, tutti i capricci della moda si affollano o si ostentano" (De Amicis, *Ricordi* 40).

4 "L'Esposizione è diventata un enorme ateneo internazionale che ci dà per venti soldi tutto lo scibile umano" (De Amicis, *Ricordi* 78).

5 Throughout the city "la strada diventa piazza, il marciapiede diventa strada, la bottega diventa museo; il caffè, teatro; l'eleganza, fasto; lo splendore, sfolgorìo; la vita, febbre" (39; the streets turn into piazzas, the sidewalks into roads, workshops into museums; coffee shops into theatres; elegance into luxury; brightness into glare; life into fever). De Amicis describes the pavilions as a similar site teeming with activities: "potete telegrafare a casa, scrivere le vostre lettere, fare il bagno, prendere di tanto in tanto una scossetta elettrica, farvi pesare, portare, fotografare, profumare, curare" (De Amicis, *Ricordi* 77; one can telegraph home, write letters, take a bath, get an electric shock every once in a while, be weighed, carried, photographed, perfumed, healed).

6 "Vuoi un pendolo che ti faccia vento? Un orologio fatto con un girasole, da cui esca un ragno ad acchiappare una mosca? Un mobile che ti si trasformi sotto le mani, a tuo piacere, in bigliardo, in scrivania, in scacchiera e in

tavola da mangiare? Una barca vera con remi e timone, da portar sotto il braccio al lago di Como? Un portamonete che tiri delle pistolettate? La carta dell'Europa in un fazzoletto? Un paio di stivaletti di squame di pesce? Un letto di ceralacca? Una poltrona di cristallo? Un violino di maiolica? Un velocipede a vapore? Qui c'é tutto." (De Amicis, *Ricordi* 71; Do you want a pendulum that acts like a fan? A clock made of a sunflower, where a spider goes out to catch a fly? A piece of furniture that can be transformed at will into a billiard table, a desk, a chess-board, or a dining table? A real boat with oars and helm, which you can carry under your arm back to Lake Como? A change purse that can shoot? A map of Europe on a tissue? A pair of boots made of fish scales? A bed made of sealing wax? An armchair in crystal? A violin in majolica? A steam-operated velocipede? Here we have everything.)

7 The original reads: "bella e tremenda peccatrice" (De Amicis, *Ricordi* 178). "É questa la grande Parigi? Se un terremoto fa crollare tutte le vetrine e una pioggia ardente cancella tutte le dorature, che cosa ci resta?" (167).

8 "Un popolo di statue candide, uno sfolgorìo diffuso di cristalli, un luccichio di sete e di musaici, un riso di colori e di forme, davanti a cui tutti i visi si rischiarano, tutti i cuori s'allargano, e tutte le bocche dicono: - Italia - prima che gli occhi ne abbiano letto l'annunzio" (De Amicis, *Ricordi* 61).

9 See my essay "An Italian in Paris" for a more detailed analysis of De Amicis's *Ricordi di Parigi.*

10 In reconstructing the early developments of Italian industrialism, I rely on Luciano Cafagna's essay "La rivoluzione industriale in Italia, 1830–1900," in Giorgio Mori's volume *L'industrializzazione in Italia (1861–1900)*, 57–73; Luciano Segreto's essay "Storia d'Italia e storia dell'industria," in *Storia d'Italia: Annali; L'industria*, vol. 15, 7–86; and the volumes by Patrizio Bianchi (*La rincorsa frenata*, 2013) and Renato Giannetti and Michelangelo Vasta (*Storia dell'impresa industriale italiana*, 2005).

11 Forced currency was adopted in 1866 as a way to cover the national debts accumulated from the wars of independence, by avoiding the equivalence of lira with the gold standard. Forced currency led to partial limitation of foreign imports (indirectly protecting Italian products), yet also to Italy's isolation in the context of European trade (Coppa 162).

12 "To speak of Italian industry before 1878 means to speak of the textile industry" (Carreras 203). The Italian textile industries were concentrated in Lombardy and Veneto: silk in Como, linen in Cremona, cotton in Gallarate (since the foundation of Cantoni in 1855), and wool in Schio (since the foundation of Lanificio Rossi in 1818). Tobacco and food represented 24.3 per cent of national industry in 1878 (Carreras 202).

Among the food companies starting in this period, it is worth mentioning Olio Sasso (1860), Filippo Berio (1867), Cirio (1875), and Barilla (1877).

13 Between 1861 and 1880 the Italian railway system grew from 2,000 to 6,200 kilometres. Over this period the state also constructed harbours (Genoa and Naples) and tunnels (Brennero in 1867, Frejus in 1871, and Gottardo in 1882). See Giuntini (551–617) for a detailed account of Italy's investments in infrastructures at the end of the nineteenth century.

14 After the Casati law of 1859 had refounded the Italian school system, the Coppino law of 1877 imposed three years of mandatory education in response to Italy's high rates of illiteracy (75 per cent in 1861, and 62 per cent in 1881; Forgacs 18).

15 Buscioni's book *Esposizioni e 'stile nazionale' (1861–1925)* presents a systematic account of Italy's participation in international and national expositions and highlights the link between the design of the façades for Italian pavilions at these events and the creation of a national style (1–32).

16 From 1878 to 1918 the Italian cultural industry gradually expanded from a relatively small market – bound to a limited audience (because of Italy's high illiteracy rates) and to dependence on foreign cultural materials – to a mass audience during the First World War (Forgacs 30–52).

17 In the late nineteenth century a day of work amounted to roughly twelve to fifteen hours (Merli 101).

18 Verga's novella "Rosso Malpelo" ("Evil Hair") and Pirandello's novella "Ciaula scopre la luna" ("Crow Discovers the Moon") denounce the tragic situation of minors working in Sicilian sulphur mines.

19 In 1893 Italy did not have a central bank but rather six banks (related to former pre-unification states) that were empowered to print money. One of them, the Banca Romana, controlled by the Papal States until the seizure of Rome, had been secretly printing a quantity of money greater than the allotted sixty million lira (by illegally authenticating banknotes printed in London, with the seal of the Papal States and a date preceding 1870) as a way to fund the political campaigns of Italian deputees. Italy's National Central Bank was created in 1894 as a result of the Banca Romana scandal.

20 In the absence of sufficient domestic demand and large surpluses for investments, "Italian industrialization needed to be 'artificially' driven by state intervention and bank credit. This created a close interpenetration of financial and industrial capitals" (Forgacs 30).

21 New important companies that started in Italy included Marelli (1891, magnets), Lavazza (1895, coffee), Fiat (1899, cars), Gazzoni (1901, Idrolitina for sparkling water), Perugina (1907, a branch of Buitoni, chocolates), Olivetti (1908, typewriters), and Caproni (1908, aeroplanes).

22 The history of the engineering corps during the First World War represents "an important chapter in Italian design owing to its high level of invention and flexibility" (Vercelloni 29).

23 The original reads: "Isolina si era buttata sul divano di cretonne giallina, a fiori rossi, molto duro, dalla spalliera diritta: guardava distrattamente il salotto. Vi erano quattro poltroncine coperte di stoffa […] stavano attorno a una tavola rotonda, dal marmo bianco. […] Poi: sei sedie di legno nero, dal colore smorto, che sembravano sempre impolverate, una mensola coperta di marmo bigio, su cui stavano sei tazze di porcellana bianca, la caffettiera, la zuccheriera; due scatole da confetti, vuote, vecchie, una di raso verde pallido, l'altra di paglia, a nappine: un piattino di frutta artificiali, anche questo in marmo, dipinte vivacemente, il fico, il pomo, la pesca, la pera e un grappoletto di ciliegie-un tavolino da giuoco, coperto di panno verde […] all'unica finestra le tendine di velo ricamato molto trasparenti, molto strette, colle bende di cretonne. Innanzi al divano un piccolo tappeto. Era tutto" (Serao, *Virtù* 13–14).

24 "Nella piccola anticamera […] aveva messo un grande cofano da nozze, nero antico, delicatamente scolpito, su cui era disteso un lungo e sottile cuscino di seta rossa e gialla; tre o quattro sedie di legno bruno, cupo, scolpito e un tavolino eguale; […] dopo veniva il salotto, che aveva un grande balcone sulla piazza, un salotto largo e luminoso, sempre pieno di sole; ma certe tende antiche, di un lampasso roseo e verdigno molto chiaro e ungrande pezzo di merletto antico giallastro, innanzi al balcone, mitigavano la luce […]. Mancava *sapientemente* qualunque mobile di legno, non un tavolino agli angoli duri, non uno sgabello; il velluto, la seta, il raso nascondevano qualunque traccia di durezza. In certi leggierissimi vaselli opalini dei giacinti rosei, carnicini, violetti, bianchi, lilla pallidissimi; sopra un divano, da un vaso giapponese, una rosa si era sfogliata, come di languore. Dei cuscini di piume, larghi, di seta rossa, rosea, scarlatta, porporina, rosa secca, in tutte le gradazioni del rosso, dal seno della rosa bianca sino al tetro color vinoso, erano ammucchiati in un angolo; se ne poteva formare un sedile, un letto, un trono." (Serao, *La conquista* 205–6; emphasis mine; In the small lobby […] he had placed a great nuptial coffer, ancient black, delicately sculpted, upon which lay a long, thin pillow of red and yellow silk; three or four chairs and a small table of dark-brown engraved wood; […] then came the living room, which had a great balcony overlooking the piazza, a large and bright living room, always lit by the sun; its light, however, was softened by some ancient curtains, made of a pink and light-green lampas fabric, and by a great piece of ancient yellow crochet, placed by the balcony. […] Every piece of wooden furniture had been *skilfully* removed from the room. Not a small table with sharp corners,

not a stool. Velvet instead, and silk, and satin, hid any trace of hardness. Pink, coral, violet, white, and pale lilac hyacinths lay in some light Opalini vases; from a Japanese vase, above the couch, a rose dropped a petal, out of languor. A few large feather pillows, made of red, pinkish, scarlet, purple, and rose silk were amassed in a corner by degrees of red, ranging from the white centre of a rose to the dark colour of wine; one could form out of them a seat, a bed, a throne.)

25 The original reads: "Perfettissimo teatro"; "virtualità afrodisiaca latente"; "avevano per lui acquistato qualche cosa della sua sensibilità [...] non soltanto erano testimoni de' suoi amori, de' suoi piaceri, delle sue tristezze, ma eran partecipi" (D'Annunzio, *Piacere* 17–18).

26 Upon her entrance into Sperelli's Palazzo Zuccari, Elena wears "a cape of Carmelite fabric, with empire-style sleeves, cut in large puffs on top and flat on the buttoned wrist" (un mantello di panno Carmélite, con maniche nello stile dell'Impero tagliate dall'alto in larghi sgonfi, spianate e abbottonate al polso), and with "an immense collar of blue fox, which covered her entire body without diminishing the grace of her slenderness" (un immenso bavero di volpe azzurra per unica guarnitura copriva tutta la persona senza toglierle la grazia della snellezza; 20). Sperelli's preparation for the dance at Palazzo Farnese is described as follows: "egli andò a vestirsi, nella camera ottagonale ch'era, in verità, il più elegante e comodo spogliatoio desiderabile per un giovine signore moderno. Vestendosi, aveva una infinità di minute cure della sua persona. Sopra un grande sarcofago romano, trasformato con molto gusto in una tavola per abbigliamento, erano disposti in ordine fazzoletti di batista, guanti da ballo, gli astucci delle sigarette, le fiale delle essenze, e cinque o sei gardenie fresche in piccoli vasi di porcellana azzurra." (D'Annunzio, *Piacere* 73; he went to get dressed in the octagonal room, which in fact was the most elegant and comfortable dressing room that a young modern man could desire. As he dressed himself, he took great care in an infinite number of ways. Upon a great Roman sarcophagus, turned with much taste into a dressing table, in order were placed *batiste* tissues, dancing gloves, cigarette cases, phials of essences, and five or six fresh gardenias in small vases of blue porcelain.)

27 As documented in Paola Sorge's book *D'Annunzio e la magia della moda*, D'Annunzio established relationships with renowned contemporary tailors like Marta Palmer (who designed clothes for his lovers), Edwige Bastianini Bellmann (who executed D'Annunzio's drawings of cloaks and shirts), and Biki (who designed his underclothing).

28 For an analysis of the appearance of outmoded items in late-nineteenth-century literature, see Orlando's volume *Gli oggetti desueti nelle immagini della letteratura: Rovine, reliquie, rarità, robaccia, luoghi inabitati e tesori nascosti.*

29 The original reads: "Due pilette d'acqua santa d'argento *cesellato,* che il tempo aveva *ossidate* [...] un fascio di *lumen-cristi* [...] *scrostati, sgretolati, contorti,* [...] i ritratti al dagherrotipo di lui e della mamma quand'erano sposi, quasi completamente *svaniti* [...] una quantità di fogli *ingialliti,* che contenevano le poesie giovanili del babbo" (Colombi, *Matrimonio* 7–8; emphasis mine).

30 The original reads: "Topaie, materassi, vasellame / lucerne, ceste, mobili: ciarpame / reietto così caro alla mia Musa" (verses 154–6). During the 1910s and 1920s the poetics of junk transfigured into a literary form, as seen, in different ways, in the fragment-centred collections by Clemente Rebora, *Frammenti lirici* (*Lyrical Fragments,* 1914); Giovanni Boine, *Frantumi* (*Smithereens,* 1917); Giuseppe Ungaretti, *L'Allegria di naufragi* (*Joy of Shipwrecks,* 1919); Camillo Sbarbaro, *Trucioli* (*Shavings,* 1920); and Eugenio Montale, originally *Rottami* (*Scraps*), then *Ossi di seppia* (*Cuttlefish Bones,* 1925).

31 At the time, Pietro Selvatico's *Storia estetico-critica delle arti del disegno in Italia* (*Aesthetic and Critical History of the Arts of Design in Italy,* 1852) constituted the sole and earliest source on an independent concept of "design" in Italy.

32 The original reads: "Questa pubblicazione vorrebbe riescire *bella,* ma più che bella *utile*" (*Arte,* no. 1, January 1890, 1; emphasis mine).

33 "Questo periodico, mentre guarderà con vivo interessamento all'avvenire, non trascurerà il passato, poiché la conoscenza del passato riesce indispensabile a preparare le forze della fantasia e del raziocinio così per le nuove e sottili ricerche sulla natura, come per le nuove e fiere battaglie contro la pigrizia o il pregiudizio di chi s'afferra tenacemente alle tradizioni." (*Arte,* no. 2, February 1901, 13–14; While observing the future with great interest, this publication will not neglect the past, for knowledge of the past is indispensible to training the forces of creativity and reasoning for both new, subtle investigations of nature, and novel, proud fights against either laziness or the prejudice of those who tenaciously cling to tradition.)

34 Sommaruga (1867–1917), a student of Camillo Boito, was the main promoter of Liberty architecture in Italy. He also designed the Italian pavilion at the Universal Exposition of Saint Louis in 1904 and the monumental Grand Hotel Campo dei Fiori over Varese in 1912 as Italy's first complex to be built above a height of one kilometre.

35 "It was during the High Renaissance that bourgeois culture first displayed a self-awareness about the fashioning of the human subject as a controllable process" (Paulicelli, "Fashion" 284). Before then, explicit attention to clothing was given in the figure of Griselda in the final novella of Boccaccio's *Decameron.* For more information about Rosa Genoni see Paulicelli's book *La moda è una cosa seria: Su Rosa Genoni.*

36 Marinetti strategically waited until February 1909, in order for the news of Messina's earthquake of December 1908 to fade, before launching Futurism

in *Le figaro*. His ability as a cultural entrepreneur also appeared in *Poesia* in his capacity to self-advertise (by sending subscribers Christmas gifts wrapped in the journal's paper), to launch reportages (on free verse, on Italian women), and to attract both famous and emerging intellectuals (Pascoli, Jammes, Laforgue, Oriani, Moréas, Claudel, Gozzano, Yeats, Neera, Govoni, Rubino, de Unamuno, Palazzeschi).

37 In the manifesto "Destruction of Syntax" of 1913, Marinetti explicitly indicates that the new possibilities offered by modern technologies alter human sensibility, and he lists all the modifications produced by them: increase in the pace of life, horror of anything old and love of the new, horror of the quiet life, destruction of a sense of the *hereafter*, removal of limitations, precise knowledge of things, semi-equality between men and women, greater sexual freedom, a rethinking of patriotism and war, a new awareness of financial matters, a new kingdom of machines, an idea of records, abolition of distances, and a new awareness of the world.

2 Timepieces and Italian Modern Times

1 The Venice congress ended with a scientific failure, as confirmed by contemporary newspapers: "nelle riunioni generali del Congresso nulla si è compiuto di serio e d'importante. La scienza non ha progredito di un passo" (*Corriere della sera*, 27 September 1881; in the general sessions nothing serious or important was accomplished. Science did not progress one inch). For a complete account of the third International Geographic Conference in Venice see George Wheeler's report of the event, published in Washington in 1885.

2 See Adolphe Hirsch and Theodore Von Oppolzer's volume *Unification des longitudes par l'adoption d'un méridien initial unique* for a comprehensive account of the International Geodesic Conference of Rome in 1883.

3 The Bay of Assab had been a possession of the Rubattino Company since 1869. Italy's purchase of the outpost in 1882, which compensated for the loss of Tunisia to France (right before the international conference of Berlin in 1881) and laid the foundation for the military occupation of Massawa in 1884, was the nation's first act of colonial policy.

4 The original reads: "Una festa solenne in cui l'Italia afferma la propria esistenza ed il proprio valore e, mostrando ciò che ha fatto, dà una sicura promessa di quanto darà in avvenire" (*Corriere della sera*, 4 May 1881). Celebrating the country's achievements twenty years after its unification and redeeming the flop of its first exposition in Florence in 1861, the National Exposition of Milan in 1881 exposed Italians to the thrill of modernity by its display of industrial products, its unprecedented attractions (including

electric illumination that stretched from Piazza Duomo to Corso Venezia), and its celebration of progress (sealed by the success of the *Excelsior* ballet).

5 See Howse (*Greenwich Time* 116–71) for a complete reconstruction of the historical process leading to the establishment of the common hour.

6 The conference proposed that governments select Greenwich as the initial meridian "for the reason that that meridian fulfills [...] all the conditions wished for by science and because being at present the best known of all, it offers the most chances of being generally accepted" (Dolan). At the same time, the conference indicated that the adoption of Greenwich would be a positive step towards Great Britain's acceptance of the 1875 Convention du Mètre, and the global unification of weights and measures.

7 The original reads: "Il giorno universale per altro non impedirà ad ogni paese di contare il tempo vero conforme alla rispettiva longitudine, ma servirà nelle relazioni internazionali d'ogni genere" (*Corriere della sera*, 22 October 1884).

8 In the issue of 1–2 November 1893, *Corriere della sera* presented the passage to European central time as a small sacrifice to the common hour: "Dalla scorsa mezzanotte tutti gli orologi pubblici vennero fatti avanzare di 10 minuti per ridurre l'ora sul Meridiano dell'Europa Centrale. Perciò i lettori che trovassero in ritardo il loro orologio non ne facciano colpa a questo. Sono dieci minuti primi e 4 secondi che si devono *sacrificare* all'ora unica." (Last night at midnight all public clocks were moved ahead by ten minutes in order to conform to the hour upon the Meridian of Central Europe. Therefore, readers who find their watches to be ten minutes behind should not blame them for the delay. Those are ten minutes and four seconds that we must *sacrifice* to the common hour; emphasis mine).

9 On 28 May 1916, *Corriere della sera* depicted as inconsequential the appearance of daylight savings time, justified by the pragmatic need to save on combustible fuels and conform Italy to its war allies – "la vita civile in tutte le sue manifestazioni non sarà per nulla turbata" (civic life in all its manifestations will not be disturbed). At the same time, it also presented that time as unmistakably fictional, in the exposed paradox of "legal aging," or in that of running backwards against the real flow of time: "Avverrà soltanto questo: che un viaggiatore partito ad esempio alle 23.50 del 3 giugno, dopo 20 minuti di marcia autentica, avrà viaggiato per un'ora e venti" (Only this will happen: that a traveller leaving, for example, at 11:50 p.m. on 3 June, after 20 minutes of actual ride, will find himself having travelled for an hour and twenty minutes).

10 Until 1866 each city measured time at its own meridian, and day was divided according to the French hour (calculated on a clockwise twelve-

hour dial, starting from midnight). After Rome and Milan had agreed on a common hour, based on the Pontifical Meridian of the Collegio Romano, Italy moved its prime meridian from the papal observatories to the geodesic station of Monte Mario in 1870. Sicily and Sardinia kept calculating their own time according to the meridians of Palermo and Cagliari.

11 According to the *ora italica,* in the case of a 5:00 p.m. sunset, 6:00 p.m. would be hour one, 11:00 p.m. hour six, 5:00 a.m. hour twelve, etc. In addition to Paolo Uccello's clock for Santa Maria del Fiore in Florence (1443), clocks indicating Italic hours appear also in Sandro Botticelli's *St. Augustine in His Studium* (Church of Ognissanti, Florence, ca. 1480) and in Canaletto's *Il campo San Giacometto verso Rialto* (Venice, 1725). For information about Paolo Uccello's clock and the Italic hour see Bigi's book *L'orologio nel Duomo di Firenze: L'unico al mondo che segna l'ora italica.*

12 The original reads: "Il tempo *vero,* indicato dai quadranti solari [...] il tempo medio calcolato secondo un *sole fittizio*" (*Illustrazione italiana,* 27 September 1883, emphasis mine).

13 The arbitrariness of Greenwich as the beginning of the earth's rotation is underlined by *Corriere della sera* in the ironic question posed by the reporter (Luigi Stefanoni) to the members of the Rome conference of 1883: "quand'è che comincia la rotazione?" (2–3 November 1883; when does the rotation start?)

14 The idea of creating a secularized temporal order is also witnessed by the parallel launching in Paris, in 1884, of a competition to create a new secular calendar as an alternative to the Gregorian one (established in 1582). Gaston Armelin won with the proposal of a perpetual calendar, made of years counted in regular trimesters of one thirty-one-day month and two thirty-day months each.

15 In *Clocks and the Cosmos* Macey argues that the British industrial revolution of the seventeenth century was made possible by the increasing reliability of mechanical timekeeping and the beginning of the serial production of clocks.

16 The original reads: "Pei bisogni scientifici e pel servizio delle grandi comunicazioni, quali sarebbero le ferrovie, le linee marittime, i telegrafi e le poste [...] senza pregiudizio delle ore locali o nazionali, che continuerebbero ad essere usate nella vita civile" (Stefanoni, *Corriere della sera,* 2–3 November 1883).

17 The original reads: "Modelli di orologi, con due quadranti opposti, l'uno indicante l'ora locale, l'altro l'ora universale" (Stefanoni, *Corriere della sera,* 2–3 November 1883).

18 While indicating with their sound the start of a new day at sunset (*ora italica*), bells also displayed with their "suono che uguale, che blando cade,

/ come una voce che persuade" ("L'ora di Barga," verses 5–6; *Tutte le poesie* 610; sound, which mildly falls / as a persuasive voice) a public order (both secular and religious), regulating and giving meaning to social life.

19 Giosuè Carducci, "Mezzogiorno alpino": "Nel gran cerchio de l'alpi, su 'l granito / Squallido e scialbo, su' ghiacciai candenti, / Regna sereno intenso ed infinito / Nel suo grande silenzio il *mezzodì.* / Pini ed abeti senza aura di venti / Si drizzano nel sol che gli penètra, / Sola garrisce in picciol suon di cetra / L'acqua che tenue tra i sassi fluì." (*Poesie* 1001, emphasis mine; in the grand circle of the Alps, upon the squalid and bare granite, upon the glaring glaciers *noon* reigns serene, intense, and infinite in its majestic silence. In absence of wind, pine trees and firs stand firm in the penetrating sunlight. Only the water, tenuously flowing amidst the rocks, chirps in its soft sound of a cithara.)

20 De Amicis similarly describes the intensification of life produced by the Universal Exposition of Paris in1878 with equal enthusiasm and anxiety: "si prova subito un raddoppiamento d'attività fisica per effetto del raddoppiamento di valore del tempo, e l'orologio, fino allora sprezzato, assume la direzione della vita" (*Ricordi di Parigi* 161; one immediately perceives a redoubling of physical activity caused by the redoubling of the value of time, and the watch, previously neglected, assumes the direction of life).

21 The original reads: "Non puoi credere che è di terribile non aver l'orologio quando s'ha l'amante! Si sbaglia già sempre l'ora. Arrivi, è troppo presto, non vi è: è una morte lenta. Arrivi tardi, è passato un quarto d'ora, per un altro quarto d'ora egli ti porta il broncio, gli uomini si seccano di aspettare. Se da lui, ogni cinque minuti gli domandi: che ora sarà? Quello s'irrita di questa domanda. A casa ritorni sempre in ritardo, con una cera sbalordita che è un miracolo non ti tradisca. Dio mio, che farei per avere un orologio!" (Serao, *Virtù* 39).

22 Luigi Capuana's shory story "Le esitanze" (1905, in *Coscienze*; "Hesitancies") retrieves Serao's image of the wrist-watch as an item of female fashion and an indispensable tool in adulterous affairs, by presenting the female protagonist vainly looking at herself in the mirror and constantly checking the hour on her "orologino," while preparing for a secret appointment with a lover.

23 "In the museum of curious objects, the clock, born to measure time, holds the space of an enigma in which is represented the existential techno-drama of a culture, an enigma that is displayed onto a metaphoric object that, in miniature, reproduces the order and disorder of the cosmos" (Bonito 9).

24 The progression of the hours accompanies Sperelli's expectation of the hour of Elena's arrival: "l'orologio della Trinità de' Monti suonò le tre e

mezzo. Mancava mezz'ora" (D'Annunzio, *Piacere* 6; the clock of Trinità de' Monti sounded three-thirty. Half an hour to go); "il momento si approssimava. L'orologio della Trinità de' Monti suonò le tre e tre quarti" (13; the moment drew closer. The clock of Trinità de' Monti sounded three forty-five); "mancavano due o tre minuti all'ora" (15; two or three minutes left to the hour); "l'orologio battè le quattro" (15; the clock struck four); "eran quasi le cinque meno un quarto [...] Elena entrò" (19; it was almost a quarter to five [...] Elena entered).

25 The original reads: "Un rigattiere [...] lo acquistò come roba fuori d'uso, non come orologio [...] Si paragonò al suo vecchio orologio di Vienna e si accorse che anche lui era un oggetto fuori d'uso" (*Demetrio Pianelli* 216).

26 In 1920, Federigo Tozzi would similarly equate clocks to individual life in the short story "Gli orologi" (1920, "Clocks"), in which seven old wall-clocks, stored in the house as memories of a past time and constantly wound by the protagonist, Bernardo Lotti, follow him in his death as they progressively stop – "non essendo più caricati, ad uno per volta gli orologi si fermarono [...] L'ultimo [...] orologio si fermò quando il Lotti era già stato messo al camposanto: il giorno dopo" (*Le novelle* 665; no longer wound, one by one the clocks stopped [...] The last one stopped when Lotti had already been taken to the cemetery, the day after) – and are sold as useless antiques.

27 The double-edged temporality visualized by Gozzano's image of a stuck, ticking clock is also implied in the poet's famous interrogation in the opening lines of "La signorina Felicita" – "a quest'ora che fai? Tosti il caffè?" (verse 8; *Opere* 188; what are you doing right this minute? Are you roasting coffee?) – which deliberately causes the contingent hour ("quest'ora") to intersect with the stilled instant eternalized by poetry.

28 The original reads: "La maggior parte del tempo la mia vita è vuota, ordinaria, comune, piena di uggia inespressa" (Papini, "Orologio" 190–1). "In questi istanti, così fuggenti, io vivo assai più cose che in tutto il tempo che corre tra un passaggio e l'altro" (192).

29 "Da quella immobilità feroce / compresi che quella / doveva esser l'ora / inesorabilmente! / tutti i giorni io doveva / a quell'ora morire?" (Palazzeschi, "L'orologio," verses 44–9; *Tutte le poesie* 266; I understood from that ferocious immobility that, inexorably, that must be the hour! Did I have to die every day at that time?) "Orologio, guarda mi getto! / E faccio l'atto. Ah! Ó sentito uno scatto! / Sei stato tu, tu che ài segnata già l'ora / Ài creduto che fosse quella! / Ahahahahahah! [...] Ora sono io che comando / sono io che darò l'ora a te, Ora!" (Palazzeschi, "L'orologio," verses 117–22, 125–6; *Tutte le poesie* 268; Now, clock, watch me do it! And I do. Ah! I heard a tick! It

was you, you who marked that hour! You thought that was it! Hahahahaha! Now I give the orders, I will give the hour to you, Now!)

30 This first wrist-watch model has a couple of antecedents: the gift (a bracelet embedding a small timekeeper) from the Duke of Leicester to Queen Elizabeth I in 1571, and the gift (a bracelet, with a watch, produced by Nitot) from the empress Josephine to the daughter of Maximilan Joseph of Bavaria on the occasion of her marriage to Napoleon in Munich in 1806 (Cobolli Gigli 9).

31 Dante refers to the new technology of mechanical clocks in *Paradiso* X (verses 139–48), where he uses the *svegliatore* to describe the harmony of the souls with the time of liturgy, and in *Paradiso* XXIV (verses 13–18), where he refers to the mechanical clock's gear transmission as a metaphor for the dancing crowns of the blessed souls.

32 "The traditional relationship between Temperance and the clock clearly carries over into Protestant ethics" (Macey 37). This explains the expansion of the clock-making industry in Protestant countries, leading London and Geneva to be, in the eighteenth century, "the two greatest centers of clock- and watch-making in Europe" (Cipolla 70).

33 Not by chance, La Mettrie compares the gears of the body to those of a clock: "je me tromp point; le corps humain est une horloge, mais immense, et construite avec tante d'artifice et d'habilité" (I am not mistaken; the human body is an immense clock, built with much artifice and skill; quoted in Macey 86).

34 Expanding from their previous traditions, the most popular foreign brands in Italy were Vacheron Constantin (Paris, 1755), Breguet (Paris, 1775), Girard Perregaux (Geneva, 1791), Longines (Geneva, 1832), Patek Philippe (Geneva, 1839), Cartier (Paris, 1847), Tissot (Geneva, 1853), Tag Heuer (Switzerland, 1860), Junghans (Germany, 1861), Movado (Geneva, 1881), Hamilton (United States, 1892), Omega (Switzerland, 1894), and Rolex (London, 1908).

35 In his travel book *Olanda* (*Netherlands,* 1876) De Amicis skeptically observes the effects of industrialism on Dutch society, stating, "In questa società ogni cosa è classificata, prestabilita, misurata; la vita scorre come una macchina agisce; l'uomo si move come un automa; il regolamento tien luogo di volontà e *l'orologio governa il pensiero*" (244–5, emphasis mine; in this society everything is classified, pre-established, measured; life runs as a machine does; man moves as an automaton; rules replace will and *the clock governs thought*).

36 "Lunedì, ore 8 si alza; ore 8,30, doppia il portone dell'Ufficio; ore 9 evacua (anche per la domenica); ore 9,30, legge in Ufficio il giornale; ore 10,15 conferisce col superiore; ore 11 evade pratiche; ore 11,45, si spolvera, si liscia, minge; ore 14,20, doppia il portone dell'Ufficio; ore 14,30–16,30

silenzio assoluto, respirazione regolare, occupazione ignota; ore 17, evade pratiche; ore 17,45, si spolvera, si liscia, minge; ore 18–19 guarnisce il centro della città, inventaria le vetrine. Calcola approssimativamente il calore sessuale delle donne che incontra per strada; ore 19–20, cena; ore 21 si corica (il sabato a questa voce si aggiunge 'compie il dovere coniugale')." (Jahier 22; Monday, 8 a.m., he wakes up; 8:30 a.m., walks in the office door; 9 a.m., defecates [even on Sundays]; 9:30 a.m., reads the newspaper in his office; 10:15 a.m., talks to his boss; 11 a.m., processes files; 11:45 a.m., dusts himself off, smoothes his clothing, urinates; 2:20 p.m., walks in the office door; 2:30–4:30 p.m., absolute silence, regular breathing, unknown occupation; 5 p.m., processes files; 5:45 p.m., dusts himself off, smoothes his clothing, urinates; 6–7 p.m. checks out the downtown, makes an inventory of shop windows. Approximately calculates the sexual heat of the women he meets on the street; 7–8 p.m., dines; 9 p.m., lies on his bed [on Saturdays, the entry "fulfils his marital duties" is added].)

37 See "The Evolution of the Bulgari Style" for information about the development of the maison's aesthetic research; Boscaini for information about Bulgari and the city of Rome; and Triossi for a catalogue of jewels.

38 As documented in a poster advertisement for Veglia designed by Maga (Achille Mauzan) in 1924 – featuring the bright and overlapping emergence of a rooster and an alarm clock over a dark background – the brand (immensely successful after the Second World War) prolonged the cultural narration of Borletti, locating horology at the intersection of mechanical timekeeping (the alarm clock) and "natural" timekeeping (the rooster), and configuring Italian-made modernity as the fusion of industrial and "human" time.

39 La Rinascente aimed at an exclusive, high-class customer. At the same time Borletti also created UPIM – Unico Prezzo Italiano Milano (The Only Italian Price Milan) – to reach a mid-range clientele.

40 "In 1919 it was Senatore Borletti who left Venice on his boat to carry to the poet-soldier in Fiume the millions collected to fund the liberation of the Istrian city by D'Annunzio's *legionaries*. D'Annunzio returned this generous friendship with book or poem dedications and with gifts of his original manuscripts" (Introna 40).

41 Senatore Borletti also became a senator in the Italian parliament during Fascism, and in 1930 he became the president of Italy's largest manufacturer of rayon, SNIA-Viscosa, providing all the uniforms for the Italian army and the Fascist officers.

42 See Taroni's book *Bergson, Einstein e il tempo: La filosofia della durata bergsoniana nel dibattito sulla teoria della relatività* for more information on the Einstein-Bergson debate.

43 In his *Time and Free Will,* Bergson uses the clock to explain the discontinuity and continuity of time: "When I follow with my eyes on the dial of a clock the movement of the hand which corresponds to the oscillations of the pendulum, I do not measure duration, as seems to be thought; I merely count simultaneities, which is very different. Outside of me, in space, there is never more than a single position of the hand and the pendulum, for nothing is left of the past positions. Within myself a process of organization or interpenetration of conscious states is going on, which constitutes true duration. It is because I endure in this way that I picture to myself what I call the past oscillations of the pendulum at the same time as I perceive the present oscillation" (108).

44 The original reads: "Perché misurava il tempo quell'orologio? A chi segnava le ore? Tutto era morto e vano" (Pirandello, "Leonora addio!" 1079; in *Novelle per un anno*).

45 The original reads: "Il sole [...] non è neanche buono da regolare gli orologi! [...] E il suo orologio, infatti, sul cui quadrante aveva scritto con inchiostro rosso: *Solis mendaces arguit horas,* non era regolato col tempo solare" (Pirandello, "Pallottoline!" 996; in *Novelle per un anno*).

46 In Aldo Palazzeschi's novel *:riflessi* (*Two Points Mirrored,* 1908) the sun is described as a "sick dancer in front of an indifferent audience" (come una povera danzatrice tisica dinanzi a un pubblico indifferente; 12) and bereft of any warmth, in its inability to cope with the new time of profit.

47 In a similar way, Marinetti saw in the sun a metaphor of passéism and vehemently attacked it in "Nuova religione morale della velocità" ("The New Ethical Religion of Speed"): "We have to persecute, lash, and torture all those who inveigh against speed. The guilt of those flophouse cities weighs heavily, where the sun takes up its residence, slows to a standstill, and moves no more. Who can really believe that the sun will be gone this evening? Come one, get a move on! It's not possible! It has settled down here. The piazzas are lakes of stagnant heat. The streets are rivers of languid fire. For the moment, there's no way of getting through them. There's no escape! Flood of sunlight [...] Solar guillotines above each door. God help anyone with a thought in his head" (*Critical Writings* 254; *Teoria* 131).

48 The image of a stuck clock recurs also in De Chirico's poem "L'ora inquietante" ("The Disquieting Hour," 1918), in which the painter sketches a world devoid of humans and presents himself (like the spectator of the *Enigma*) as the only remaining voice, still interrogating or searching for immortality in the post-metaphysical age (verses 12–18; *Il meccanismo* 54).

49 In *L'intelligenza della folla* (1903, *The Intelligence of the Crowds*) Scipio Sighele clearly articulates this feeling of lost superiority: "Gli italiani sono –

individualmente – dotati di una innegabile superiorità su molti altri popoli, e nondimeno – tutti insieme – formano una nazione che, se è stata una volta la prima, non è purtroppo più la prima nel mondo moderno" (6; Italians are – individually – endowed with an undeniable superiority towards many other peoples, yet – collectively – form a nation that, though it once was, is unfortunately no longer the first in the modern world).

50 Religious modernism has four stages: textual criticism (with Ernesto Bonaiuti's diffusion of Loisy's condemned book *L'évangile et l'église*); philosophical influence (by Tyrrell and LeRoy); political action (with Romolo Murri's foundation of the Lega democratica nazionale); and heresy (with the establishment of the anti-Modernist oath).

51 De Chirico's tension between duality and synthesis is also part of his Nietzschean heritage. Like Nietzsche, De Chirico "pursued art as a means of knowing possible truths" and "explored knowingly the Apollonian and the Dionysian duality, seeking relentlessly a means of their impossible unification" (Walker 7).

52 The *Doctrine of Fascism* was published in 1932 under the entry "Fascismo" in the *Enciclopedia italiana di scienze, lettere ed arti* (847–84; vol. 14). Gentile wrote the first part, "Fundamental Ideas," but the credit for the whole text was given to Mussolini.

3 Industrial Photographs and the Fictional Vision

1 In his speech Paolo Mantegazza defined photography as "a good and democratic art, since it grants everyone the right to preserve the appearance of their loved ones" (quoted in Ceccuti 306).

2 The parallel establishment of the Società Fotografica Italiana in May 1889 and the Società Dante Alighieri in July 1889 suggests a common endeavour to codify and spread Italy's culture through its visual and linguistic heritage.

3 During the 1899 Exposition of Photography, organized in Florence by SFI, Alinari was recognized as the first business in Italy's photographic industry, with the manufacture of twenty-five thousand plates of Italian artworks. During the twentieth century Alinari images constituted roughly 10 per cent of the illustrations of the entire Italian publishing industry (Benedetti 121–51).

4 Quintavalle explicitly references Vittorio Alinari in relation to photographic pictorialism, even though he admits that "this relation between Vittorio Alinari and the *pictorialist* research does not seem to be a theme that scholarship has investigated until now" (468).

5 "The ultimate origins of photography lie in the fifteenth-century invention of linear perspective" (Galassi 12). From this intuition of a tri-dimensional space by Leonardo, Brunelleschi, and Alberti, the path towards the invention of a tool for perfect perspective leads to the inventions of the camera obscura – used in painting by Albrecht Dürer and Anthony van Dyck, and perfected by Giovan Battista Della Porta (*De refractione*, 1593), Athanasius Kircher (*Ars magna lucis et umbrae*, 1646), and Marc'Antonio Cellio (*Descrizione di un nuovo modo di trasportare qualsiasi figura disegnata in carta mediante i raggi solari*, 1686) – and later of William Hyde Wollaston's camera lucida (1807).

6 I rely on Becchetti's book *Fotografi e fotografia in Italia* (9–48) and Bertelli's essay "La fedeltà incostante" for a reconstruction of the different techniques and the cultural impact of photography in Italy during the nineteenth century.

7 Marvel over Daguerre's invention continued throughout the nineteenth century, as documented in the astonishment of both Pope Leo XIII, who praised, in his 1877 Latin poem "Ars photographica" ("Photographic Art"), the "mira virtus ingenii / novumque monstrum!" (*Carmina* 35; marvellous virtue and the new prodigy of human intellect!), and D'Annunzio, who defined it as "una magia novissima" (quoted in Rocchini 78; a most novel magic).

8 Many pioneering photographers worked in Rome: John Henry Parker, Robert MacPherson, Eugène Piot, Frederic Flachéron, Pompeo Molins, Giorgio Sommer, Giacomo Caneva, Leopoldo Alinari, Gioacchino Altobelli, Stefano Lais, Ludovico Tuminello, Tommaso Cuccioni, Antonio D'Alessandri, Vittorio della Rovere, and Lorenzo Suscipj.

9 Following in the footsteps of the Italian politician and three-time finance minister Quintino Sella, his nephew, Vittorio Sella, completed a photographic charting of the Alps, the Caucasus, the Saint Elias range in Alaska, the Ruwenzori range in Africa, and the Himalayas at the turn of the twentieth century. A catalogue of these pictures is contained in Ansel Adams's volume *Vittorio Sella, Mountaineer and Photographer* (1999).

10 "The coupling of evidence and photography in the second half of the nineteenth century was bound up with the emergence of new institutions and new practices of observation and record-keeping" (Tagg 5). Photography was fundamental in the development of disciplinary institutions like prisons, asylums, hospitals, schools, and factories.

11 The photographic taxonomy of society ultimately expressed instances of social surveillance (against social mobility, crime, or military desertion) and moral containment (even in the spread of pornography, which was used as a way to tame or frame the illicit).

12 Photographs of Rome and Pompeii represent, in Freudian terms, "on the one hand, the time of accumulation, continuity, duration, contemplation, diffusion, overload – yet fragmentary; on the other, a time of seizure, cutting, the instant, rupture, uniqueness – yet totalizing" (Dubois 273).

13 The mystification of Risorgimento intrinsically led to its fetishization, as seen in the "relics" of Italy's heroes stored in the Vittoriano monument: Maroncelli's hairs, Garibaldi's bone fragments, Mazzini's glasses, and Cavour's death mask.

14 After *Camera obscura* (Milan, 1863–7) by Ottavio Baratti, and *Rivista fotografica universale* (*Universal Photographic Journal*, 1870–6) by Antonio Montagna, new photographic magazines were established from the 1890s: Luigi Gioppi's *Il dilettante di fotografia* (*The Amateur Photographer*, Milan, 1890), Rodolfo Namias's *Il progresso fotografico* (*Photographic Progress*, Milan, 1894), and Tancredi Zanghieri's *Il corriere fotografico* (*The Photographic Courier*, Piacenza, 1904). After the publication of Giovanni Muffone's handbook *Come il sole dipinge* (*How the Sun Paints*, 1886), other photographers and editors wrote manuals of photography, including Luigi Gioppi's *La fotografia secondo i processi moderni* (*Photography according to Modern Processes*, Hoepli, 1891); Carlo Brogi's *Il ritratto in fotografia: Appunti pratici per chi posa* (*The Portrait in Photography: Notes for Those Who Pose*, 1895); and Tancredi Zanghieri's *La fotografia turistica* (*Tourist Photography*, 1908). For more information about contemporary writings on photography, see Zannier (*Cultura fotografica* 131–68) and the second volume of Bertelli and Bollati's *L'immagine fotografica.*

15 As Pierre Bourdieu notes, although they seemingly favoured individual improvisation, nothing was "more regulated and conventional than photographic practice and amateur photographs" (7).

16 See Maffioli's *I macchiaoli e la fotografia* for a more complete account of the influence of photography on the *macchiaioli* painting.

17 In the "Manifesto of Futurist Painters" (11 February 1910; Rainey 62–4), Boccioni, Carrà, Russolo, Balla, and Severini labelled the divisionist painters Previati and Segantini and the contemporary sculptor Medardo Rosso as true modern artists. In the "Technical Manifesto of Futurist Painting" (11 April 1910; Rainey 64–6), they refer to divisionism as the key element of modern painting, explicitly stating: "Painting cannot exist today without *divisionism.* This is not a technical *device* that can be methodically learned and applied at will. Divisionism, for the modern painter, must be an *innate complementariness,* which we deem essential and necessary" (quoted in Rainey 66).

18 See Nemiz's *Capuana, Verga, De Roberto: Fotografi* for a catalogue of photographs taken by the *veristi* writers. With regard to Verga, Mutti (15–

22) gives a detailed report of his photographic practice, referenced in his epistolary exchange with Capuana (who started taking pictures in 1863) and first recorded in 1878 (as indicated by the first dated image in his archive). See Minghelli's essay "L'occhio di Verga" for a broader cultural account of the relationship between verismo and photography.

19 Verga theorizes this equivalence in the dedicatory letter to Salvatore Farina, preceding the novella "L'amante di Gramigna" ("Gramigna's Lover," 1880; in *Novelle* 157).

20 Unlike Capuana, who published his pictures, Verga did not photograph for public use, but only for private purposes.

21 In the early twentieth century the same photographic dialectics of staged absence and imaginative fiction informed Giovanni Papini's mid-life autobiography *Un uomo finito* (*The Failure,* 1913), which he introduces with a vision of half a photograph, of himself as a child, and similarly constructs a binary space: of loss (of his childhood) and of creative self-fashioning (in the beginning of his autobiography in the present).

22 See Pietromarchi's *Un occhio di riguardo* for a catalogue of Primoli's photographic work.

23 Nunes Vais photographed royals (King Vittorio Emanuele III, Queen Elena, and Queen Margherita), writers (Aleramo, D'Annunzio, De Amicis, Marinetti, Moretti, Negri, Palazzeschi, Papini, Serao, Trilussa, Vivanti), artists (Boccioni, Carrá, Sartorio, Soffici), intellectuals (Croce, Gentile, Marconi, Martini, Ojetti, Panzini, Salvemini), composers (Mascagni, Puccini), actors (De Sica, Fregoli, Petrolini), actresses (Borelli, Duse, Gramatica, Menichelli), politicians (Giolitti, Murri, Mussolini, Orlando, Turati), and middle-class Florentines. The artistry of Nunes Vais is praised by D'Annunzio: "the device, which at first was only apt for the brutal representation of reality, has now become, in his hands, an instrument of infinite poetic refinement" (quoted in Rocchini 78). See Vannucci's *Mario Nunes Vais: Gentiluomo fotografo* for a catalogue of his portraits.

24 See Miraglia's *Culture fotografiche e società a Torino* for a comprehensive account of the evolution of photographic culture in the city.

25 See Tucker (*The Modernist Still Life – Photographed*), Weiermair (*The Nature of Still Life: From Fox Talbot to the Present Day*), and Martineau (*Still Life in Photography*) for more information about the origins and development of the photographic still life.

26 See Costantini's *La fotografia artistica* for a more extensive account of the journal's development and content.

27 "Fu lui a donarmi quel libro, / ricordi? che narra siccome, amando senza fortuna, / un tale si uccida per una, per una che aveva il mio nome"

("L'Amica," verses 96–8; it was he who gave me that book, remember? which narrates how, unlucky in love, some guy took his life over someone, someone who had my name).

28 Carlotta's photograph (1850) is dated eleven years after the invention of photography (1839). In the same way, Gozzano's poem (1907) is eleven years after Lumière's invention of cinema (1896).

29 The complementary characters of Carlotta and Speranza find a synthesis in Gozzano's figure of Signorina Felicita. The poem "La signorina Felicita ovvero la Felicità" is structured around the same narrative elements: the vacation house (Canavese, Belgirate), the accumulated objects (in the attic, in the *salotto*), the reported conversations inside, the games in the garden, the woman's love for a poet, and the ambiguous protagonist, living in the present, yet already cast in the past.

30 The original reads: "Tragedia d'Oreste in un teatrino di marionette [...] uno strappo nel cielo di carta del teatrino [...] Oreste, insomma, diventerebbe Amleto" (Pirandello, *Il fu Mattia Pascal* 136).

31 D'Annunzio also branded himself through photography, as seen in his partnership with Mario Nunes Vais and as documented in Miraglia's essay "D'Annunzio e la fotografia."

32 The author alludes in the poem to Goethe, Foscolo, and Leopardi. He ironically quotes Leopardi's poems (in *Canti*), first, "La quiete dopo la tempesta" ("The Calm after the Storm") in the awareness of the past *affanno* (anxiety), after the exam – "Han fatto l'esame più egregio di tutta la classe. Che affanno / passato terribile!" (Verses 25–6; They completed the best exam in their class. What terrible anxiety has passed!) – and, second, the "Canto notturno" ("Night Song") in his ironic questions to the moon – "Romantica Luna [...] / non sorta sei da una stampa del *Novelliere Illustrato*? / Vedesti le case deserte di Parisina la bella?" (Verses 85–9; Romantic moon [...] aren't you from a print of *Novelliere Illustrato*? Did you by any chance see the deserted houses of Parisina the beautiful?)

33 Umberto Boccioni dismissed Bragaglia and his photodynamism as "una presuntuosa inutilità che danneggia le nostre aspirazioni di liberazione della riproduzione schematica e successiva della statica e del moto" (letter to Sprovieri, 4 September 1913, quoted in Marra 36; a presumptuous pointlessness that damages our aspirations to liberation in the schematic and subsequent reproduction of stasis and motion).

34 In 1911 Ambrosio obtained the film rights to all of D'Annunzio's literary works (Armenante).

35 "No other European country rivaled Italy in the scope of its overflowing inter-textual adaptions. [...] There was a disproportionate relationship

between the modest investment and the range of ambition of filmmakers' cultural objectives" (Brunetta 31).

36 See De Gregorio's essay "Nascita e morte dell'Ambrosio film" for more information about Ambrosio Films. With regard to the broader development of the cinematographic industry in its early period see Bernardini's *Cinema muto italiano*, vol. 2, *Industria e organizzazione dello spettacolo 1905-1909* and Nicoli's *The Rise and Fall of the Italian Film Industry*.

37 The following study of official and amateur photography during the First World War refers to my essay "Presente! The Latent Memory of Italy's Great War in Its Photographic Portraits" (University of Toronto Press, 2014), 149–80.

38 The Army Supreme Command prohibited pictures, sketches, and drawings by "those without special authorization" (*Norme per i corrispondenti di guerra*, July 1917; part 2, section 1) and scrutinized "all written correspondance of a journalistic character, both postal and telegraphic" (part 1, section 2, 13; quoted in Renzi 144, 146).

39 After the establishment in 1916 of the Ufficio storiografico per la mobilitazione, which systematically filed soldiers' profiles and portraits, photo-portraits of soldiers in uniform were meant to assemble a larger picture of Italy in war as "tiles of an extended national monument" (Caffarena 119).

40 In his war memoir, *Trincee* (*Trenches*, 1924), Carlo Salsa noticed the saturation of war stories in the public sphere: "Penso anch'io oggi che, sulla guerra, si sia scritto e parlato a sazietà. […] Si sono udite poi troppe narrazioni […]. Siamo giunti così ad un punto di saturazione: l'argomento della guerra è passato agli atti come una pratica sdrucita, e, a riparlarne, c'è da far inorridire le belle signore golose di letteratura alla moda." (Salsa 15; I, too, think that today enough has been written and said about war. […] Too many narrations have been heard […]. We have reached, then, a point of saturation: the topic of war has been processed as an overused file, and, just in talking about it, the fashionable ladies in search of fancy new literature are horrified.)

41 War portraits constituted, in the aftermath of war, not just a politically relevant affirmation of the soldier's participation in the conflict but also, in their visual exhibition of private sorrow, a subtle form of "censorship and taming of mourning itself" (Dolci and Janz 44).

4 Bicycles and the Moving Body of the Nation

1 The original reads: "Quest'anno avevo ideato per le mie prossime vacanze la gita in velocipede da Milano a Chicago e ritorno […] ma c'è di mezzo il

mare […]. Datemi un biglietto da 500 lire o prosciugatemi il mare; ed io vi farò vedere l'utilità pratica del bicicletto con l'andata e ritorno in due mesi circa da Milano alla grande Esposizione mondiale di Chicago" (*Corriere della sera*, 7 July 1893).

2 The Fasci dei Lavoratori (Workers' League), founded in Sicily in 1891 in connection with peasantry and anarchism, preceded the foundation of Filippo Turati's Partito dei Lavoratori (Workers' Party) in 1892. During the summer of 1893 the Sicilian fasces arranged a series of strikes, which soon turned into an organized revolt against heavy taxation by the state. Although Giolitti had adopted a tolerant policy towards the turmoil, the new prime minister, Francesco Crispi, who took office in December 1893, violently repressed and outlawed the movement out of fear that it could expand to the whole nation.

3 As documented in Fiorentino's *Gli Stati Uniti e l'Italia alla fine del XIX secolo,* the concept of *americanismo* applied to both ecclesiastical and social spheres. As outlined in Confessore's book *L'americanismo cattolico in Italia,* it referred to the modern religious trends coming from the United States, later condemned as heresies in Pope Leo XIII's encyclical letter *Testem benevolentiae nostrae* of 1899. Against the backdrop of a widespread cultural bias towards America, it also related to the broader European concerns about the growing international and industrial power of the United States (overtly manifested in the Spanish-American War of 1898).

4 These words also title Rossi's volume *L'anarchico delle due ruote,* which constitutes the most reliable source for documenting the life and feats of Luigi Masetti.

5 The original reads: "Il momento è dei velocipedi. Padroni delle vie, padroni dei giornali, padroni dell'interesse pubblico. Passerà questa moda?... Passerà il fanatismo; e rimarrà l'uso dilettevole, igienico, soprattutto utile" (*Illustrazione,* 9 July 1893, 19).

6 In 1901 the journalist Ottone Brentari produced an important piece for *La rivista mensile del Touring Club Italiano* (no. 1), describing Masetti's eighteen-thousand-kilometre trip across Africa, Europe, and Russia, as well as reporting the cyclist's encounter with Tolstoy.

7 Explicitly directed towards mechanics and operators in the bicycle industry, some later publications in Milan included *L'industria della bicicletta: Giornale tecnico per l'industria e il commercio dei velocipedi* (*The Bicycle Industry: Technical Journal for the Industry and Commerce of Velocipedes,* 1895); and *Il progresso ciclistico: Rivista mensile illustrata tecnica e sportiva di ciclismo e automobilismo* (*Cycling Progress: Monthly Illustrated Technical and Sport Magazine of Cycling and Motoring,* 1896–7). As indicated in the opening issue of *L'industria della*

bicicletta ("Il nostro scopo," 17 March 1895), these publications shared a common goal: "connecting those who have a direct or an indirect interest in the cycling industry, through well-devised advertising – both expositive and narrative – news, and technical reports" (mettere in diretta relazione fra loro quanti hanno interesse diretto o indiretto all'industria ciclistica, per mezzo di ben intesa réclame, sia espositiva che narrativa, di notizie e ragguagli tecnici).

8 Paolo Mantegazza vehemently opposed the use of bicycles by women ("Eve's daughters can never ride a velocipede if they want to reconcile hygiene and morality"; quoted in Marchesini et al. 70), and Cesare Lombroso wrote a famous 1900 essay in *La nuova antologia* about cycling and crime, "Il ciclismo nel delitto" ("Cycling in Crime").

9 The equation of bicycles with a less rigid feminine code (in clothing and behaviour) is a common theme in fin de siècle imagery. In a poster for the magazine *La bicicletta* (post-1894 to 1898) the illustrator, Aleardo Villa, was the first to depict a modern woman riding her bicycle, décolleté, wearing an elegant hat, and exposing her knees and socks by her tied-up skirt (Panzeri). In addition, the cover of Maurice Leblanc's 1898 novel *Voici des ailes!* (*It Gives You Wings!*) featured a pedalling and flying woman with a bare chest.

10 In the novel *Il santo* (1905) Fogazzaro hints at the hierarchy's aversion to cycling in the questions posed by young students to the protagonist Benedetto with regard to the position of the Church in the modern world: "Sarebb'egli stato disposto a farsi propugnatore di una riforma della Chiesa? [...] Era democratico cristiano? [...] Gli piaceva che fosse proibito ai cardinali di uscire a piedi e ai preti di andare in bicicletta?" (212; would he be ready to promulgate a reform in the Church? [...] Was he a Christian democrat? [...] Did he like that cardinals were forbidden to take a walk and priests to ride a bicycle?)

11 After the first organized races (in Padua in 1869, and the Milan-Turin race from 1876) and the building of the first velodromes (the Umberto I was built in Turin in 1890), many cycling clubs were founded after the model of the Florence Veloce Club (1870) and then became associated, in 1884, as part of the Unione Velocipedistica Italiana (UVI).

12 Several contemporary publications intended to explain the practice of cycling and to sponsor its growing market. The era's most important promotional books were *Il velocipedismo (storia, igiene e pratica)* (*Velocipedism: History, Hygiene, Practice*, 1894) by V. Monaco; *La pratica del velocipede e la tecnica dell'allenamento* (*A Practical Guide to Velocipedes and Training Techniques*, Florence, 1895) by A. Roster and A. Orlandini; *La bicicletta* (*The Bicycle*, Turin, 1896) by S. Stella; and *La bicicletta* (*The Bicycle*, Livorno, 1897) by A. Lombardini.

13 After their early beginnings (with the presentation in Paris of the *célérifère* by Count Mède de Sivrac in 1791) bicycles underwent a process of slow transformation. Throughout the nineteenth century the *vélocifère,* conceived as a support for fast walking, had rapidly evolved, into the so-called *draisienne* (with the introduction of the handlebars in 1817 by Ludwig Drais von Sauerbronn, allowing directioning); into a tricycle (with the introduction of two back wheels by Gourdoux in 1819, allowing stability); into the *michaudine* (after MacMillan's invention of the pedals in 1838 and Michaux's introduction of a transmission chain in 1855, allowing a quicker, balanced motion); and into the *grand-bi* (with the assembling of an oversized forewheel in the 1870s, allowing increased speed). See Marchesini et al. (18–28) and Ormezzano (28–39) for a complete genealogy of the bicycle.

14 Production moved from 30,000 items in 1896 to 200,000 in 1900, to 605,000 in 1910, and to 1,364,000 in 1919 (Marchesini 90).

15 See Mari (133–58) for information about Bianchi's marketing strategies, and Marchesini et al. (32–3) for Bianchi's evolution into a car manufacturer.

16 The original reads: "La Bianchi che ha imperato nelle pacifiche competizioni dello sport contribuisce ora validamente all'avanzata gloriosa delle forze italiane" (quoted in Antonio Gentile 38).

17 See Bardelli's *L'Italia viaggia* for an analytical documentation of the Touring Club Italiano, and Pivato's *Il Touring Club Italiano* for a synthetic overview of its history and activities.

18 *Sicilia 1898* (*Sicily, 1898*), subtitled *Note di una passeggiata ciclistica,* by Bertarelli, is a diary of his tour of Sicily, an ethnographic instrument of discovery, and an example of his deliberate attempt to transform cycling into a pedagogical tool for the cultural and geographical education of Italians.

19 In addition to the magazine the TCI published maps for its subscribers (*Planimetrie e profili ciclistici d'Italia* in 1896; *Carta d'Italia del TCI al 250.000,* with polychrome map indicators and stylizations of attractions, in 1906; *Carta della Tripolitania e della Cirenaica,* in 1912, after the Libyan war), as well as several guides: *Guida-itinerario dell'Italia e di alcune strade delle regioni limitrofe* in 1895 (replacing, with its 370 itineraries, the more common *Baedeker*); the seven-volume *Guida d'Italia* in 1912; *Guida d'Italia per gli stranieri* in 1922.

20 Favoured by the medical discoveries regarding the good influence of seawater on health in the 1860s and by the decreased travel time from London to Rome (from four weeks in 1830 to fifty-five hours in 1870), the British were the first to populate seaside localities like Sanremo, Bordighera, and Viareggio and to invest in Italy's tourist infrastructures, with the construction in 1908 of the Excelsior in Venice-Lido and the Grand Hotel in Rimini (Pivato, *Touring* 10–22).

21 Boccioni signed only three of the magazine covers in 1908 (January, March, and April). Given the magazine's investment in both Boccioni and Malerba, and based on the clear attribution to Malerba of the issues from July to December (Breda), it is reasonable to surmise that Boccioni authored the first six issues of 1908.

22 The cover advertisement for Bianchi in the June 1908 issue further confirms the magazine's endeavour to create a visual common ground across different productive areas and brands. By staging the rapid encounter on a mountain road of a car (passing a slower, horse-drawn carriage on the way up) and a bicycle (effortlessly descending down the road) in one snapshot, the advertisement visualizes the connections of the mechanical industry with tyre manufacturing and represents Bianchi's transition from a line focused mainly on bicycles to the manufacture of cars, by transferring onto the automobile the cycling imagery of elegance, outdoor living, and domination over nature.

23 "Era una scena ammirabile e insieme spaventevole, il vedere quella piccola nave, guidata dal suo audace capitano, sfidare la rabbia di uno dei più vasti oceani!... Sembrava un moscerino che lottasse contro un titano, ma quel moscerino non aveva paura e non dava indietro [...] come fosse deciso a vincere o morire." (Salgari 46; It was at once an admirable and frightening sight, seeing that small boat, led by its bold captain, defying the wrath of one of the vastest oceans!... It seemed like a gnat fighting against a titan, yet that gnat was not afraid and did not withdraw [...] as if he were resolved to either win or die.)

24 The original reads: "La bicicletta siamo ancora noi che vinciamo lo spazio e il tempo" (Oriani 87). Federigo Tozzi would also recall the model of Oriani's trip from Faenza to Tuscany and back, in the novella "Un'osteria" ("A Tavern"), staging the deliberate bicycle trip through Emilia Romagna of the protagonist and his friend Giulio (*Novelle* 265).

25 After Foscolo's translation of Sterne's *Sentimental Journey,* the genre of a plotless narration (or anti-narration), made of unstructured anecdotes and humorous episodes, achieved great success in Italy with Collodi's *Romanzo in vapore* (*Novel on a Steam Engine,* 1856), Yorick's *Su e giù per Firenze* (*Up and Down Florence,* 1877), and Pirandello's 1908 essay on *umorismo* (humour). See Mazzacurati's volume for a detailed study of the "Sterne effect" on Italian literature.

26 I rely on Weidlich (*Ciclismo e letteratura*), Bertellini (*Scrittori della bicicletta*), Pedroni ("Poesia ciclistica delle origini"), Barsella ("Bicicletta: Il mito e la poesia"), and Bosi Maramotti ("La bicicletta e la letteratura") to reconstruct the relationship between bicycles and poetry.

27 Excerpts from Betteloni's "Canto dei Ciclisti": "Avanti! Avanti! Rapidi / precipitando a volo / noi divoriam lo spazio / radendo appena il suolo / ed irruente palpita / pieno d'ebbrezza il cor" (verses 1–6); "e correr, correr, correre / lo sconfinato pian / e salir monti e scendere / ne 'l divin sole immersi / cento ammirar spettacoli / di Natura diversi" (verses 17–22); "fansi d'acciaio i muscoli / ne l'esercizio ardito" (verses 25–6; *Poesie edite ed inedite* 487).

28 The original reads: "Corro così la solitaria landa […] e vivo e volo!" ("Sole d'inverno" ["Winter Sun"], verses 23, 26; *Rime* 509).

29 The quotation continues: "Virgilio cantò il cavallo, Monti il pallone, Carducci il vapore, molti la nave, nessuno ancora la bicicletta; eppure né il cavallo, né il pallone, né il vapore, né la nave resero all'uomo più facile il *trasportarsi* ovunque una qualche necessità lo richiami, lasciandolo più *signore di se stesso.*" (Oriani 43–4, emphasis mine; Virgil celebrated the horse, Monti the hot air balloon, Carducci the steamboat, many the ocean liner, yet no one the bicycle; nonetheless, no horse, no hot air balloon, no steamboat, no ocean liner ever made it easier for man to *self-transport* wherever necessity calls him, leaving him more *master of himself.*)

30 The originals read: "Ogni volta che per una via di città, […] io vedo questa macchina fuggente su cui sta accovacciato un uomo, non posso frenare un moto di paura e di disgusto" (Serao); "un arrotino impazzito" (Carducci; quoted in Samaritani).

31 The original reads: "Un attimo […] un battito […] un palpito" (verses 3, 7, 11); "Mia terra, mia labile strada / sei tu che trascorri o son io?" (verses 27–8); "impeto d'ala" (verse 31); "l'ebbrezza del giorno" (verse 32; Pascoli, *Tutte le poesie* 584–5).

32 Remo Ceserani's volume *Treni di carta* offers a detailed account of the emergence of the train symbol in Italian literature since the mid-nineteenth century.

33 The paronomasia of *ascesa* and *accesa* is repeated twice in the poem: "condussi nell'*ascesa* / la bicicletta *accesa* d'un gran mazzo di rose" (verses 17–18); "discenderai al Niente […] queste pensavo cose, guidando nell'*ascesa* / la bicicletta *accesa* d'un gran mazzo di rose" (verses 57, 61–2, emphasis mine).

34 The originals read: "Donna e bicicletta si somigliano" (Oriani 60); "la bicicletta è più seduttrice della donna; la sua velocità diventa una carezza alla quale è impossibile resistere" (58); "basta la punta di uno spillo a sgonfiare la corazza di una bicicletta e quella di una parola a vuotare il cuore di una donna" (60).

35 As witnessed by the journalist Argo, in *Il ciclo* (19–20 May 1894), bicycles (in the feminine) relate to pleasure and to an affirmed male control over

it: "chiamar desidero / Il bicicletto ancora bicicletta. [...] / Per le signore sia la desinenza / Ed il genere sia sempre maschile / Ma resti a noi però di conseguenza / Il piacere di montare il femminile." (Quoted in Samaritani; I still wish to call the *bicicletto bicicletta.* [...] / To the ladies the ending and to the gentlemen always the gender. Consequently, however, may we retain the pleasure of mounting the feminine.)

36 An example of this topos appears in Govoni's "La fiera" ("The Fair"): "Sull'erba della darsena intrecciammo / le nostre impolverate biciclette / come in gelosa lotta due caprette. / Sul loro esempio, muti, ci avvinghiammo." (Verses 7–12, *Poesie* 73; Upon the grass of the river bank we intertwined our dusty bicycles like two little goats in a jealous fight. Following their example, in silence, we clung to each other.)

37 An example from Argia Sbolenfi's poetry: "mi avvolge un'onda di piacere sovrano / quando vengo stringendo il trionfale / manubrio in mano / io son beata allor che fra le gambe / sento il rigido ordigno e in quegli istanti / tendo le coscie e l'agitar d'entrambe / lo spinge avanti" ("In bicicletta" ["On the Bicycle"], verses 18–24; *Rime di Argia Sbolenfi* 79–80; a wave of sovereign pleasure overwhelms me when I come holding the triumphant handlebars in my hand. I am blessed as soon as I feel the rigid device between my legs, and in those moments I tighten my thighs, and their agitation moves it forward.)

38 The originals read: "Abituata a volare, lei, a correre, a correre, in treno, in automobile, in ferrovia, in bicicletta, su i piroscafi. Correre, vivere!" (Pirandello, "Mondo di carta" 486); "perpetua irrequietezza di perplessità" (484); "si sentiva soffocare in quel mondo di carta" (486); "Ma che! [...] – Io ci sono stata, sa? E le so dire che non è com'è detto qua!" (486); "Non voglio farle offesa. Ma mi colora tutto diversamente, capisce? E io ho bisogno che nulla mi sia alterato; che ogni cosa mi rimanga tal quale" (484); "M'importa un corno che lei c'è stata! È com'è detto là, e basta! Dev'essere così, e basta!" (486); "Niente lì si doveva toccare. Era così, e basta. Il suo mondo. Il suo mondo di carta. Tutto il suo mondo" (486).

39 The original reads: "Ogni volta che, non visto, potevo esaminare a mio comodo una bicicletta appoggiata a un muro, mi sentivo forzato, come dall'attrazione di un frutto proibito, ad afferrarla, a palparla, a metterla ritta e in moto" (De Amicis, "La tentazione" 76–7).

40 Some critics point to cycling as the source of an improvement in De Amicis's style: "se ella pedalasse, ella riuscirebbe più stringatamente sintetico nell'espressione del suo pensiero" (De Amicis, "La tentazione" 76; if you pedalled, you would acquire a more concisely synthetic expression of your thought).

41 The original reads: "Diventai un biciclista del cuscino. Nel sonno [...] avevo il pieno e vivissimo inganno della sensazione della corsa. Ah, finalmente! E ci voleva tanto a decidersi!" (83). As the quotation continues, De Amicis describes the enjoyment of pedalling, by adopting the typical imagery of cycling poetry: "È davvero il senso delizioso dello *scioglimento da ogni legame* molesto della vita, della *libertà*, dell'oblio, della *dominazione dello spazio*, della *fuga verso l'infinito*. Questo fendere l'aria senza quasi sentire il contatto della terra dà veramente l'illusione d'esser portati via da *due grandi ali invisibili* [...] questo *volo* che mi fa parere intorno tutte le altre creature umane torpide, sonnolente, schiave, che muta tutto attorno a me ad ogni istante, che *mi toglie il concetto del tempo, che m'inebria d'aria, di luce e di freschezza*, che mi fa pensare a lampi e a visioni, che mi fa *fremere, sorridere, palpitare e sognare. Questa è una vita nuova, una voluttà sovrumana, un rapimento celeste*." (It is truly the sweet perception of *throwing off any annoying bondage* in life, of *freedom*, of oblivion, of *domination over space*, of *escaping towards infinity*. This splitting of the air without even feeling any contact with the ground truly gives the illusion of being transported by two large *invisible wings* [...] this *flight* that makes every human creature around me seem torpid, sleepy, enslaved, that transforms everything around me in every instant, that *removes me from the concept of time*, that *inebriates me of air, light, and freshness*, that makes me think in lightning visions, that makes me *quiver, smile, palpitate, and dream. This is a new life, a super-human pleasure, a celestial bliss*" ("La tentazione" 83, emphasis mine).

42 The originals read: "La grande battaglia tra i giganti della strada"; "un epico spettacolo"; "una lotta titanica" (*Gazzetta dello sport*, 12, 24, and 26 May 1909).

43 "Il più meraviglioso dei viaggi" (Morasso, "Mentre"); "la più grande prova ciclistica non solo d'Italia, ma di Europa" (Morasso, "Il giro d'Italia").

44 "Straordinaria apoteosi della bicicletta"; "un trionfo innumerevole, sterminato, maestoso della piccola macchina, dell'umile congegno di acciaio" (Morasso, "La fine").

45 "In quello spettacolo supremamente bello di migliaia e migliaia di persone fuse in un solo sentimento e unico, vibrava intenso l'animo fremente del nostro popolo generoso" (*La gazzetta dello sport*, 1 June); "la folla stava dappertutto, enorme [...] la via formicolava di ciclisti, una processione interminabile. [...] Fu una valanga di uomini e di macchine che invase la strada" (*Corriere della sera*, 31 May); "Si sono viste a Bologna, a Chieti, a Napoli, a Roma città intere deliranti agli arrivi [...] irrompere come fiumane al traguardo per stringere da vicino il vincitore, per gridare al suo orecchio un bravo e un evviva" (Morasso, "Mentre").

46 Luigi Ganna is indeed "very strong and inflexible, like a powerful and well-regulated machine that always achieves the highest performance, without alteration, indefinitely, to the extreme" (fortissimo e inflessibile come una poderosa e ben regolata macchina che lavora al massimo rendimento sempre, senza alterarsi, indefinitamente, fino all'estremo; Morasso, "Le ultime tappe").

47 The first significant acceptance of international sport as an element of global modernity in Italian media can be traced to the episode of Dorando Pietri's victory and unfair disqualification in the marathon of the London Olympic Games in 1908. The relevance attributed to the event, however, came as a reflection of its enthusiastic coverage by Arthur Conan Doyle in the *Daily Telegraph*, which offered Italians an occasion to celebrate their national pride.

48 For more information on De Coubertin and the development of the modern Olympic Games, see A. Lombardo (*Pierre de Coubertin: Saggio storico sulle Olimpiadi moderne, 1880–1914*), Ioli (*Letteratura e sport: Per una storia delle Olimpiadi*), Young (*The Modern Olympics: A Struggle for Revival*), Smith (*Olympics in Athens 1896: The Invention of the Modern Olympic Games*), and Roche (*Mega-events and Modernity: Olympics and Expos in the Growth of Global Culture*).

49 The tension between Catholic and national gymnastics societies – the lively educational system of the Church (which used physical education as a healthy outlet for sinful inclinations) and the supposedly neutral state system (which openly boycotted Catholic associations) – mirrored the divide between the Vatican (which forbad Catholics from participating in Italy's political life) and the Liberal state (which was still incapable of tolerating different voices; Fabrizio 27–65).

50 From 1908 to 1911, and 1915 to 1919, Morasso was also the director of the magazine *Motori cicli e sports* (*Motoring, Cycling, and Sports*), dedicated to cycling, car and boat racing, and aeronautics.

51 The originals read: "Sitibondo delirio di onnipotenza" (94); "correre per correre sempre più in fretta" (*La nuova arma* 4).

52 In Morasso's thought, the "machine" spurred the bold protagonism of a new human being and ultimately envisioned a new model of knowledge in movement (as opposed to the one coming from the past and books), which was conceived as a naturalistic vitalism and a perennial conquest in the present, as pointed out in an excerpt from the book that foresees the later rhetoric of Futurism: "Al fuoco il libro, non vogliamo sottrarci al suo gioco arcaico, vogliamo essere noi, vogliamo uscire dai musei, dalle accademie, da tutti i luoghi rinchiusi, bui e silenziosi, da tutti i depositi di anticaglie e

di muffa, da tutti i recinti del passato, [...] vogliamo correre fra i nostri compagni che lavorano e che amano, vogliamo correre là dove si opera, dove si lotta, dove si crea, dove si vive, oppure dove nell'occhio della donna desiderata o sulla vetta agognata si contempla una infinita conquista!" (*La nuova arma* 158; In the fire with the book, we don't want to give in to its archaic game, we want to be us, we want to get out of museums, of academies, of all enclosed, dark, and silent places, of all deposits of antique junk and mould, of all the fences of the past, [...] we want to run among our companions who work and love, we want to run there where one acts, fights, creates, lives, or where, in the eye of the desired woman or upon the coveted summit, one contemplates an infinite conquest!)

53 "Non siamo più uomini come eravamo prima, come sono tutti gli altri, noi siamo terribilmente armati e gli altri sono inermi, noi siamo esseri nuovi fortissimi di una specie ignota, centauri di carne e di ferro, di ruote e di membra. Ci sale al cervello l'ebbrezza orgogliosa del nostro nuovo potere." (*La nuova arma* 54; We are no longer men as we were before, as all the others are. We are powerfully armed and the others are defenceless, we are strong new beings of an unknown species, centaurs of flesh and iron, wheels and muscles. The proud inebriation of our new power overwhelms us.)

54 Gian Emilio Malerba's poster advertisement for Cicli Stucchi of 1903 ("The Art of Cycling") similarly depicts the visual union of human and technological elements through the vision of a naked figure (a woman) lightly taking flight while pedalling a unicycle. Malerba's advertisement shares some elements with Codognato's representation of 1910, namely nudity, the back view of the figures, the overlapping of pedals and wings, and the juxtaposition of the body with the bicycle. In contrast, Codognato shifts from a female to a male subject and redefines bicycles in relation to the body, emphasizing muscular strength over sensual grace.

55 Campana's poem, the first ever about the Giro d'Italia, was initially dedicated to Marinetti. The original version was written in 1913 and published abridged in *Canti Orfici* as "Il traguardo." The original text is quoted: "Dall'alta ripida china precipite / come movente nel caos d'un turbine / come un movente grido del turbine / come il nocchiero del cuore insaziato. // Bolgia di roccia alpestre: grida di turbe rideste / vita primeva di turbe in ebbrezze: / un bronzeo corpo dal turbine / si dona alla terra con lancio leggero. // Oscilla di vertigine il silenzio dentro la muta catastrofe di rocce ardente d'intorno. / Tu balzi anelante fuggente fuggente nel palpito indomo / un grido fremente dai mille che rugge e scompare con te / balza una turba in caccia si snoda s'annoda una turba / vola una turba in caccia Dionisos Dionisos Dionisos." (Campana, in *Il più lungo*

giorno, 1973; From the high, precipitous slope, like moving in the chaos of a whirlwind, like a moving scream of the whirlwind, like the helmsman with an unquenched heart. Ditch of alpine rocks; screams of reawakened crowds, primordial life of inebriated crowds; a bronze body is given to the earth from the whirlwind with a light leap. The silence oscillates in vertigo in the muted catastrophe of rocks burning all around. You leap, yearning, escaping, escaping in the untamed palpitation, a quivering scream from the thousands, which roars and disappears with you, a crowd hunts, unleashing an energy, the hunting crowd flies, *Dionisos Dionisos Dionisos.*)

56 The documentary, which focused on the second Giro d'Italia, was the first work released by Comerio Films, founded in 1910 by the renowned photographer Luca Comerio.

57 The original reads: "Le piste, le gare atletiche, le corse ci esaltano! Il traguardo è per noi il meraviglioso simbolo della modernità" (quoted in Salaris 106).

58 Alongside Darwin and Nietzsche, the contemporary diatribe on the cultural value of the body engaged those who conceived of it deterministically – as an object in Marxist materialism (considering the soul and mind as functions of the brain) or in Lombroso's positivism (identifying in corporal signs, like fingerprints or skulls, the evidence of identity or the predisposition to crime) – and those who observed it experientially – as a storage place of "bodily memory" in Bergson (*Matter and memory* 197), and as the source of intersubjectivity in Husserl.

59 Casini comments: "the most intimate depths, the most authentic essence of the personality is in the body. The body constitutes the centre of man, and conscience is its by-product. The most authentic wisdom is the wisdom that emanates from the body; it is the wisdom that comes from the immanent rationality of the body" (304).

60 In his *Battista al Giro d'Italia* (1929) the writer Achille Campanile would say "scrivere pedalando" (23). In more recent times Augé synthetically defined the new intellectual self-awareness of the *homo pedalans* with the expressions "je suis ce que je découvre" (29; I am what I discover) and "je pedale donc je suis" (86; I pedal therefore I am).

61 The number of bicycles circulating in Italy was 2,224,000 as of 1924 (860,000 more than in 1919), and 4,935,000 as of 1938 (Marchesini 90).

5 Gramophones, Radio, and the New Languages of Sound

1 Born in Naples in 1873, Caruso made his first operatic appearance at the age of twenty-two, on 15 March 1895, at the Teatro Nuovo in Naples, singing

in Domenico Morelli's opera *L'amico Francesco*; his first appearance at La Scala in Milan took place on 26 December 1900, in Puccini's *La Bohème* directed by Arturo Toscanini. The ten arias of his 1902 recorded performance were "E lucevan le stelle" *(Tosca,* Puccini); "Una furtiva lagrima" (*L'elisir d'amore,* Donizetti); "Celeste Aida" (*Aida,* Verdi); "Questa o quella" (*Rigoletto,* Verdi); "En fermant les yeux" (*Manon,* Massenet); "Giunto sul passo estremo" (*Mefistofele,* Boito); "Apri la tua finestra" (*Iris,* Mascagni); "Studenti, udite" (*Germania,* Franchetti); "Ah, vieni qui … no, non chiuder gli occhi vaghi" (*Germania,* Franchetti); and "Dai campi, dai prati" (*Mefistofele,* Boito) (Scott 57).

2 "The music historians still debate whether Caruso created the record industry or the record industry created Caruso" (Gargano 54).

3 U.S. President Roosevelt wrote to the King of England (Sewall 22):

His Majesty, Edward VII. London, Eng.

In taking advantage of the wonderful triumph of scientific research and ingenuity which has been achieved in perfecting a system of wireless telegraphy, I extend on behalf of the American People most cordial greetings and good wishes to you and to all the people of the British Empire.

THEODORE ROOSEVELT
Wellfleet, Mass., Jan. 19, 1903.

King Edward VII replied with the following message:

Sandringham, Jan. 19, 1903.

The President, White House, Washington, America.

I thank you most sincerely for the kind message which I have just received from you, through Marconi's trans-Atlantic wireless telegraphy. I sincerely reciprocate in the name of the British Empire the cordial greetings and friendly sentiment expressed by you on behalf of the American Nation, and I heartily wish you and your country every possible prosperity.

EDWARD R. and I

4 The 1899 message was sent from Wimereux, France (near Boulogne), to South Foreland Lighthouse, England (near Dover). The 1901 message was sent from Poldhu, Cornwall, to Signal Hill, St John's, Canada, and consisted of a wireless signal in Morse code of the letter *S.* After this major achievement Marconi built a new station in Glace Bay, Nova Scotia (on 17 December 1902), and in South Wellfleet, on the Cape Cod shore (on 18 January 1903) (Hong 60–80).

5 See Birch's volume *Marconi Radio Pioneer* for more information about the life of Marconi, his experiments with wireless communication, and the evolution of his company.

6 "For recording, one yells into the microphone and turns the crank. The cylinder rotates and the stylus makes contact with the foil. As sound waves move the diaphragm in and out, the stylus makes corresponding indentations in the foil. The sound wave is encoded as a time-locked impression of varying depths in the helical track around the metal foil on the cylinder. For playback, one rewinds the cylinder with the crank and places the stylus of the speaker in contact with the beginning of the groove in the foil. When the crank is turned, out comes the recorded sound. One of the early tests was to record and playback 'Mary had a little lamb …' The early versions were capable of recording about ten seconds of sound" (R. Weber 151).

7 "If the fifteenth century had invented print for a more rapid communication of the written language, the nineteenth century found a way to enclose and reproduce word, man's principle means of communication" (Kron 13).

8 In a 1887 article published in *Scientific American* Edison presented his new device by listing its old applications: "the new phonograph is to be used for taking dictation, for taking testimony in court, for reporting speeches, for the reproduction of vocal music, for teaching languages […], for civil and military orders […], for the distribution of the songs of great singers, sermons and speeches, the words of great men and women" (quoted in Kittler 78).

9 Even though Edison, in 1888, had travelled to Europe to record the voices of Gladstone and Bismarck, or Brahms playing *Hungarian Rhapsody* in Vienna, "until the mid-Nineties Europe remained on the periphery of phonographic affairs and depended solely on exports from American factories" (Gelatt 101). Within this context, the fact that Owen combined the production of gramophones with that of typewriters indicates the equivalence between sound recording devices and the broader category of automatic writing.

10 Barraud's painting represented "a double message of 'fidelity,' entrusted both to the little dog (faithful by definition) and to the gramophone (whose technical perfection had been confirmed by the dog's attention to the perfectly recognizable voice of his master)" (Curci 7).

11 The Anglo-Italian Commerce Company "relied for its operatic repertoire on members of Milan's several opera houses" and produced recordings of Caruso. "No one has been able to say with assurance just when Caruso recorded his three A.I.C.C. cylinders; most experts incline to late 1900 or early 1901" (Gelatt 104).

12 SAIF, or simply Fonotipia, was established by the Anglo-French composer Baron Frederic d'Erlanger, and it also produced, with serious recordings of male and female celebrities, Fregoli and Neapolitan songs (Andrews 11–12).

13 "Il fonografo ha detronizzato il telefono. Di quest'invenzione *L'illustrazione* fu la prima a parlare in Italia; del fonografo s'è già parlato di molto fra noi, ma saremo primi a darne il disegno. [...] Sentiamo la nostra meraviglia, sentiamo la gioia della nostra sorpresa e mandiamo i mi rallegro della maggiore ammirazione possibile al felice e perseverante inventore, il signor Thomas A. Edison, di Menlo Park." (*Illustrazione italiana,* 6 January 1878; The phonograph overthrew the telephone. *Illustrazione* was the first in Italy to talk about this invention; much has already been said about the phonograph, but we will be the first to provide a drawing of it. [...] We feel our marvel, we feel the joy of our surprise, and we send our congratulations, filled with as much admiration as possible, to the happy and persevering inventor Thomas A. Edison from Menlo Park.)

14 "Non piove più perché hanno messo quel maledetto filo del telegrafo, che si tira tutta la pioggia, e se la porta via [...] il telegrafo portava le notizie da un luogo all'altro; questo succedeva perché dentro il filo ci era un certo succo come nel tralcio della vite, e allo stesso modo si tirava la pioggia dalle nuvole, e se la portava lontano, dove ce n'era più di bisogno." (Verga, *I Malavoglia* 72–3; It no longer rains, because someone put up that cursed telegraph wire, which pulls all the rain and carries it away [...] the telegraph carried news from one place to another; this happened because inside the wire there was a certain juice, like in the shoot of the vine, and in the same way it pulled the rain away from the clouds and carried it far away, where there was more need of it.)

15 The original reads: "I fili del telegrafo s'incroceranno come un'immensa tela di ragno sopra i tetti della città rumorosa" (De Amicis, *Costantinopoli* 151).

16 By the mid-1900s "manufacturers enclosed the gramophone entirely within a cabinet which became important as a piece of furniture" (Melville-Mason 117).

17 The mass production of pianos started in the eighteenth century with the German Silbermann and continued in the nineteenth century with the British Broadwood and the American Steinway (M. Weber 117–23).

18 "The introduction of the phonograph and the ready availability of Caruso's record made it easy for the next generation to copy much of his vocal style and yet at the same time ignore this salient feature of his art: the perfection of the relationship between his legato and portamento" (Scott 169–70).

19 The original reads: “Una vecchia scatola armonica, con l’anima di metallo costrutta d’un pettine d’acciaio i cui denti vibravano al girare d’un cilindro irto di punte” (D’Annunzio, *Forse che sì* 704).

20 The original reads: “Chi può dire perché una vecchia cosa meccanica ci diventi amica, si leghi a qualche fibra della nostra vita intima, e non possiamo mai separarcene senza temere di far morire un po’ di noi stessi? [...] Era l’*aura* dei miei sogni puerili. Anche ora non posso udirla senza un certo ondeggiamento del cuore. Per nessuna cosa ho avuto il sentimento della proprietà come per questo scarabillo” (D’Annunzio, *Forse che sì* 705, emphasis mine).

21 The original reads: “Coloro che usano oggi del telegrafo, *del telefono e del grammofono*, del treno, della bicicletta, della motocicletta, dell’automobile, del transatlantico, del dirigibile, dell’aeroplano, del cinematografo, del grande quotidiano (sintesi di una giornata del mondo) non pensano che queste diverse forme di comunicazione, di trasporto e d’informazione esercitano sulla loro psiche una decisiva influenza” (Marinetti, *Teoria* 65–6, emphasis mine).

22 The original reads: “nulla è più interessante per un poeta futurista che l’agitarsi della tastiera di un pianoforte meccanico” (Marinetti, *Teoria* 51).

23 “Bisogno di sapere ciò che fanno i loro contemporanei di ogni parte del mondo [...] di comunicare con tutti i popoli della terra [...] di fissare ad ogni istante i nostri rapporti con tutta l’umanità” (Marinetti, *Teoria* 69).

24 By the mid-nineteenth century Ricordi had operative branches in Naples (1864), Florence (1865), Rome (1871), and London (1878).

25 Grandi magazzini italiani E. & A. Mele opened in 1889 in Naples, under the influence of the Parisian department stores Lafayette and Le Bon Marché, and closed in 1932 after the financial crisis of 1929.

26 Benjamin comments on the impact of lithography on modern industrial societies in his *Work of Art in the Age of Its Technological Reproducibility*: “Lithography marked a fundamentally new stage in the technology of reproduction. This much more direct process [...] first made it possible for graphic art to market its products not only in large numbers, as previously, but in daily changing variations. Lithography enables graphic art to provide an illustrated accompaniment to everyday life” (20).

27 Over the following years Campari also commissioned works from such talents as Leopoldo Metlicovitz, Franz Laskoff, Plinio Codognato, Leonetto Cappiello, Achille Mauzan, Marcello Nizzoli, and Fortunato Depero.

28 Even though the Italian Impresa Generale d’Affissioni was founded in 1881, Italy’s *cartellonismo* really started in imitation of the French style in the mid-1890s, as confirmed by the art critic Vittorio Pica, writing in 1895 on the

French journal *La Plume*: "The Italian posters [...] do not arouse the same interest as the French, English, or American ones. Nevertheless, in recent times, a great advancement is evident in taste, conception, and print run of colour images, especially in the Ricordi and Treves workshops" (quoted in De Iulio 13).

29 Jules Chéret is the father of the modern poster. His most famous works, which inspired Seurat and Toulouse-Lautrec among others, are *La biche au bois*, *Le bal Valentino*, *La folie Bergère*, *Moulin rouge*, and *Eldorado*. Alphonse Mucha, imitated by Hohenstein, created in *Gismonda* (1894) a visual archetype of the reassuring female character, naked or dressed in a tunic, adorned with flowers, and surrounded by a halo of light.

30 "The works of Hohenstein or Marcello Dudovich, Leopoldo Metlicovitz or Leonetto Cappiello were to become symbols of the Italian 1900s, just as the blown glass of Murano is a trademark of the 1920s and '30s, the first and second Futurist movements, the Metaphysical school in painting and Rationalism in architecture symbolise the years between the two World Wars, cinematic Neo-realism is the symbol of the post-war years, and design the art of the 1950s and '60s" (Mughini ix).

31 While Puccini was composing *Tosca* in 1899, Metlicovitz designed a set of souvenir postcards (each representing a different scene of the opera) as a way to launch Ricordi's new line of production. His postcards were presented at the first International Exposition of Illustrated Postcards of Venice in 1899 (Ferraris 194). *Tosca* premiered at the Teatro Costanzi in Rome on 14 January 1900 and at La Scala in Milan on 17 March 1900. Hohenstein choreographed the opera and designed the set.

32 Leonetto Cappiello replicated this technique in his famous poster for Chocolat Klaus of 1903. By untying the bond between the character and the product, he turned the advertising representation into "an image-brand" (Martinelli 43) made to astonish and be remembered.

33 "The poster ad, as Davide Campari always wanted it, often suggesting it personally to the artist, is a music, and, like music, it requires hearing. [...] Colour, then, needs to be seen in order to be turned into music" (Cenzato, quoted in "Cordial Campari").

34 Aldo Mazza (1880–1964) was a Milanese painter, caricaturist, and illustrator. He exposed his divisionist paintings at the Biennale of Venice (1906) and the National Exposition of Turin (1911). He collaborated with the humorous magazine *Guerin Meschino* and designed posters for such companies as Biciclette Bianchi, Articoli per la fotografia Ganzini, La Rinascente, and Credito Italiano.

35 Pratella performed *Inno alla vita* (*Hymn to Life*) in 1913. After *Inno alla vita* he responded to his critics by attacking aesthetic pleasure in an article (published on 15 May 1913 in *Lacerba*) entitled "Contro il 'grazioso'" ("Against 'Gracefulness'"). Russolo premiered *Risveglio di una città* (*The City Reawakens*, parallel to Boccioni's painting *The City Rises*), *Si pranza sulla terrazza del Kursaal* (*Lunch on the Kursaal Terrace*), and *Convegno di automobili e aeroplani* (*Convention of Cars and Aeroplanes*) in two controversial *serate futuriste* of 1914 (in Milan on 21 April and in Genoa on 20 May), after which brawls ensued.

36 As in other countries, radio diffusion started in Italy in the early 1920s. Founded in 1923, Società Italiana Radio Audizioni Circolari (SIRAC) and the Società Anonima Radiofono were unable to obtain concessions by the government, so the minister of communications, Costanzo Ciano, invited the companies in 1924 to find a common agreement. After the establishment of Unione Radiofonica Italiana on 27 August 1924, the government agreed to fund it. For more information about the birth and evolution of radio broadcasting in Italy, see Monteleone's volume *Storia della radio e della televisione in Italia: Un secolo di costume, società e politica.*

37 Campbell relates D'Annunzio's blindness to the development of acoustic languages that are de-territorialized ("acoustic perception is de-territorialized when eyesight is lost") and capable of bodily penetration ("D'Annunzio's near-blindness [...] is made the means by which acoustic events penetrate the body"; 46).

38 The original reads: "Sintesi di infinite azioni simultanee," "immensificazione dello spazio," "amplificazione e trasfigurazione di vibrazioni" (Marinetti, "La radia").

39 The original reads: "Queste mie liriche radiofoniche sono espressioni adatte per la trasmissione a distanza. L'ascoltatore non è più unicamente raccolto in un salotto silenzioso e romantico, ma si trova ovunque: per strada, nei caffè, in aeroplano, sui ponti di una nave, in mille atmosfere diverse. Quindi il carattere della lirica radiofonica deve essere spaziale, volitivo, sonoro, inaspettato, magico. In una parola la poesia radiofonica da me inventata dev'essere l'espressione lirica di un purissimo stato d'animo" (Depero, *Liriche radiofoniche* 8).

40 The original, taken from the poem "New York nuova Babele" ("New York New Babel") reads: "milioni di minuscola umanitá affaccendata dentro questa ciclopica, metrocubica, meccanopoli. Dentro questa babelimmensa manicomiofficina, dentro questo montagneppipedo DINAMONDO" (Depero, *Liriche radiofoniche* 70).

6 Cigarettes and Smoke: The Modern Lightness of Being

1 The original reads: “Occhi moltiplicati, a tiro lungo, che scaricano i loro sguardi come le palle del nostro fucile a ripetizione” (Marinetti, *La battaglia di Tripoli* 11).

2 The desert, theatre of the war in Libya, is descibed as enclosed by a “soffitto di nuvole bituminose” (Marinetti, *La battaglia di Tripoli* 34; ceiling of bituminous clouds) and a “tenera corolla di fumo” (43; tender corolla of smoke).

3 The original reads: “Un enorme sigaro che le grosse labbra del mare fumano ora, tranquillamente” (Marinetti, *La battaglia di Tripoli* 56); “un’immensa nuvola gessosa” (52).

4 Marinetti captured the experience of his first flight in “The New Ethical Religion of Speed” (1916) in the same terms: “When I flew for the first time with aviator Bielovucic, I felt my breast opening up like a great hole into which the vast skyscape was delightfully plunging, smooth, cool, streaming down” (*Critical Writings* 257; *Teoria* 130).

5 In the “Manifesto of Futurism” Marinetti aspires to sing “of the pulsating, nightly ardour of arsenals, and shipyards, *ablaze* with their violent electric moons; of railway stations, voraciously devouring *smoke-belching* serpents; of workshops hanging from the *clouds* by their twisted threads of *smoke*; of bridges which, like giant gymnasts, bestride the rivers, flashing in the sunlight like gleaming knives; of intrepid *steamships that sniff out* the horizon; of *broad-breasted* locomotives, champing on their wheels like enormous steel horses, bridled with pipes; and of the lissom flight of the aeroplane, whose propeller flutters like a flag in the wind, seeming to applaud, like a crowd excited” (*Critical Writings* 14; *Teoria* 11; emphasis mine).

6 The original reads: ““la mano che scrive sembra staccarsi dal corpo e si prolunga in libertà assai lungi dal cervello, che, anch’esso in qualche modo staccato dal corpo e divenuto *aereo*, guarda dall’alto, con terribile lucidità, le frasi inattese che escono dalla penna” (Marinetti, *Teoria* 56; emphasis mine).

7 Christopher Columbus first wrote about tobacco in his diary on 6 November 1492 in reference to the Indigenous populations of the Indies: “halcaron los dos cristianos por el camino mucha gente que atravesaba a sus pueblos, mujeres y hombres con un tizón en la mano, yerbas para tomar sus sahumerios que acostumbraban” (quoted in Monti 11; The two Christians found on their way many people crossing their villages, men and women with a charred stick in hand, and herbs to produce the aromatic smoke to which they were accustomed).

8 Tobacco was first brought to Spain (where the physician Nicolas Monardes praised it as a curative plant for cold and arthritis), then to France (where the ambassador Jean Nicot gave it as a gift to Queen Catherine de Medici in 1560, praising its virtues against ulcers and asthma), Italy (where Cardinal Prospero di Santa Croce created the smoking hour as a remedy for boredom), England (where Sir Walter Raleigh offered it to Queen Elizabeth I in 1565 as a cure for plague), Germany (1570), and the Ottoman Empire (1580) (Gately 39–44).

9 The Dutch iconography of pleasure and vanitas in smoking was recovered two centuries later by Vincent van Gogh in *Self-Portrait with Pipe* (1886) and *Skull with Cigarette* (1885).

10 Ashes alluded to the visual language of still-life painting, in which the representation of dirt, dust, and powder indicated both the ephemerality of consumption and the eternal remnant of the present. As Grazioli points out, "if dust is the prime metaphor of consumption and destruction, it is nonetheless also a metaphor of the indestructible remnant, both in the sense of what remains after all possible fragmentation, and of what cannot be further consumed, what cannot be de facto eliminated, and thus eternal in this sense" (5).

11 King James I, a famous opponent of tobacco, wrote *A Counterblaste to Tobacco* and signed a decree in 1604, increasing the tax on tobacco by 4000% (Pollard 39–42). Pope Urban VIII promulgated a note of excommunication in 1642 against smokers in the diocese of Seville (later lifted by the inveterate smoker and snuffer Pope Benedict XIII). Shah Abbas of Persia punished tobacco snuffers by cutting their noses, and in Turkey, Amurat IV punished smokers with the death penalty (Burns 33–40).

12 The poem, set in a smoke-filled tavern in Pisa, is a dialogue between a smoking doctor and a sceptical university student on the pleasures of tobacco. The doctor persuades the student by indicating the varieties and advantages of tobacco: as a medicine for solitude and as a remedy for tedious academic classes. In the poem the doctor presents the nineteenth century as "il secolo dei fumi / il secol delle macchine a vapore" (Guadagnoli, *Il tabacco*; verse 50).

13 Palmesi states that "the earth's surface was entirely wrapped in the fascination of the dense, terrible, and fatal cloud" (10; la superficie della terra si trovava tutta avvolta nel fascino della densa, terribile e micidiale nube), and Scalzi confirms that "we are wrapped in the fascination of the prodigious cloud which, in less than three centuries, managed to cover the entire surface of the earth" (11; ci troviamo avvolti nel fascino della prodigiosa nube che in men di tre secoli è giunta a cuoprire la superficie di tutta la terra).

14 The original reads: "Meglio a chi 'l senso smarrì de l'essere / meglio quest'ombra, questa caligine: / Io voglio io voglio adagiarmi / in un tedio che duri infinito" (Carducci, *Poesie* 879; verses 57–60).

15 The ambivalent representation of industrial smoke clouds persisted into the early twentieth century. On the one hand, in Umberto Boccioni's canvas *Officine di Porta Romana* (*Factories at Porta Romana,* 1908), smoke represents the urban conquest of the countryside (mirrored in the turmoil of peasants who are converging on the factories and turning into factory workers) and the vital energy that is transforming Milan. On the other hand, in Mario Sironi's canvas *Periferia* (*Suburbs,* 1922), smoke from a chimney configures the gray contours of urban life, and the by-product of a human-deprived society that is moved by the automatism of industrial machinery.

16 The original reads: "Il fumo saliva nell'aria *tingendosi* ai raggi quasi orizzontali del sole; le tappezzerie *s'armonizzavano* in un colore caldo e festoso. L'aroma del tè *si mesceva* all'odor del tobacco" (*Piacere* 236, emphasis mine).

17 "Dalle finestre prive di tende entrava lo splendore rossastro del tramonto, entravano tutti gli strepiti della via sottoposta. Alcuni uomini staccavano ancòra qualche tappezzeria dalle pareti, scoprendo il parato di carta a fiorami volgari, su cui erano visibili qua e là i buchi e gli strappi. Alcuni altri toglievano i tappeti e li arrotolavano, suscitando un *polverio denso* che riluceva ne' raggi. Un di costoro canticchiava una canzone impudica. E il *polverio* misto al fumo delle pipe si levava sino al soffitto." (*Piacere* 357, emphasis mine; From the windows bereft of curtains, the inflamed splendour of sunset and all the clamour of life on the street entered the room. Some men were still detaching wallpaper from the wall, uncovering the unexceptional flowery decorations, upon which holes and tears were laid bare here and there. Some others removed and rolled up the carpets, creating a dense *cloud of dust,* which shone in the sunlight. One of them sang an obscene song. And the *cloud of dust,* mixed with the smoke of pipes, rose to the ceiling.)

18 The original reads: "Roma diventa la città delle demolizioni. La gran polvere delle ruine si leva da tutti i punti dell'urbe e si va disperdendo a questi dolci soli maggesi [...] ma dalle rovine sorgerà e risplenderà la nuova Roma, la Roma nitida, spaziosa e salutare" (D'Annunzio, *Scritti giornalistici* 311).

19 After the Egyptian bombing had destroyed the Turks' pipes, Ottoman soldiers found a way to smoke by filling with tobacco (instead of gunpowder) the paper tubes that were used to light cannons, thereby inventing cigarettes. Ottoman cigarettes spread to Western Europe by way of English and French soldiers fighting with Turks in the Crimean War of 1855 (Shechter 30).

20 The original reads: "D'improvviso il mio piede s'inumidì e il mio capo arse. Io nacqui" (D'Annunzio, "Autobiografia" 593). "Le forze mi mancavano. In

non vidi più nulla, non udii più nulla. [...] le mie povere membra si sparsero tra la cenere fredda" (596).

21 In *Il fu Mattia Pascal* (1904) Pirandello retrieves D'Annunzio's imagery of smoking (in his case a cigar) in reference to a collective "I" and deliberately placing in opposition the rhetoric of fluidity and rhetoric of solidification. He comments on Rome's impossibility of becoming modern, through the contrasting associations of the old city to a basin of holy water and of the modern city to an ash-tray: "I papi ne avevano fatto – a loro modo s'intende – un'acquasantiera; noi italiani ne abbiamo fatto, a modo nostro, un portacenere. D'ogni paese siamo venuti qua a scuotervi la cenere del nostro sigaro, che è poi il simbolo della frivolezza di questa miserrima vita nostra e dell'amaro e velenoso piacere che essa ci dà." (116; The popes had made of it – in their own way – a holy water font; we Italians had made of it – in our own way – an ash-tray. We came here from every village to shake the ashes off our cigars, which are then the symbol of the frivolousness of this miserable life of ours, as well as of the bitter and poisonous pleasure that it gives us.)

22 "There is no doubt that the explosion of interest in cigarette smoking was furthered by the birth of modern advertising. [...] Advertising imagery became the place where twentieth-century fantasies about the power of smoking to provide status, desirability or pleasure were defined. With every purchase, consumers were offered an intense yet vague, magical yet repeatable promise: they too could be like the smokers illustrated in the advertisements" (Gilman and Xun 20–1).

23 Represented in contemporary painting – Édouard Manet's *The Plum* (1887) or Vincent van Gogh's *A Woman in the Café le Tambourin* (1888) – cigarettes become a symbolic tool "where struggle over women's liberation took place" (Gilman and Xun 22).

24 For an annual account of the Italian state's monopoly of tobacco in the early years (1884–1900) see pages 73–111 of the volume issued by the ministry of finance, *Cenni storico-statistici sul monopolio del tabacco in Italia* (Rome, 1900).

25 *Il tabacco* (*Tobacco*), published in Rome since 15 January 1897, was officially *Organo dell'industria e del commercio del tabacco* (*Instrument of the Tobacco Industry and Trade*). Its technical and industrial connotation separated it from other tobacco-related magazines of the nineteenth century that linked smoke to humour. Among them it is worth mentioning the satirical magazines *La pipa* (*The Pipe*; Genoa, 1864), *Il sigaro* (*The Cigar*; Sassari, 1874), *La sigaretta* (*The Cigarette*; Genoa, 1889), and *Il resto al sigaro* (*The Change to the Cigar*; Florence, 1885).

26 Serao allows women to smoke cigarettes but not in excess. Serao states that "every once in a while, in the countryside, while travelling, in cheerful company, a lady can smoke a cigarette without marring the poetry of her image" (*Saper vivere* 179; ogni tanto, in campagna, in viaggio, in una gaia brigata, una signora può fumare una sigaretta senza che la poesia della sua immagine ne sia offuscata).

27 The link between cigarettes and sensuality will be further staged both in advertising (as exemplified by Marcello Dudovich's poster for *Sigarette Macedonia* in the 1920s) and in early cinema (as seen in the common smoking representation of the Carmen-like divas Francesca Bertini and Lyda Borelli).

28 The smoking sheriff of Puccini's *La fanciulla del West* is considered "the prototype of those film characters later played by Humphrey Bogart, James Dean and Clint Eastwood" (Hutcheon and Hutcheon 230).

29 The original reads: "quando avremo quarant'anni, altri uomini più giovani e più validi di noi, ci gettino pure nel cestino, come manoscritti inutili – Noi lo desideriamo!" (Marinetti, *Teoria* 13).

30 The original reads: "impossibilità assoluta di arrestarsi e di ripetersi" (Marinetti, *Teoria* 81). "Il teatro di Varietà è il solo che utilizzi la collaborazione del pubblico. Questo non vi rimane statico come uno stupido *voyeur*, ma partecipa rumorosamente all'azione, cantando anch'esso, accompagnando l'orchestra, comunicando con motti imprevisti e dialoghi bizzarri cogli attori [...] Il teatro di Varietà utilizza il fumo dei sigari e delle sigarette per fondere l'atmosfera del pubblico con quella del palcoscenico" (Marinetti, *Teoria* 83).

31 Smoke expressed the "taste for action for action's sake" (Poggioli 40), as hinted in Marinetti's comment on Cangiullo's "Fumatori IIa Classe," in "The Geometrical Splendour," praising its capacity to render "the long, monotonous flights of fancy and the outward spreading of smoke-boredom experienced on a long train journey, with this TYPOGRAPHICALLY DESIGNED IMAGE" (*Critical Writings* 138; *Teoria* 103).

32 The original reads: "Turrrrbine rivoluzione di calcinacccci fumo acre acido solfidrico gaz ammoniacale odor di bruciato" (*Zang Tumb Tumb* 621).

33 The original reads: "Sbriciolarsi bianchezza di specchi solletico dei tentacoli del legno dispersione di granelli particelle slegate sottili finissime dadi sputati dal bossolo della legge Dispersione di 40 milioni di miliardi molecole-fuggiaschi-polverosi senza gambe senza testa senza braccia" (*Zang Tumb Tumb* 632).

34 The original reads: "noi distruggiamo sistematicamente l'io letterario perché si *sparpagli nella vibrazione universale,* e giungiamo ad esprimere l'infinitamente piccolo e le agitazioni molecolari. Es: fulmineo agitarsi di

molecole nel buco prodotto da un obice (ultima parte di *Forte Chiettam-Têpé*, nel mio *Zang tumb tuum*" (*Teoria* 100, emphasis mine).

35 The original reads: "i suoi differenti impulsi direttivi, le sue forze di compressione, di dilatazione, di coesione e di disgregazione, le sue torme di molecole in massa o i suoi turbini di elettroni" (*Teoria* 50).

36 "Essere solo padrone del SOLE avere il proprio azzurro ricino agilità monopolio del cielo avere sotto i piedi pianure pianure pianure pianure di nuvole nuvole nuvole nuvole **VRRRRRRRR** morbidezze di pellicce gasose NAVIGAZIONE DI MONTAGNE MALLEABILI deformazione incorporarsi strati strati strati di nuvole disfa-cimento reincarnarsi analisi di cumuli contorsioni smembramento brandelli condensazione **SINTESI** sofferenza e rimorso del-l'**UNITÀ** prodigalità di forme concorrenza di masse fusione anatomie di nuvole sbrandellamento sfilacciamento a tutta velocità **SCULTURE FLUIDE** Vapore acqueo Polvere di ghiaccio Nelle orecchie 1300 m. Termometro 0, 2 Identità di 3 miliardi piccole onde di vento Concentriche = 3 miliardi mattonelle d'ovatta Pavimento geometrico" (*Zang Tumb Tumb* 602–3).

37 Fernand Léger offers a visual counterpart to Marinetti's image in the contemporary Cubist painting *The Smokers* (1911) by significantly alternating the analogical succession of blank spots and moments of vision from an elevated perspective.

38 The original reads: "Ài veduto come lo abbiamo impolverato? Non si capiva più che cosa fosse. —Quando siamo stati vicini mi sembrava di averlo veduto scomparire —Scomparire? —Sicuro, anche a me. —Ma quello non era un uomo sapete! —Che cos'era sentiamo? —Sembrava una nuvola. —Lo abbiamo ricoperto di polvere, una nuvola sembriamo noi caro mio, in questa porca strada! —No no, l'ò veduto prima che la strada fosse invasa dalla polvere, è un uomo di fumo!" (*Il codice di Perelà* 138).

39 In the original novel of 1911 Alloro dies because of his aspiration "to remain on fire without burning" (*Il codice di Perelà* 161; rimanere sul fuoco senza bruciare).

40 The original reads: "Il condannato, in mezzo, si avanza agilissimo, leggero, quasi saltellando ad ogni passo, mentre il corteo s'incammina pesantemente, lento, con cadenza funebre" (*Il codice di Perelà* 347).

41 In Palazzeschi's first novel, *:riflessi* (1908), the fire of a straw stack and the smoke – which "grew bigger and bigger" and turned into "a large cloud" (59; una grossa nube) – identifies the poet's diversity in his will to vivify an immobile life (symbolized by his dusty home, untouched after his mother's suicide) and in his homosexuality.

42 "À l'extrême bout de la rangée de baraques, comme si, honteux, il s'était exilé lui-même de toutes ces splendeurs, je vis un pauvre saltimbanque,

voûté, caduc, adossé contre un des poteaux de sa cahute. [...] Partout la joie, le gain, la débauche; partout la certitude du pain pour les lendemains; partout *l'explosion frénétique de la vitalité.* Ici la misère absolue, la misère affublée, pour comble d'horreur, de haillons comiques, où la nécessité, bien plus que l'art, avait introduit le contraste. [...] Obsédé par cette vision, je cherchai à analyser ma soudaine douleur, et je me dis: Je viens de voir l'image du vieil homme de lettres qui a survécu à la génération dont il fut le brillant amuseur; du vieux poète sans amis, sans famille, sans enfants, dégradé par sa misère et par l'ingratitude publique." (Baudelaire, "Le vieux saltimbanque," 80–2, emphasis mine; At the farthest end of the rows of booths, as if he had banished himself in shame from all those splendid sights, I noticed an old acrobat, hunched and decrepit, a ruin of a man, who was leaning against one of the posts of his hut. [...] Everywhere else there was nothing but exuberance and prosperity and debauchery; everywhere the certainty of daily bread for the days to come; everywhere the *uninhibited overflow of vital energy.* But here was poverty at its nadir, and to complete the horror of it, a poverty dressed to kill in the rags and tatters of farce, making a contrast dictated more by destitution than by art. [...] Obsessed by the whole scene, I tried to analyse my sudden painful feelings, and I said to myself, I have just seen the very image of some old man of letters who has outlived his own generation, which once he entertained so brilliantly: the old poet without any friends or family or children, degraded by poverty and public ingratitude.)

43 The *fumismo* of Italian Futurist theatre, derived from "that attitude which was called *fumisterie* from Laforgue's *Pierrot fumiste,*" is intimately related to humour but also, as Poggioli suggests, to "another attitude, which takes the name of *funambolismo*" (164).

44 Palazzeschi draws the model for his laughing yet melancholic man of smoke from the French models of smoking Pierrots by Jules Laforgue and Albert Giraud. Laforgue's *Pierrot fumiste* (*Pierrot the Smoker,* 1882) narrates the story of the mask's unsuccessful marriage to Colombinette, from their wedding day to the trial brought against him by his wife because of his unwillingness to procreate. Laforgue relates Pierrot's depiction as a *fumiste* to his lightness, his *spleenetique* irony (suspended over the void of impotence and dismissal), his preference to metamorphosis over reproduction, and of his dancing final escape on a train, "leger et ricanant, dansant dans son compartiment á chaque station" (107; light and snickering, dancing in his compartment at each station). Giraud's *Pierrot lunaire: Rondels bergamasques* (*Lunar Pierrot: Rondels from Bergamo* 1884), later adapted to melodrama by Schoenberg in 1912, deliberately relates Pierrot to the repertoire of the commedia dell'arte by equating his style to its improvised rondels,

by identifying himself with Harlequin as "le Pierrot Bergamasque" ("Pierrot from Bergamo"; Richter 26), and by multiplying his character into infinite transformations during the poems. Giraud, however, interprets Pierrot in a new modern key, characterizing him with the new elements of his lunatic pallor (making him *lunaire*), of his dissonant viola – as in "La sérénade de Pierrot" ("Pierrot's Serenade"; Richter 12) – and of his tragic laughter – as in "Suicide," where "en sa robe de Lune blanche / Pierrot rit son rire sanglant" (verses 1–2, Richter 36; in his white-moon dress Pierrot laughs his bloody laughter).

45 From Soffici's poem: "stringo il volante con mano d'aria, / premo la valvola con la scarpa di cielo; / Frrrrrr frrrrrr frrrrrr, affogo nel turchino ghimè, / mangio triangoli di turchino di mammola, / fette d'azzurro; / ingioio bocks di turchino cobalto […] impennamento erotico fra i pavoni reali delle nuvole." (Verses 18–23, 32, in *BÏF§ ZF+18 Simultaneità e chimismi lirici*; I hold the steering wheel with an air hand, I press the valve with a sky shoe; Frrrrrrr frrrrrr frrrrrr, I sink into the turquoise fabric, I eat triangles of turquoise violet, slices of blues; I swallow pitchers of turquoise cobalt […] erotic rearing up amidst the royal peacocks of the clouds.)

46 Following the request of General Pershing who had demanded tens of thousands of cigarettes for soldiers, the "Smokes for Soldiers Fund" (run by the *New York Sun* and supported by President Wilson) and the Young Men's Christian Association "shipped more than 12 million dollars' worth of tobacco products to European battlefields in America's year and a half of war" (Burns 158).

47 "Tirammo fuori, chi la sigaretta, chi la pipa, e ci mettemmo a fumare e a motteggiare. La tempesta delle cannonate, degli urli, dei rombi, dei sibili continuava. Continuasse pure; noi ridevamo intanto per l'ultima volta, trasfigurati in una sorta di luce tragica che ci rendeva grandi." (Soffici, *Kobilek* 171; Some took out cigarettes, some pipes, and we started smoking and making fun of each other. The storm of cannon shots, screams, rumbles, and whistles went on. Let it continue, too; meanwhile, we kept laughing for the last time, transfigured in a sort of tragic light that made us great.)

48 The original reads: "Ci sarà un'esplosione enorme che nessuno udrà e la terra ritornata alla forma di nebulosa errerà nei cieli priva di parassiti e di malattie" (Svevo, *Coscienza* 474).

49 A student of Freud and pioneer of psychoanalysis in Italy, Weiss edited the entries "Freud" and "Psicoanalisi" for the *Enciclopedia italiana*; promoted Freud's theories in the XVII Congress of the Società Italiana di Freniatria in Trieste in 1925; authored the first manual in Italian, *Elementi di psicoanalisi* (*Elements of Psychoanalysis*) in 1930; and founded the Società Psicoanalitica Italiana in 1932 (Pavanello).

50 In 1908, the year of Lombroso's death, the first two essays on Freud's thought were published in Italy. Luigi Baroncini published "Il fondamento e il meccanismo della psico-analisi" in *Rivista di Psicologia,* and Gustavo Modena published "Psicopatologia ed etiologia dei fenomeni psiconevrotici: Contributo alla dottrina di Freud" in *Rivista sperimentale di freniatria.*

51 In spite of the ecclesiastical resistance, Agostino Gemelli, former counsellor of the army's supreme commander Luigi Cadorna during the First World War, and author of the pioneering book *Il nostro soldato: Saggi di psicologia militare* (*Our Soldier: Essays in Military Psychology,* 1917), developed a parallel psychological thought related to the neo-Thomist project of a "Christian spiritualism" (Lombardo and Foschi 64), which was launched through his journal *Archivio italiano di psicologia* (*Italian Psychology Archive,* founded in 1919) and later evolved into the foundation of the Catholic University of the Sacred Heart in Milan in 1921.

52 For a more detailed analysis of the relationship between Fascism and psychoanalysis see Zapperi's volume *Freud e Mussolini: La psicanalisi in Italia durante il regime fascista.*

53 The original reads: "«15.4.1890 ore 4½. Muore mio padre. U.S.» per chi non lo sapesse quelle due ultime lettere non significano *United States* ma ultima sigaretta" (91).

54 "Oggi, 2 febbraio 1886, passo dagli studii di legge a quelli di chimica. Ultima sigaretta" (*Coscienza* 71; Today, 2 February 1886, I move from the study of law to that of chemistry. Last cigarette).

7 Toys, Clothes, Furniture, and the Aesthetic Power of Play

1 For a reconstruction of the symbolic impact of the national caravan accompanying the unknown soldier from the warfront (Aquileia) to the Vittoriano monument (Rome) in 1921, see Laura Wittman's book *The Tomb of the Unknown Soldier, Modern Mourning, and the Reinvention of the Mystical Body* (University of Toronto Press, 2011).

2 *Balli plastici* premiered at the Teatro dei Piccoli in Rome on 14 April 1918. Gilbert Clavel wrote the play, Fortunato Depero realized the sets and the wooden marionettes, and Alfredo Casella composed the music. The spectacle was composed of five short actions – "I pagliacci," "L'uomo coi baffi," "I selvaggi," "Ombre," "L'orso azzurro" ("Clowns," "The Man with the Moustache," "Savages," "Shadows," "The Blue Bear") – combining mimics and musical accompaniment (Scudiero, *Depero* 16).

3 The publication of several pedagogical essays between the seventeenth and nineteenth centuries led to the rediscovery of childhood as a critical age in the formation of the modern citizen. Among them it is worth mentioning

François Fénelon's *Treatise on the Education of Daughters* (1687), John Locke's *Some Thoughts Concerning Education* (1693), Jean-Jacques Rousseau's *Emile or on Education* (1762), and Friedrich Schiller's letters *On the Aesthetic Education of Man* (1794). By virtue of these studies, play was redefined as a decisive factor in children's development, and toys were revalued as tools of knowledge, empiric forms of acquiring experience, and creative engines of mankind.

4 For the subsequent analysis I rely on the categories outlined by the sociologist Roger Caillois in *Les jeux et les hommes*, 1958 (*Man, Play, and Games*, 1961), defining the four facets of play's enjoyment: *ágon* (competition), linked to merit and risk management; *alea* (gambling), connected with denial of work and fatalistic surrender to destiny; *mimicry*, associated with travesty and fiction; and *ílinx* (from the Greek word for "whirlpool"), related to the pursuit of the extreme (tight-rope walking or car racing) and to the mastering of vertigo.

5 By the late-nineteenth century, fencing duels in Italy still constituted a heavily codified social ritual of diatribe solving. In literature, duelling enacted a symbolical surrogate of virtue (in front of the rapidly disappearing aristocracy) and a dazzling contest to "merit" victory in the political or amorous field. Ceccarelli states that at the time "everyone duelled. Noblemen, ministers, doctors, artists, socialists, republicans, liberals, conservatives; and then D'Annunzio [...], Serao's husband Edoardo Scarfoglio, the Futurist Marinetti, and also Bontempelli and Ungaretti, who engaged in a sword duel in Pirandello's house. [...] Women, too, duelled, and even priests. [...] It was the First World War, with its charge of brutal modernization, that exhausted the pace and fashion of sword fights."

6 D'Annunzio's association of play with death relates to the extreme pursuit of man's divinity (leading Giulio to crash on the ground, and Paolo to achieve the sky's extreme heights) and to a burning erotic tension, as confirmed in the novel's opening scene, which stages the intermingling of the sexual desires of Paolo and his lover Isabella while their race car dauntingly flies down an ancient road.

7 Fogazzaro's séance, organized by Professor Gilardoni to evoke the spirit of Luisa's dead daughter Maria, presents a self-deceiving mental projection and a wishful imagination. Pirandello's séance, organized by the philosopher Paleari to evoke the spirit of the dead musician Max Oliz, stages an ironic mockery of bourgeois society. Svevo's séance, organized at the Malfenti house, enacts an ambiguous social interplay, allowing Zeno to defy bourgeois codes by jokingly evoking Guido's name through the movements of the

table, yet trapping him into the terrible misunderstanding of his declaration of love made to Augusta instead of Ada. The modernist topos of the *seduta spiritica* will significantly re-appear, after the Second World War, in Federico Fellini's movie *La dolce vita* (1959). For more information about the impact of séance on Italian culture see Cigliana's book *La seduta spiritica* (2007).

8 Gambling refers to an unconscious state of mind, leading De Marchi's Cesarino Pianelli to addiction (to high fashion and stylish parties), to financial and moral ruin (as he pays his debts with his co-workers' money), and eventually to suicide. Borgese's figure of the war veteran Rubè follows the same degenerating path.

9 Baudelaire's originals read: "vie en miniature"; "désir"; "première initiation à l'art"; "joujou vivants"; and "acteurs dans le grand drame de la vie" ("La morale" 681).

10 In Ancient Greece, funeral games, Olympic rituals, tragic representations, or theatrical plays constituted a holy time, separated from ordinary life and giving meaning to it by manifesting man's struggle for perfection (contest) and his ecstatic contact with divinity (through dance or Bacchic inebriation).

11 During the Middle Ages, play related to the carnival attitude of jesting, buffoonery, and religious parody. According to Bakhtin, play outlined "a second world and a second life outside officialdom," offering, in the medieval period, a "nonofficial, extraecclesiastical and extrapolitical aspect of the world, of man, and of human relations" (6), and, in the masquerades of Renaissance courts, a "temporary suspension, both ideal and real, of hierarchical rank," allowing "a special type of communication impossible in everyday life" (10).

12 Buffalo Bill's spectacular "Wild West" show, re-enacting the American conquest of the West (through a mix of acrobatics, tamed animals, disguised characters, and fiction), had an immense success in Italy in the tours of 1890 (five shows) and 1906 (119 shows over thirty-four cities) and contributed to the generation of a cultural and social myth of America. See Bussoni's *Buffalo Bill in Italia* for a historical reconstruction of his myth in Italy.

13 In "The Painter of Modern Life" Baudelaire compares the artist to a child who "sees everything in a state of newness," and adds that "genius is nothing more nor less than *childhood recovered* at will (8, emphasis mine).

14 The originals read: "Una vocina sottile sottile" (*Adventures* 82); "Ho pensato di fabbricarmi da me un bel burattino di legno: ma un burattino maraviglioso, che sappia ballare, tirare di scherma e fare i salti mortali" (88); "Geppetto, che era povero e non aveva in tasca nemmeno un centesimo, gli fece allora un vestituccio di carta fiorita, un paio di scarpe di scorza d'albero e un berrettino di midolla di pane" (132).

15 In the manifesto “The Futurist Reconstruction of the Universe” Balla and Depero called for a renewed attention to materials in the production of objects, and systematized elements in detailed lists: “strands of wire, cotton, wool, silk of every thickness and coloured glass, tissue paper, celluloid, metal netting, every sort of transparent and highly coloured material. Fabrics, mirrors, sheets of metal, coloured tin foil, every sort of gaudy substance” (198).

16 Pirandello confirms the longevity of the artisanal local business by choosing a Sicilian artisan of marionettes, Saverio Càrzara, as the protagonist of his novella “La Paura del sonno” (“Fear of Sleep”), published in *Roma letteraria* on 25 March 1900.

17 In the late nineteenth century toys in Italy still represented an element of economic disparity and an exclusive imported amusement for the élites. De Amicis emblematically documents this in *Cuore* (*Heart,* 1886) through the gesture of its protagonist, Enrico, in sharing and eventually leaving his train to a poor friend (Precossi) after his father’s suggestion: “a Precossi piace il tuo treno. Egli non ha giocattoli. Non ti suggerisce nulla il tuo cuore?” (103; Precossi likes your train. He doesn’t have toys. Isn’t your heart telling you anything?)

18 Modern achievements are presented in *Il corriere dei piccoli* with a fantasizing tone, which prefigures and echoes the excited discourse of Futurist manifestos. Blériot’s aeroplane flight is described as follows: “Il cielo delle grandi città sarà solcato in tutti i sensi da veicoli aerei, come ora il loro suolo è percorso per ogni verso da carri, tramvais, carrozze, biciclette, automobili” (8 August 1907; The skies of big cities will be criss-crossed in all directions by aerial vehicles, just as the earth is now criss-crossed in every direction by wagons, tramways, coaches, bicycles, and cars). Marconi’s wireless telegraphy is presented in these terms: “dei pali, dei fili, qualche macchina, e il nostro pensiero vola oggi lontano centinaia e centinaia di chilometri, con la celerità del fulmine. [...] È grazie a questa velocità sbalorditiva che noi possiamo comunicare anche coi popoli più lontani e mandare rapidamente ai fratelli, ai parenti, agli amici, il nostro saluto, la nostra parola.” (7 March 1909; some posts, some wires, some machinery, and our thought today flies far, hundreds and hundreds of kilometres away, with the swiftness of lightning. [...] It is by virtue of this mesmerizing rapidity that we can communicate with even the farthest peoples and quickly send our greetings, our words, to brothers, relatives, and friends.)

19 The most important Italian representative of Dadaism was Julius Evola. After his brief association with Marinetti’s Futurism, and his avant-garde period (deemed as too commercial), Evola moved into esotericism in the

1920s by gradually developing his interests in the occult, Oriental studies, and hallucinogenic drugs. In 1927 he founded the Ur Group, aiming at a revival of ancient Roman paganism in contemporary Fascist culture (Schnapp, *Modernitalia* 23–52).

20 The original reads: "la bocca non era ancora finita di fare, che cominciò subito a ridere e a canzonarlo" (*Adventures* 98).

21 Puppetry and humour – in satirical magazines (often titled after the puppet masks Harlequin and Pulcinella) or in itinerant literature (as in Yorick's *Storia dei burattini* [*History of Puppets,* 1864] or Carlo Collodi's *Viaggio per l'Italia di Giannettino* [*Giannettino's Trip around Italy,* 1880]) – represented for adults "an antidote to the conservativism and traditionalism of bourgeois culture" (Segel 37).

22 First represented in Turin, on 15 January 1909 as *La donna è mobile* (*Woman Is Fickle*) a few weeks before the publication of the "Manifesto of Futurism," and vehemently booed by the audience (while Marinetti provokingly praised its disapproval), the play represents the story of a bored bourgeois couple – opposing but joining the American engineer John Wilson and his sensual Egyptian wife, Mary – as they decide to make love in the company of the husband's dancing dolls (*poupées*).

23 The vision of a perfect man-machine had been idealized with the emergence of the automaton in the seventeenth century and the publication of La Mettrie's *L'Homme-machine* (1747), a book that "represents the archetype of twentieth century, machine-age manifestos, including Marinetti's futurism and English vorticism" (Tiffany 61).

24 The same vision appears in Charlie Chaplin's movie *Modern Times* (1936) and in the episode of "The Sorcerer's Apprentice" in Walt Disney's *Fantasia* (1940).

25 Before Depero, the illustrator Aleardo Villa had imagined a similar point of connection between humour (as a form of destabilizing deformation) and fashion (as a site of endless creativity) in a pioneering advertisement for the satirical magazine *La settimana umoristica.* In the representation of a clownish figure of a woman cutting the illustrated pages of the magazine over her dress, and in the detail of her paper hat, topped with a brush, a pen, and a feather, Villa visualized the humorous and subversive transformation of clothing into a new support for art, as well as foreshadowed the impact of satirical writing and painting on the formation of fashion as a new plastic art.

26 "La moda è un'arte come l'architettura e come la musica. Una veste femminile genialmente ideata e ben portata ha lo stesso valore di un affresco di Michelangiolo o di una Madonna del Tiziano." (Quoted in Braun 40; Fashion is as much an art as architecture and music. A brilliantly

designed and well-worn feminine gown has the same value as a fresco by Michelangelo or a Madonna by Titian.)

27 "I gioielli e le stoffe dolci al tatto distruggono nel maschio l'assaporamento tattile della carne femminile. [...] Il maschio perde a poco a poco il senso potente della carne femminile e lo rimpiazza con una sensibilità indecisa e tutta artificiale, che corrisponde soltanto alla sete, ai velluti, ai gioielli, alle pellicce." (Marinetti, "Contro il lusso femminile," point 6, quoted in Spackman, *Fascist Virilities* 10; Jewels and fabrics that are sweet to the touch destroy in the male the tactile tasting of the female body. [...] Little by little, the male loses the powerful sense of the female body and replaces it with an uncertain and artificial sensibility, which only corresponds to thirst, velvets, jewels, and fur coats.)

28 In parallel with women's fashion, the female literature of the late 1920s also constituted a less censored and less constrained space to critique the regime. With regard to the uniformity of men's clothing see Riello 50–6.

29 Maria Monaci Gallenga's innovation consisted in the decomposition of ancient decorative themes and in the reassembling of their single elements in simpler patterns. Thanks to this technique she could not only reinterpret old models in a modern key but also acquire a vast repertoire of moulds (Raimondi).

30 For information about Schiaparelli's life see Secrest's volume *Elsa Schiaparelli: A Biography* (2014).

31 The Latin tradition of the Lares and the Italian puppet theatre constituted the cultural background of the Italian idea "of the house as a theatrical place, a scenographic site, where objects are actors who interact with inhabitants" (Annicchiarico and Branzi 44).

32 Along the lines of Antonio Sant'Elia's "Manifesto of Futurist Architecture" (1914), "The Futurist Reconstruction of the Universe" would also have an impact on urban design, as documented in Piero Portaluppi's playful drawings, published in *Aedilitia I* (1924) and *Aedilitia II* (1930), imaginatively defining a new fictional space of modern cities (Schnapp, *Modernitalia* 53–76).

33 With the commissioning of posters from Dudovich, Cappiello, and Nizzoli, Davide Campari defined "the three pillars of Campari's communication: a recognizable logo, a distinctive colour, and a quality product" (Gemmellaro).

34 The original reads: "arte viva, moltiplicata, e non isolata e sepolta nei musei – arte libera d'ogni freno accademico – arte gioconda – spavalda – esilarante – ottimista – arte di difficile sintesi dove l'artista è alle prese con l'autentica creazione" (quoted in Belli, *Depero* 150).

8 The Industrial Laboratory of Italian Modernity

1 “Nella casa all’italiana non ci è grande distinzione di architettura tra esterno ed interno […]. Da noi l’architettura di fuori penetra nell’interno […] la casa all’italiana è di fuori e di dentro senza complicazioni, accoglie suppellettili e belle opere d’arte e vuole ordine e spazio fra di esse, e non folla o miscuglio. Giunge ad essere ricca con i modi della grandezza, non con quelli soli della preziosità. Il suo disegno non discende dalle sole esigenze materiali del vivere, essa non è soltanto una ‘*machine à habiter.*’ Il cosiddetto ‘comfort’ non è nella casa italiana solo nella rispondenza delle cose alle necessità e ai bisogni, ai comodi della nostra vita ed alla organizzazione dei servizi. Codesto suo ‘comfort’ è in qualcosa di superiore, esso è nel darci con l’architettura una misura per i nostri stessi pensieri, nel darci con la sua semplicità una salute per i nostri costumi, nel darci con la sua larga accoglienza il senso di una vita confidente e numerosa, ed è infine, per quel suo facile e lieto e ornato aprirsi fuori a comunicare con la natura, nell’invito che la casa all’italiana offre al nostro spirito di ricrearsi in riposanti visioni di pace, nel che consiste, nel pieno senso della parola italiano, il *conforto.*” (Ponti, “La casa all’italiana” 10; In the Italian-style house there is no architectural distinction between exterior and interior […]. In our style, exterior architecture penetrates the interior […] the Italian-style house is both outdoor and indoor without complications, gathers together daily objects and beautiful artwork, and arranges them with order and space, without overcrowding or jumbling them. It achieves richness through greatness, not only through luxury. Its design does not exclusively flow from thc material needs of life, since it does not represent solely a ‘*machine à habiter.*’ The so-called comfort of the Italian-style house does not only mean correspondence of things to necessities and needs, to a comfortable dimension of our life, and to the organization of services. Such ‘comfort’ is rather related to something superior and consists in giving us a measure for our thoughts through architecture, a healthy balance for our habits through simplicity, and the sense of a confident and prolific life through abundant welcoming. The full meaning of the Italian word *conforto* consists, finally, in the Italian-style house’s easy, glad, and ornate outward opening to communicate with nature, as well as in its invitation to our spirit to recreate in relaxing visions of peace.)

2 “L’industria dunque fa stile […] l’industria è la maniera del nostro tempo. […] Qual è la formula dell’industria? È la riproduzione esatta e numerosa di un modello perfetto.” (Ponti, “Arte e industria” 134; Industry, then, makes style […] industry is the manner of our time. […] What is the

formula of industry? It is the exact and numerous reproduction of a perfect model.) Artists have the task to make "authentic and vital a production that otherwise would just be, in the best case, a perfected imitation" (autentica e vitale una produzione che altrimenti sarebbe nel migliore dei casi una imitazione perfezionata). The artist then "makes a production typical and characteristic" (rende tipica e caratteristica una produzione) and creates authenticity through "a signature production" (produzione d'autore) (134).

3 The house "should not be fashionable, since it should not go out of fashion [...] our house – wanted, built, furnished with loving understanding of its material and ethical functions – will be the worthy dwelling of Man and will represent not just the traces of ephemeral and subsequent trends, but rather the testimony of our intelligence, of our life, of our culture, and of the nobility of the things we love." (Ponti, "Casa di moda" 11; non deve essere di moda, perché non deve passare di moda [...] voluta, costruita, arredata con amorosa comprensione di queste sue funzioni materiali ed etiche, la nostra abitazione sarà la vera nostra casa, sarà la dignitosa dimora dell'Uomo e rappresenterà non le tracce di mode caduche e successive ma la testimonianza della nostra intelligenza, della nostra vita, della nostra cultura e della nobiltà delle cose che amiamo.)

4 "È opinione comune che modernità significhi mortificazione della tradizione. La tradizione è mortificata invece dai pigri profittatori di essa. Da qui la decadenza delle arti, proprio per opera inconsapevole di fedeli alla tradizione, che poi la vanno a ficcare, gelosi a comodo loro, anche dove (come in tante attività e tecniche d'oggi) tradizione non esiste, o dove (come in certe nostrane industrializzazioni che datano dalla fine dell'ottocento, vedi tappeti) la tradizione sarebbe l'imitazione o contraffazione di modelli esotici. [...] Non è questo snaturamento da temere, è da temere solo la decadenza di quelle viventi energie che creano via via gli elementi della grande tradizione [...] Le forze che operano nella tradizione sono occulte, di volta in volta le individuiamo anche dove non ci apparvero presenti: ma esse operano attraverso i più *vivi*: la tradizione è fatta *solo di autenticità*." (Ponti, "Morte e vita" 133, emphasis mine; It is commonly held that modernity means mortification of tradition. Tradition is instead mortified by its lazy profiteers. From here comes the decadence of the arts, by the unconscious work of those very people who claim to be faithful to tradition, who, in their own self-defined jealousy, stick tradition even where it does not apply [as in many techniques and activities of today], or where it would only be imitation or counterfeit of exotic models [as in certain industrial productions of ours, for example carpets, dating from the late nineteenth century]. [...] We do not need to fear this distortion, but rather the decay of

those living energies that gradually built up the elements of great tradition […] The forces that operate in tradition are latent, and we locate them from time to time even where they do not seem present to us, yet they keep operating in those who are most lively. *Tradition is only made of authenticity.*)

5 Ponti's experience at Richard-Ginori represented the first case of a "continuing relationship between an industry and a designer [*progettista*] extraneous to the factory system" (Bosoni, "Per una 'profezia'" 199). See Matteoni (18–31) for a history of Ponti's relationship with ceramics, and La Pietra (16–41) for a catalogue of Ponti's works for Richard-Ginori.

6 As well as Ponti's ceramics, the first Biennale of Decorative Arts of Monza in 1923 featured the work of glassmakers Vittorio Zecchin and Paolo Venini, who were reinterpreting the Renaissance tradition of Murano glasses.

7 During the 1920s "the figure of the artist/craftsman was replaced by that of the architect/designer" (De Guttry and Maino, "From Artist-Artisans" 50).

8 "In Italy the car remained, at least to end of the 1920s, an item of luxury and leisure" (Paolini 212).

9 For more information about the origins and development of the Italian aviation industry see Paolo Ferrari's volume *L'aeronautica italiana: Una storia del novecento.*

10 "In the Fascist twenty-year period [*ventennio*] the decorative arts was the only sphere in which an authentic and real free will survived. Starting from the Liberty period, they became an extraordinary creative workshop, a laboratory whose experimentations were unmatched before the twentieth century. The forms of the future were born in this period" (Cogeval 16).

11 Mussolini had little interest in the artistic life of the country and entrusted the Jewish journalist Margherita Sarfatti (his mentor and lover) with responsibility for the figurative arts. Sarfatti even wrote Mussolini's official biography, *Dux*, which appeared in Italy in 1926. The book was eventually translated into eighteen languages and helped to shape the international image of the Fascist dictator as Italy's saviour and the heir to the Caesars of Ancient Rome.

12 "While historians have established that Italian design has its roots in the Rationalist movement, it is equally well substantiated that these roots draw their sustenance equally from the exploration of classical forms reinterpreted in metaphysical terms by certain artists in the Novecento group […] and from the pursuits of a part of the Second Futurism movement, as expressed by such disruptive and multifaceted personalities as Fortunato Depero" (Bosoni, *Modo italiano* 144). For a study of the connections between Italian design and the rationalist movement see Emilio Ambasz's seminal volume *Italy: The New Domestic Landscape* and Bruno Zevi's essay "Gruppo 7: The

Rise and Fall of Italian Rationalism." For more information about the relationship between Italian design and the second Futurism movement, as expressed in the work of Giacomo Balla, see Ara Merjian's essay "A Future by Design: Giacomo Balla and the Domestication of Transcendence."

13 See Lees-Maffei and Fallan (3–4) for a complete list of works dealing with the historiography of Italian design.

14 Sparke confirms that "the achievements of the Renaissance had the effect of applying a stylistic brake to the evolution of Italian material culture in the late nineteenth and early twentieth centuries" ("A Modern Identity" 266).

15 "The project of Italian industrial design in the early twentieth century oscillates between theoretical research and typological deviation" (Vercelloni 28).

16 Branzi locates the origins of Italy's "intimate vocation to follow a process of mediation" in the Renaissance, an age in which "radically different realities, like Faith and Reason, Paganism and Christianity, Politics and Morals, could live together within an artificial and elastic style system, capable, over the next four centuries, of containing (without breaking) the powerful thrusts produced by its intrinsic and incurable contradictions" (18).

17 In their *ars combinatoria,* in their visual quest for unrepeatable variation (in lighting and composition), and in their self-transformation (as they were previously covered with paint or devoid of scale), objects represented for Morandi not only the "point of departure to explore abstraction" (Abramowicz 13) but also the mental site for the construction of an architectural space and of a perennial state of self-overcoming.

18 A singular example of the creativity born from the scarcity of materials is the design of Alfonso Bialetti's aluminium Moka coffee maker in 1933. In the essay "The Romance of Caffeine and Aluminum" Schnapp offers a detailed account of the genealogy and cultural meanings of this product.

Works Cited

Abramowicz, Janet. *Giorgio Morandi: The Art of Silence.* New Haven, CT: Yale University Press, 2004. Print.

Adams, Ansel, et al. *Vittorio Sella, Mountaineer and Photographer: The Years 1879–1909.* New York: Aperture, 1999. Print.

Adorno, Theodor. *Philosophy of Modern Music.* Trans. Anne Mitchell and Wesley Blomster. New York: Seabury Press, 1973. Print.

Albanese, Giuseppe. "'I mobili a sorpresa parlanti e paroliberi': Il contributo di Francesco Cangiullo alla definizione dell'arredo futurista." *Il Cerchio,* vol. 15, no. 69–70, January–March 2009, pp. xvi–xxv. Periodical.

Aleramo, Sibilla. *Momenti: Liriche.* Florence: Bemporad, 1921. Print.

Alighieri, Dante. *La divina commedia.* Ed. Anna Maria Chiavacci Leonardi. Milan: Mondadori, 2007. Print.

Alinari, Vittorio, editor. *Il Decamerone illustrato da Tito Lessi.* Florence: Alinari, 1909–15. Print.

– *La divina commedia: Novamente illustrata da artisti italiani.* Florence: Alinari, 1902–3. Print.

– *Firenze e contorni, catalogo n. 1: Riproduzioni fotografiche pubblicate per cura dei Fratelli Alinari fotografi-editori.* Florence: Tip. Barbera, 1891. Print.

– *Il paesaggio italico nella "Divina Commedia."* Florence: Alinari, 1921. Print.

Alunni, Umberto. *La radio in soffitta.* Raleigh, NC: Lulu Press, 2014. Print.

Amatori, Franco. "La grande impresa." *Storia d'Italia: Annali; L'industria,* vol. 15. Turin: Einaudi, 1999, pp. 691–756. Print.

Ambasz, Emilio. *Italy: The New Domestic Landscape; Achievements and Problems of Italian Design.* New York: Museum of Modern Art, 1972. Print.

Ambrosio, Arturo, director. *Pagliacci.* Società Anonima Ambrosio, 1907. Film.

L'amico Francesco. Composed by Domenico Morelli. Perf. Enrico Caruso. Teatro Nuovo, Naples, 15 March 1895. Performance.

Anderson, Benedict. *Imagined Communities: Reflections on the Origins and Spread of Nationalism.* New York: Verso, 1983. Print.

Andrews, Frank. *A Fonotipia Fragmentia: A History of the Società Italiana di Fonotipia, Milano, 1903–1948.* UK: Historic Singers Trust, 2002. Print.

Annicchiarico, Silvana, and Andrea Branzi. *Che cosa è il design italiano? Le sette ossessioni del design italiano.* Milan: Electa, 2008. Print.

Archetti, Marcello. *Ordine ritmo misura: Le rappresentazioni culturali del tempo.* Bergamo: Moretti & Vitali, 1992. Print.

Archivio italiano di psicologia. Turin: Laboratorio di psicologia sperimentale della R. Università, 1921–35. Periodical.

Arendt, Hannah. *The Human Condition.* Chicago: University of Chicago Press, 1958. Print.

Arisi, Francesco. *Il tabacco masticato, o fumato: Trattenimenti ditirambici.* Milan: 1725. Print.

Armenante, Lucia. "Film Ambrosio." *Enciclopedia del cinema. Treccani,* web. Accessed 2 February 2017.

L'arte decorativa moderna. Dir. Enrico Thovez. Turin: 1902. Periodical.

Arte italiana decorativa e industriale. Dir. Camillo Boito. Bergamo: Istituto Italiano di arti grafiche, 1890–1911. Periodical.

"The Art of Cycling in Italy." *Italian Ways.* Web. Accessed 8 February 2017.

Ashplant, T.G., et al. "The Politics of War Memory and Commemoration: Contexts, Structures and Dynamics." *The Politics of War Memory and Commemoration,* ed. T.G. Ashplant, Graham Dawson, and Michael Roper. New York: Routledge, 2000, pp. 3–86. Print.

Attfield, Judy. *Wild Things: The Material Culture of Everyday Life.* New York: Berg, 2000. Print.

Augé, Marc. *Eloge de la bicyclette.* Paris: Manuels Payot, 2008. Print.

Avanzi, Beatrice. "'Forse è l'arte il solo incantesimo concesso all'uomo': Pittura italiana 1900–1940." *Una dolce vita? Dal liberty al design italiano 1900–1940,* ed. Guy Cogeval and Beatrice Avanzi. Milan: Skira, 2015, pp. 87–95. Print.

Bachelard, Gaston. *The Psychoanalysis of Fire.* Boston: Beacon Press, 1964. Print.

Baia Curioni, Stefano. *Mercanti dell'opera: Storie di casa Ricordi.* Milan: Il Saggiatore, 2011. Print.

Bakhtin, Mikhail. *Rabelais and His World.* Trans. Helene Iswolsky. Cambridge, MA: MIT Press, 1968. Print.

Baldissone, Giusi. Introduction. *Opere,* by Guido Gozzano, ed. Giusi Baldissone. Turin: UTET, 2006. Print.

Balla, Giacomo. "Futurist Manifesto of Men's Clothing." *Futurist Manifestos,* ed. Umbro Apollonio. Boston: Art Works, 2001, pp. 132–4. Print.

– “Il vestito antineutrale.” Milan: Direzione del Movimento Futurista, 11 September 1914. Print.

Balla, Giacomo, and Fortunato Depero. “The Futurist Reconstruction of the Universe.” *Futurist Manifestos*, ed. Umbro Apollonio. Boston: Art Works, 2001, pp. 197–9. Print.

Balli plastici. Written by Gilbert Clavel. Set design by Fortunato Depero. Music by Alfredo Casella. Teatro dei Piccoli, Rome, 14 April 1918. Performance.

Ballo Excelsior. Written by Luigi Manzotti. Music, Romualdo Marenco. Teatro La Scala, Milan, 11 January 1881. Performance.

Baranski, Zygmunt, and Rebecca West. *The Cambridge Companion to Modern Italian Culture.* Cambridge: Cambridge University Press, 2001. Print.

Bardelli, Daniele. *L'Italia viaggia: Il Touring Club, la nazione e la modernità (1894–1927).* Rome: Bulzoni, 2004. Print.

Barisone, Silvia. “Futurist Ceramics.” *Italian Futurism, 1909–1944: Reconstructing the Universe,* ed. Vivien Greene. New York: Guggenheim Museum, 2014, pp. 287–94. Print.

Baroncini, Daniela. “Orologio.” *Oggetti della letteratura italiana,* ed. Gian Mario Anselmi and Gino Ruozzi. Rome: Carocci, 2008, pp. 119–26. Print.

Baroncini, Luigi. “Il fondamento e il meccanismo della psico-analisi.” *Rivista di Psicologia,* vol. 1, no. 4 January–February 1908, pp. 211–32. Periodical.

Barsella, Suzanna. “Bicicletta: Il mito e la poesia.” *Italica,* vol. 76, no. 1, Spring 1999, pp. 70–97. Periodical.

Barthes, Roland. *Camera Lucida: Reflections on Photography.* New York: Hill & Wang, 1981. Print.

– *What is Sport?* New Haven, CT: Yale University Press, 2007. Print.

Bartky, Ian R. *One Time Fits All: The Campaigns for Global Uniformity.* Stanford, CA: Stanford University Press, 2007. Print.

Baruffaldi, Girolamo. *Tabaccheide.* Ferrara: 1714. Print.

Bassetti, Remo. *Storia e storie dello sport in Italia: Dall'unità a oggi.* Venice: Marsilio, 1999. Print.

Baudelaire, Charles. “La morale du joujou.” *Oeuvres complètes,* by Charles Baudelaire, ed. Y.-G. Le Dantec. Paris: Bibliothèque de la Pléiade Gallimard, 1961, pp. 681–7. Print.

– *The Painter of Modern Life and Other Essays.* Ed. Jonathan Mayne. New York: Phaidon, 1995. Print.

– “Le vieux saltimbanque.” *Paris Blues: The Poems in Prose.* Trans. Francis Scarfe. London: Anvil Press Poetry, 2012, pp. 78–83. Print.

Baudrillard, Jean. *The System of Objects.* Trans. James Benedict. New York: Verso, 1996. Print.

Bazin, André. "Ontologie de l'image photographique." *Qu'est-ce que le cinéma?*, by André Bazin. Paris: Éditions du Cerf, 1958. Print.

Becchetti, Piero. *Fotografi e fotografia in Italia 1839–1880.* Rome: Quasar, 1978. Print.

Belli, Gabriella. *Depero: Dal Futurismo alla Casa d'Arte.* Milan: Charta, 1994. Print.

– "Gilbert Clavel, Fortunato Depero, and Balli Plastici: Rome and Capri, 1917–1918." *Italian Futurism, 1909–1944: Reconstructing the Universe*, ed. Vivien Greene. New York: Guggenheim Museum, 2014, pp. 191–210. Print.

Benedetti, Amedeo. "La fortuna delle immagini Alinari nella grande editoria italiana del Novecento." *Culture del Testo e del Documento*, vol. 12, no. 36, 2011, pp. 121–51. Print.

Ben Ghiat, Ruth. *Fascist Modernities: Italy, 1922–1945.* Berkeley: University of California Press, 2001. Print.

Benjamin, Walter. *The Arcades Project.* Trans. Howard Eiland and Kevin McLaughlin. Cambridge, MA: Belknap Press of Harvard University Press, 1999. Print.

– *Illuminations: Essays and Reflections.* Ed. Hannah Arendt. Trans. Harry Zohn. New York: Schocken Books, 1968. Print.

– *The Origin of German Tragic Drama.* New York: Verso, 1985. Print.

– *The Work of Art in the Age of Its Technological Reproducibility and Other Writings on Media.* Ed. Michael Jennings, Brigid Doherty, and Thomas Y. Levin. Trans. Edmund Jephcott, et al. Cambridge, MA: Belknapp Press of Harvard University Press, 2008. Print.

Bergson, Henri. *Matter and Memory.* New York: Cosimo Classics, 2007. Print.

– *Time and Free Will: An Essay on the Immediate Data of Consciousness.* London: George Allen & Unwin, 1959. Print.

Berman, Marshall. *All That Is Solid Melts into Air: The Experience of Modernity.* New York: Simon & Schuster, 1982. Print.

Bernardini, Aldo. *Cinema muto italiano.* Vol. 2, *Industria e organizzazione dello spettacolo 1905–1909.* Bari: Laterza, 1981. Print.

Bertarelli, Vittorio. *Sicilia 1898: Note di una passeggiata ciclistica.* Ed. Vittorio Cappelli. Palermo: Sellerio, 1994. Print.

Bertelli, Carlo. "La fedeltà incostante. Schede per la fotografia nella storia d'Italia fino al 1945." *Storia d'Italia: Annali 2; L'immagine fotografica (1845–1945)*, vol. 1. Ed. Carlo Bertelli and Giulio Bollati. Turin: Einaudi, 1979, pp. 59–198. Print.

Bertelli, Carlo, and Giulio Bollati. *Storia d'Italia: Annali 2; L'immagine fotografica (1845–1945).* Vol. 2. Turin: Einaudi, 1979. Print.

Bertellini, Nello. *Scrittori della bicicletta.* Florence: Vallecchi, 1985. Print.

Besio, Armando. "Quando Torino era capitale del Liberty." *Repubblica*, 11 December 2002. Web. Accessed 1 November 2017.

Betteloni, Vittorio. *Poesie edite ed inedite.* Milan: Mondadori, 1946. Print.

Bianchi, Patrizio. *La rincorsa frenata: L'industria italiana dall'unità alla crisi globale.* Bologna: Il Mulino, 2013. Print.

La bicicletta: Giornale popolare. Milan: Tip. Rivara, 1894–8. Periodical.

Bigi, Lucio. *L'orologio nel Duomo di Firenze: L'unico al mondo che segna l'ora italica.* Florence: Libreria Editrice Fiorentina, 2008. Print.

Binni, Walter. *La poetica del decadentismo.* Florence: Sansoni, 1961. Print.

Birch, Beverley. *Marconi Radio Pioneer.* Woodbridge, CT: Blackbirch Press, 2001. Print.

Blasetti, Alessandro, dir. *Vecchia guardia.* Fauno, 1935. Film.

Boatti, Giorgio. "Pedalare, pedalare: La bicicletta tra due secoli." *Biciclette: Lavoro, storie e vita quotidiana su due ruote,* ed. Guido Conti. Parma: Monte Università Parma, 2007, pp. 9–18. Print.

Boccaccio, Giovanni. *Decameron.* Ed. Amedeo Quondam, et al. Milan: Rizzoli, 2013. Print.

Bodei, Remo. *La vita delle cose.* Bari: Laterza, 2009. Print.

La Bohème. Composed by Giacomo Puccini. Dir. Arturo Toscanini. Perf. Enrico Caruso. Teatro La Scala, Milan, 26 December 1900. Performance.

Boime, Albert. *The Art of the Macchia and the Risorgimento: Representing Culture and Nationalism in Nineteenth-Century Italy.* Chicago: University of Chicago Press, 1993. Print.

Boine, Giovanni. *Frantumi.* Genoa: San Marco dei Giustiniani, 2007. Print.

Bollati, Giulio. "Note su fotografia e storia." *Storia d'Italia: Annali 2; L'immagine fotografica (1845–1945),* vol. 1, ed. Carlo Bertelli and Giulio Bollati. Turin: Einaudi, 1979, pp. 5–55. Print.

Bonito, Vitaniello. *Le parole e le ore: Gli orologi barocchi; Antologia poetica del Seicento.* Palermo: Sellerio, 1996. Print.

Bonito Oliva, Achille. "Italian Ways in Contemporary Art." *Il Modo Italiano: Italian Design and Avant-Garde in the 20th Century,* ed. Giampaolo Bosoni. Montreal: Montreal Museum of Fine Arts, 2006, pp. 91–102. Print.

Bonsaver, Guido. *Censorship and Literature in Fascist Italy.* Toronto: University of Toronto Press, 2007. Print.

Borgese, Giuseppe Antonio. *Rubè.* Milan: Mondadori, 1974. Print.

Boscaini, Lucia, et al. *Bulgari y Roma.* Catálogo de la Exposición (Museo Thyssen-Bornemisza, Madrid, 30 November 2016–26 February 2017). *Museo Thyssen.* Web. Accessed 25 January 2017.

Bosi Maramotti, Giovanna. "La bicicletta e la letteratura." *I quaderni del Cardello,* vol. 5. Ravenna: Longo Editore, 1993, pp. 119–39. Periodical.

Bosoni, Giampaolo, editor. *Il Modo Italiano: Italian Design and Avant-Garde in the 20th Century.* Montreal: Montreal Museum of Fine Arts, 2006. Print.

– "Of the *Modo Italiano* and Its Ways." *Il Modo Italiano: Italian Design and Avant-Garde in the 20th Century*, ed. Giampaolo Bosoni. Montreal: Montreal Museum of Fine Arts, 2006, pp. 15–26. Print.

– "Per una 'profezia' del design (1919–1940): La via italiana alla modernità nel disegno dell'oggetto utile." *Una dolce vita? Dal liberty al design italiano 1900–1940*, ed. Guy Cogeval and Beatrice Avanzi. Milan: Skira, 2015, pp. 197–222. Print.

Bourdieu, Pierre. *Photography: A Middle-Brow Art.* Stanford, CA: Stanford University Press, 1990. Print.

Bozzini, Giuseppe. *90 anni di turismo in Italia: 1894–1984.* Milan: Touring Club Italiano, 1984. Print.

Brandt, Allan. *The Cigarette Century: The Rise, Fall and Deadly Persistence of the Product That Defined America.* New York: Basic Books, 2007. Print.

Branzi, Andrea. *Introduzione al design italiano. Una modernità incompleta.* Milan: Baldini Castoldi Dalai Editore, 2008. Print.

Braun, Emily. "Futurist Fashion: Three Manifestos." *Art Journal*, vol. 54, no.1, Spring 1995, pp. 34–41. Periodical.

Breda, Giovanni. "Rivista Mensile del TCI. Immagini." *Rivista mensile del Touring Club Italiano e vie d'Italia.* Web. Accessed 7 February 2017.

Brogi, Carlo. *Il ritratto in fotografia: Appunti pratici per chi posa.* Florence: Landi, 1895. Print.

Brunetta, Gian Piero. *The History of Italian Cinema: A Guide to Italian Films from Its Origins to the Twenty-First Century.* Trans. Jeremy Parzen. Princeton, NJ: Princeton University Press, 2009. Print.

Buci-Glucksmann, Christine. *La folie du voir: De l'esthétique baroque.* Paris: Galilée, 1986. Print.

Bullettino della Società Fotografica Italiana. Florence: Società Fotografica Italiana, 1889–1914. Periodical.

Burns, Eric. *The Smoke of the Gods: A Social History of Tobacco.* Philadelphia: Temple University Press, 2007. Print.

Buscioni, Maria Cristina. *Esposizioni e 'stile nazionale' (1861–1925): Il linguaggio dell'architettura nei padiglioni italiani delle grandi kermesses nazionali ed internazionali.* Florence: Alinea Editrice, 1990. Print.

Bussoni, Mario. *Buffalo Bill in Italia: L'epopea del Wild West Show.* Fidenza: Mattioli, 1885, 2011. Print.

Cafagna, Luciano. "La rivoluzione industriale in Italia 1830–1900." *L'industrializzazione in Italia (1861–1900)*, ed. Giorgio Mori. Bologna: Il Mulino, 1977, pp. 57–73. Print.

Caffarena, Fabio. *Lettere dalla Grande Guerra: Scritture del quotidiano, monumenti della memoria, fonti per la storia; Il caso italiano.* Milan: Unicopli, 2005. Print.

Caillois, Roger. *Man, Play and Games.* Trans. Meyer Barash. New York: MacMillan, 1961. Print.

Camera Oscura: Rivista universale dei progressi della fotografia. Dir. Ottavio Baratti. Milan: Redaelli, 1863–7. Periodical.

Camera Work: A Photographic Quarterly. Dir. Alfred Stieglitz. New York: 1903–17. Periodical.

Camerini, Mario, director. *Gli uomini che mascalzoni...* Cines, 1932. Film.

– *The Last Days of Pompeii.* Società Anonima Ambrosio, 1913. Film.

Campana, Dino. *Il più lungo giorno.* Ed. Domenico De Robertis. Florence: Vallecchi, 1973. Print.

Campanile, Achille. *Battista al Giro d'Italia.* Milan: La Vita Felice, 1996. Print.

Campassi, Gabriella, and Maria Teresa Sega. "Uomo bianco, donna nera: L'immagine della donna nella fotografia coloniale." *Rivista di storia e critica della fotografia,* vol. 5, 1983, pp. 55–61. Periodical.

Campbell, Timothy. *Wireless Writing in the Age of Marconi.* Minneapolis: University of Minnesota Press, 2006. Print.

Cangiullo, Francesco. "Fumatori IIa classe." *Lacerba,* 1 January 1914, pp. 10–11. Periodical.

– "Il mobilio futurista. I mobili a sorpresa parlanti e paroliberi." *Roma futurista,* 22 February 1920. Periodical.

Čapek, Karel. *R.U.R.* Trans. Paul Server and Nigel Playfair. Mineola, NY: Dover, 2001. Print.

Capra Marco. "Il teatro in casa. L'idea di musica riprodotta in Italia tra Otto e Novecento." *Il suono riprodotto: Storia, tecnica e cultura di una rivoluzione del Novecento,* ed. Alessandro Rigolli and Paolo Russo. Turin: EDT, 2007, pp. 3–12. Print.

Capuana, Luigi. *Coscienze.* Catania: Fratelli Battiato, 1905. (Reference text for "Esitanze."). Print.

Cardoza, Anthony. "Making Italians? Cycling and National Identity in Italy: 1900–1950." *Journal of Modern Italian Studies,* vol. 15, no. 3, 2010, pp. 354–77. Print.

Carducci, Giosuè. *Odi Barbare.* Bologna: Zanichelli, 1877. Print.

– *Poesie di Giosuè Carducci MDCCCL–MCM.* Bologna: Zanichelli, 1905. (Reference text for "Alla stazione in una mattina d'autunno" and "Mezzogiorno alpino"). Print.

– *Rime e Ritmi.* Bologna: Zanichelli, 1903. Print.

Carmen. Composed by Georges Bizet. Opéra-Comique, Paris, 3 March 1875. Performance.

Carrà, Carlo. "The Painting of Sounds, Noises, and Smells." *Futurist Manifestos,* ed. Umbro Apollonio. Boston: Art Works, 2001, pp. 111–14. Print.

Carreras, Albert. "Un ritratto quantitativo dell'industria italiana." *Storia d'Italia: Annali; L'industria*, vol. 15, ed. Franco Amatori et al. Turin: Einaudi, 1999, pp. 179–272. Print.

Caruso, Enrico. *Ten Arias*. British Gramophone & Typewriter Co, 1902. Record.

Casini, Leonardo. *La riscoperta del corpo: Schopenhauer, Feuerbach, Nietzsche*. Rome: Edizioni Studium, 1990. Print.

Castiglione, Baldassar. *Il libro del cortegiano*. Ed. Nicola Longo. Milan: Garzanti, 2008. Print.

Ceccarelli, Filippo. "Quando il politico gettava il guanto." *Repubblica*, 4 April 2006. Web. Accessed 20 January 2014.

Ceccuti, Cosimo. "La fiorentinità degli Alinari." *Fratelli Alinari, fotografi in Firenze: 150 anni che illustrarono il mondo*, ed. Arturo Carlo Quintavalle and Monica Maffioli. Florence: Alinari, 2003, pp. 299–312. Print.

Ceen, Allan. "Cartografia dei primi meridiani." *Meridiani e longitudini a Roma*, ed. Costantino Sigismondi. Rome: Università degli studi di Roma, 2006. Print.

Cellio, Marc'Antonio. *Descrizione di un nuovo modo di trasportare qualsiasi figura disegnata in carta mediante i raggi solari*. Rome: 1686. Print.

Cenzato, Giovanni. *Campari, 1860–1960*. Milan: Edizioni Campari, 1960. Print.

Ceserani, Remo. *Treni di carta: L'immaginario in ferrovia; L'irruzione del treno nella letteratura moderna*. Turin: Bollati Boringhieri, 2002. Print.

Le chant du rossignol. Composed by Igor Stravinsky. Set design by Henri Matisse. Théâtre National de l'Opéra, Paris, 2 February 1920. Performance.

Chaplin, Charles, director. *Modern Times*. United Artists, 1936. Film.

Choate, Mark. *Emigrant Nation: The Making of Italy Abroad*. Cambridge, MA: Harvard University Press, 2008. Print.

"The Chronology." *Alinari: Cronologia*. Web. Accessed 8 February 2017.

Il ciclista. Dir. Elso Rivera. Milan: Sonzogno, 1895–6. Periodical.

Cigliana, Simona. *La seduta spiritica: Dove si racconta come e perché i fantasmi hanno invaso la modernità*. Rome: Fazi, 2007. Print.

Cipolla, Carlo. *Clocks and Culture, 1300–1700*. New York: Norton, 1978. Print.

Cobisi, Luigi. "Il Vate e Marconi: E la radio fu." *Avvenire*, 10 September 2010. Periodical.

Cobolli Gigli, Nicoletta. *Orologi: Storia, costume, collezionismo dell'orologio da polso*. Milan: Mondadori, 1986. Print.

Cogeval, Guy. "Una spensierata corsa verso l'abisso." *Una dolce vita? Dal liberty al design italiano 1900–1940*, ed. Guy Cogeval and Beatrice Avanzi. Milan: Skira, 2015, pp. 16–21. Print.

Cogeval, Guy, and Beatrice Avanzi. *Una dolce vita? Dal liberty al design italiano 1900–1940*. Milan: Skira, 2015. Print.

Cohen, Milton. *Movement, Manifesto, Melee: The Modernist Group, 1910–1914.* Lanham, MD: Lexington Books, 2004. Print.

Colletti, Lucio. "Lo stato etico in Gentile." *Giovanni Gentile: La filosofia, la politica, l'organizzazione della cultura,* ed. Maria Ida Gaeta. Venice: Marsilio, 1995, pp. 19–22. Print.

Collodi, Carlo. *The Adventures of Pinocchio. Le avventure di Pinocchio.* Trans. Nicolas Perella. Bilingual edition. Berkeley: University of California Press, 1991. Print.

– *Le avventure di Pinocchio.* Illustrated by Attilio Mussino. Florence: Bemporad, 1911. Print.

– *Un romanzo in vapore: Da Firenze a Livorno; Guida storico-umoristica.* Florence: Tipografia Mariani, 1856. Print.

– *Il viaggio per l'Italia di Giannettino: Italia settentrionale, centrale e meridionale, Sicilia e Sardegna.* Florence: Bemporad, 1918. Print.

Colombi, Marchesa (Maria Antonietta Torriani). *La gente per bene: Galateo.* Novara: Interlinea, 2000. Print.

– *Un matrimonio in provincia.* Milan: Galli, 1890. Print.

Comerio, Luca, director. *Secondo giro ciclistico d'Italia.* Comerio Films, 1910. Film.

"The Complete Works." *Giovanni Boldini.* Web. Accessed 17 March 2017.

"Concerto d'inaugurazione." Narrated by Maria Luisa Boncompagni. Unione Radiofonica Italiana, 6 October 1924. Radio.

Confessore, Ornella. *L'americanismo cattolico in Italia.* Rome: Studium, 1984. Print.

Conrad, Joseph. *The Secret Agent.* Oxford: Oxford University Press, 2008. Print.

Coppa, Frank. "Commercio estero e politica doganale nell'Italia liberale." *L'industrializzazione in Italia (1861–1900),* ed. Giorgio Mori. Bologna: Il Mulino, 1977, pp. 161–70. Print.

"Il Cordial Campari e l'arte dell'illustrazione." *Italian Ways,* 12 February 2016. Web. Accessed 8 February 2017.

Corriere dei piccoli: Supplemento illustrato del Corriere della sera. Milan, 1908–92. Periodical.

Corriere della sera. Milan. 1876– . Periodical.

Il corriere fotografico: Periodico quindicinale illustrato pei dilettanti fotografi e raccoglitori di cartoline illustrate. Dir. Tancredi Zanghieri. Piacenza: Stab. Tip. Piacentino, 1904. Periodical.

Costantini, Paolo. "L'esposizione internazionale di fotografia artistica." *Torino 1902: Le arti decorative internazionali del nuovo secolo,* ed. Rossana Bossaglia, Ezio Godoli, and Marco Rosci. Milan: Fabbri, 1994, pp. 95–105. Print.

– *La fotografia artistica, 1904–1917: Visione italiana e modernità.* Turin: Bollati Boringhieri, 1990. Print.

Cottini, Luca. "An Italian in Paris: De Amicis, the World Exposition of 1878, and the New Culture of Industrial Modernity." *Romance Studies*, vol. 33, no. 3–4, November 2015, pp. 216–29. Periodical.

– "Presente! The Latent Memory of Italy's Great War in Its Photographic Portraits." *Stillness in Motion: Italy, Photography and the Meanings of Modernity*, ed. Sally Hill and Giuliana Minghelli. University of Toronto Press: Toronto, 2014, pp. 149–80. Print.

Crepas, Nicola. "Le premesse dell'industrializzazione." *Storia d'Italia: Annali; L'industria*, vol. 15, ed. Franco Amatori, et al. Turin: Einaudi, 1999, pp. 85–179. Print.

Croce, Benedetto. *Storia d'Italia dal 1871 al 1915*. Bari: Laterza, 1928. Print.

Cronaca bizantina. Rome: Angelo Sommaruga, 1881–6. Periodical.

Curci, Roberto. "Fonografi, manifesti: La bellezza della modernità." *La voce del padrone: Manifesti e fonografi*, ed. Eugenio Manzato. Treviso: Canova, 1996, pp. 4–10. Print.

Daly, Selena. "Futurist War Noises: Coping with the Sounds of the First World War." *California Italian Studies*, special issue on "Italian Sound," vol. 4, 2013, pp. 1–16. Ed. Deanna Shemek and Arielle Saiber. http://escholarship.org/uc/item/8fx1p115. Accessed 13 February 2017.

D'Annunzio, Gabriele. "Autobiografia di una sigaretta." *Tutte le novelle*, by Gabriele D'Annunzio, ed. Annamaria Andreoli and Marina De Marco. Milan: Mondadori, 1992, pp. 591–7. Print.

– *Forse che sì forse che no. Prose di Romanzi*, vol. 2, by Gabriele D'Annunzio, ed. Annamaria Andreoli and Niva Lorenzini. Milan: Mondadori, 1989, pp. 519–887. Print.

– *Il notturno*. Trans. Stephen Sartarelli. New Haven, CT: Yale University Press, 2011. Print.

– *Il piacere. Prose di romanzi*, vol. 1, by Gabriele D'Annunzio, ed. Annamaria Andreoli and Niva Lorenzini. Milan: Mondadori, 1988–9, pp. 3–358. Print.

– *Scritti giornalistici 1889–1938*. Ed. Giorgio Zanetti. Milan: Mondadori, 1996 (vol. 1); 2003 (vol. 2). Print.

Darwin, Charles. *The Descent of Man and Selection in Relation to Sex*. New York: Burt Company, 1874. Print.

De Amicis, Edmondo. *Amore e ginnastica*. Preface by Italo Calvino. Turin: Einaudi, 1971. Print.

– *Costantinopoli*. Milan: Treves, 1888. Print.

– *Cuore*. Milan: Mondadori, 1984. Print.

– *Olanda*. Milan: Treves, 1885. Print.

– *Ricordi di Parigi*. Chieti: Solfanelli, 2012. Print.

– *Sull'oceano*. Reggio Emilia: Diabasis, 2005. Print.

– “La tentazione della bicicletta.” *Pagine Allegre di Edmondo De Amicis.* Milan: Treves, 1906. Print.

– *La vita militare.* Rome: Avagliano, 2008. Print.

De Chirico, Giorgio. *Il meccanismo del pensiero: Critica, polemica, autobiografia, 1911–1943.* Ed. Maurizio Fagiolo. Turin: Einaudi, 1985. Print.

De Fusco, Renato. *Filosofia del design.* Turin: Einaudi, 2012. Print.

– *Made in Italy: Storia del design italiano.* Rome: Laterza, 2007. Print.

Degrada, Francesco. “Il segno e il suono: Storia di un editore musicale e del suo mondo.” *Musica, musicisti, editoria: 175 anni di casa Ricordi 1808–1983,* ed. Francesco Degrada. Milan: Ricordi, 1983, pp. 9–190. Print.

De Gregorio, Domenico. “Nascita e morte dell’Ambrosio film.” *Bianco e nero,* vol. 1–2, 1963, pp. 70–6. Periodical.

De Guttry, Irene, and Maria Paola Maino. “From Artist-Artisans to Architect-Designers.” *Il Modo Italiano: Italian Design and Avant-Garde in the 20th Century,* ed. Giampiero Bosoni. Montreal: Montreal Museum of Fine Arts, 2006, pp. 41–56. Print.

– “Le straordinarie avventure delle arti decorative italiane.” *Una dolce vita? Dal liberty al design italiano 1900–1940,* ed. Guy Cogeval and Beatrice Avanzi. Milan: Skira, 2015, pp. 29–86. Print.

De Iulio, Simona. “L’immagine della réclame: La prima stagione dello spettacolo della merce nella pubblicistica italiana tra Otto e Novecento.” *Lumi di progresso: Comunicazione e persuasione alle origini della cartellonistica italiana,* ed. Alberto Abruzzese and Simona De Iulio. Treviso: Canova, 1996, pp. 13–27. Print.

Della Coletta, Cristina. *World’s Fairs Italian Style: The Great Exhibitions in Turin and Their Narrative, 1860–1915.* Toronto: University of Toronto Press, 2006. Print.

Dellanoce, Maura. “Viaggi e reportage in bici: Le cronache di Oriani e Panzini.” *Biciclette: Lavoro, storie e vita quotidiana su due ruote,* ed. Guido Conti. Parma: Monte Università Parma, 2007, pp. 19–26. Print.

Della Porta, Giovan Battista. *De refractione*. Naples: Carlino & Pace, 1593. Print.

Del Noce, Augusto. *Secolarizzazione e crisi della modernità.* Naples: Istituto suor Orsola Benincasa, Edizioni Scientifiche Italiane, 1989. Print.

De Marchi, Emilio. *Demetrio Pianelli.* Milan: Mursia, 1985. Print.

Denisoff, Dennis. “Small Change: The Consumerist Designs of the Nineteenth-Century Child.” *The Nineteenth-Century Child and Consumer Culture,* ed. Dennis Denisoff. Burlington, VT: Ashgate, 2008, pp. 1–26. Print.

Depero, Fortunato. *Depero 1913–1927.* Milan: Edizioni della Dinamo, 1927. Print.

– *Liriche radiofoniche.* Milan: Editore Morreale, 1934. Print.

– *Numero unico futurista Campari.* Paris: Éditions Jean-Michel Place, 1979. Print.
Diana, Giampietro. "La storia del tabacco in Italia: III, Dalla formazione del monopolio di stato fino alla seconda guerra mondiale." *Il tabacco,* vol. 8, 2000, pp. 91–103. Periodical.
Il dilettante di fotografia. Dir. Luigi Gioppi. Milan: Tip. Trevisini, 1890–1908. Periodical.
Disney, Walt, director. *Fantasia.* Walt Disney Productions, 1940. Film.
Dolan, Graham. "The Adoption of a Prime Meridian and the International Meridian Conference of 1884." *The Greenwich Meridian.* Web. Accessed 26 January 2017.
Dolci, Fabrizio, and Olivier Janz. *Non omnis moriar: Gli opuscoli di necrologio per i caduti italiani nella grande guerra.* Rome: Edizioni di Storia e Letteratura, 2003. Print.
La domenica del Corriere: Supplemento illustrato del Corriere della sera. Milan: 1899–1989. Periodical.
La domenica letteraria. Rome. 1882–5. Periodical.
Domus: Architettura e arredamento dell'abitazione moderna in città e in campagna. Milan: Domus, 1928– . Periodical.
La donna è mobile. By Filippo T. Marinetti. Teatro Alfieri, Turin, 15 January 1909. Performance.
Dorfles, Gillo. *Il divenire delle arti.* Turin: Einaudi, 1959. Print.
Dowd, Charles. *System of National Time and Its Application by Means of Hour and Minute Indexes to the National Railway Time-Table.* Albany, NY: Weed, Parsons, 1870. Print.
Dubois, Philippe. *L'acte photographique et autres essais.* Brussels: Nathan, 1990. Print.
Eco, Umberto. "Times." *The Story of Time,* ed. Kristen Lippincott, Umberto Eco, and E.H. Gombrich. London: Merrell Holberton, 1999, pp. 7–17. Print.
Edison, Thomas. "The Phonograph and Its Future." *North American Review,* vol. 126, no. 262, May–June 1878, pp. 527–36. Periodical.
Einstein, Albert. "On the Electrodynamics of Moving Bodies." *The Principle of Relativity.* Trans. Wilfred Perrett and George Jeffery. London: Methuen, 1923, pp. 35–65. Print.
Eisenberg, Evan. *The Recording Angel: Music, Records and Culture from Aristotle to Zappa.* New Haven, CT: Yale University Press, 2005. Print.
Emporium. Bergamo: Istituto Italiano d'Arti Grafiche, 1895–1964. Periodical.
"The Evolution of Bulgari Style." *Bulgari.* Web. Accessed 25 January 2017.
Fabi, Lucio. "La guerra nel mirino." *Storia e Dossier,* vol. 62, May 1992, pp. 56–60. Periodical.
Fabrizio, Felice. *Alle origini del movimento sportivo cattolico in Italia.* Milan: Sedizioni, 2009. Print.

Fallan, Kjetil, and Grace Lees-Maffei. Introduction. *Made in Italy: Rethinking a Century of Italian Design*, ed. Grace Lees-Maffei and Kjetil Fallan. New York: Bloomsbury, 2014, pp. 1–33. Print.

Falzone del Barbarò, Michele. *Secondo Pia: Fotografie 1886–1927*. Turin: Allemandi, 1989. Print.

La fanciulla del West. Composed by Giacomo Puccini. Dir. Arturo Toscanini. Perf. Enrico Caruso. Metropolitan Opera House, New York, 10 December 1910. Performance.

Il fanfulla della domenica. Rome, 1879–1919. Periodical.

Favaro, Adriano. "La Fotoceramica." *Provincia Treviso*. Web. Accessed 14 June 2011.

Fellini, Federico, director. *La dolce vita*. Pathé Consortium Cinema, 1959. Film.

Fénelon, François de Salignac de La Mothe. *A Treatise on the Education of Daughters*. Boston: Charles Ewer, 1820. Print.

Ferrari, Oreste. Preface. *I fiorentini fotografati da Mario Nunes Vais*, ed. Maria Teresa Contini. Florence: Centro Di, 1978. Print.

Ferrari, Paolo. *L'aeronautica italiana: Una storia del novecento*. Milan: Franco Angeli, 2004. Print.

Ferraris, Maria Pia. "Attività grafica di Ricordi." *Musica, musicisti, editoria: 175 anni di casa Ricordi 1808–1983*. Milan: Ricordi, 1983, pp. 191–230. Print.

Ferrieri, Enzo. "La radio come forza creativa." *Il Convegno*, vol. 12, no. 6, 25 June 1931, pp. 297–320. Periodical.

Fioravanti, Giorgio. "Evoluzione tecnica della stampa della musica." *Musica, musicisti, editoria: 175 anni di casa Ricordi 1808–1983*, ed. Francesco Degrada. Milan: Ricordi, 1983, pp. 260–6. Print.

Fiorentino, Davide. *Gli Stati Uniti e l'Italia alla fine del XIX secolo*. Rome: Gangemi Editore, 2010. Print.

Fleming, Sandford. *Uniform Non-local Time (Terrestrial Time)*. Ottawa: 1876. Print.

Fogazzaro, Antonio. *Piccolo mondo antico*. Rome: Newton, 2007. Print.

– *Il santo*. Reggio Emilia: Città Armoniosa, 1991. Print.

Forgacs, David. *Italian Culture in the Industrial Era, 1880–1980: Cultural Industries, Politics, and the Public*. New York: Manchester University Press, 1990. Print.

Fotografia artistica: Rivista internazionale illustrata. Dir. Annibale Cominetti. Turin: 1904–17. Periodical.

4'33". Composed by John Cage. Perf. David Tudor. Woodstock, NY, 1952. Performance.

Freud, Sigmund. "Creative Writers and Daydreaming." *The Freud Reader*, ed. Peter Gay. New York: Norton, 1989, pp. 436–43. Print.

Freund, Gisèle. *Photography and Society*. Boston: David Godine, 1980. Print.

Fumerton, Patricia. *Cultural Aesthetics: Renaissance Literature and the Practice of Social Ornament*. Chicago: University of Chicago Press, 1991. Print.

Fussell, Paul. *The Great War and Modern Memory*. New York: Oxford University Press, 1975. Print.

Galassi, Peter. *Before Photography: Painting and the Invention of Photography*. New York: Museum of Modern Art, 1981. Print.

Galison, Peter. *Einstein's Clocks, Poincaré's Maps: Empires of Time*. New York: Norton, 2004. Print.

Gangale, Daniela. "Il suono dei futuristi: La musica in *Lacerba* e altre polemiche musicali (1913–1915)." *California Italian Studies* (special issue on "Italian Sound"), vol. 4, 2013, pp. 1–30. Ed. Deanna Shemek and Arielle Saiber. http://escholarship.org/uc/item/8fx1p115. Accessed 13 February 2017.

Gargano, Pietro. *Una vita una leggenda: Enrico Caruso, il più grande tenore del mondo*. Milan: Editoriale Giorgio Mondadori, 1997. Print.

Gately, Iain. *Tobacco: A Cultural History of How an Exotic Plant Seduced Civilization*. New York: Grove Press, 2001. Print.

La gazzetta dello sport. Milan. 1896– . Periodical.

Gelatt, Roland. *The Fabulous Phonograph, 1877–1977*. New York: MacMillan, 1977. Print.

Gemelli, Agostino. *Il nostro soldato: Saggi di psicologia militare*. Milan: Treves, 1918. Print.

Gemmellaro, Claudia. "Futurismo: Fortunato Depero e il marchio Campari." *Visual Advertising*. Web. Accessed 11 January 2017.

Gentile, Antonio. *Edoardo Bianchi*. Milan: Giorgio Nada Editore, 1992. Print.

Gentile, Emilio. *L'apocalisse della modernità: La grande guerra per l'uomo nuovo*. Milan: Mondadori, 2008. Print.

– *The Sacralization of Politics in Fascist Italy*. Cambridge, MA: Harvard University Press, 1996. Print.

Gentile, Giovanni. *Dottrina del Fascismo: Idee Fondamentali*. Milan: Hoepli, 1935. Print.

– *La filosofia di Marx: Studi critici*. Pisa: E. Spoerri, 1899. Print.

– *Rosmini e Gioberti: Saggio storico sulla filosofia italiana del Risorgimento*. Pisa: Tipografia Successori Fratelli Nistri, 1898. Print.

– *Sommario di pedagogia come scienza filosofica*. Florence: Sansoni, 1954. Print.

Gerarchia. Roma: Tip. Il Popolo d'Italia, 1922–43. Periodical.

Germania. By Ugo I. Tarchetti. Dir. Arturo Toscanini. Perf. Enrico Caruso. Teatro La Scala, Milan, 11 March 1902. Performance.

Giannetti, Renato, and Michelangelo Vasta. *Storia dell'impresa industriale italiana*. Bologna: Il Mulino, 2005. Print.

Gilman, Sanders, and Zhou Xun. Introduction. *Smoke: A Global History of Smoking*, ed. Sanders Gilman and Zhou Xun. London: Reaktion Books, 2004, pp. 9–28. Print.

Ginna, Arnaldo. "Il mobilio futurista." *L'Italia futurista,* 16 December 1916. Periodical.

– *Vita futurista.* Dir. Arnaldo Ginna, 1916. Film.

Gioberti, Vincenzo. *Del primato morale e civile degli italiani.* Turin: Unione Tip. Editrice Torinese, 1920. Print.

La gioconda. By Amilcare Pomchielli. Teatro La Scala, Milan, 8 April 1876. Performance.

Gioppi, Luigi, *La fotografia secondo i processi moderni.* Milan: Hoepli, 1891. Print.

Giraud, Albert. *Pierrot lunaire: Rondels bergamasques.* Paris: Lemerre Éditeur, 1884. Print.

Giuntini, Andrea. "Nascita, sviluppo e tracollo della rete infrastrutturale." *Storia d'Italia: Annali; L'industria,* vol. 15, ed. Franco Amatori, et al. Turin: Einaudi, 1999, pp. 551–617. Print.

Gnoli, Sofia. *The Origins of Italian Fashion, 1900–1945.* London: Victoria & Albert Museum, 2014. Print.

Govoni, Corrado. *Poesie (1903–1958).* Milan: Mondadori, 2000. Print.

Gozzano, Guido. *I colloqui.* Milan: Treves, 1911. Print.

– *Opere.* Ed. Giusi Baldissone. Turin: UTET, 2006. (Reference text for "L'Amica di nonna Speranza," "Le due strade," and "La signorina Felicita."). Print.

– "I sandali della diva." *I sandali della diva: Tutte le novelle,* by Guido Gozzano. Milan: Serra e Riva Editori, 1983. Print.

– *La via del rifugio.* Turin: Streglio, 1907. Print.

Graziani, Luigi. *Lira classica: Versioni e poemetti originali.* Bologna: Zanichelli, 1931. (Reference text for "In bicycula" and "In re cyclistica Satan."). Print.

Grazioli, Elio. *La polvere nell'arte.* Milan: Bruno Mondadori, 2004. Print.

Green, Martin, and John Swan. *The Triumph of Pierrot: The Commedia dell'Arte and the Modern Imagination.* New York: MacMillan, 1986. Print.

Greene, Vivien. *Italian Futurism, 1909–1944: Reconstructing the Universe.* New York: Guggenheim Museum, 2014. Print.

Gregotti, Vittorio. *Il disegno del prodotto industriale: Italia 1860–1980.* Milan: Electa, 1982. Print.

Guadagnoli, Antonio. *Il tabacco: Sestine.* Pisa: Tipografia Nistri, 1834. Print.

Guazzoni, Enrico, director. *Quo Vadis.* Cines, 1912. Film.

Habermas, Jürgen. *The Philosophical Discourse of Modernity.* Cambridge, MA: MIT Press, 1987. Print.

Herbert, Stephen. "Set Design." *Encyclopedia of Early Cinema,* ed. Richard Abel. Routledge: New York, 2005. Print.

Hill, Sally, and Giuliana Minghelli. *Stillness in Motion: Photography and the Meanings of Modernity.* Toronto: University of Toronto Press, 2014. Print.

Hilton, Matthew. "Smoking and Sociability." *Smoke: A Global History of Smoking*, ed. Sanders Gilman and Zhou Xun. London: Reaktion Books, 2004, pp. 126–33. Print.

Hirsch, Adolphe, and Theodor Von Oppolzer. *Unification des longitudes par l'adoption d'un méridien initial unique, et introduction d'une heure universelle: Extrait des Compte rendus de la septième conférence générale de l'Association Géodésique Internationale réunie à Rome, en Octobre 1883*. Stankiewicz: Berlin, 1883. Print.

Hirsch, Robert. *Seizing the Light: A History of Photography*. Boston: McGraw-Hill, 2000. Print.

Hockemeyer, Lisa. "Manufactured Identities: Ceramics and the Making of (Made In) Italy." *Made in Italy: Rethinking a Century of Italian Design*, ed. Grace Lees-Maffei and Kjetil Fallan. New York: Bloomsbury, 2014, pp. 127–44. Print.

Hong, Sungook. *Wireless: From Marconi's Black-Box to the Audion*. Cambridge, MA: MIT Press, 2001. Print.

Horkheimer, Max, and Theodor Adorno. *Dialectic of Enlightenment*. London: Verso, 1997. Print.

Howse, Derek. "1884 and Longitude Zero." *Vistas in Astronomy*, vol. 28. New York: Pergamon Press, 1984, pp. 11–22. Periodical.

– *Greenwich Time and the Discovery of the Longitude*. New York: Oxford University Press, 1980. Print.

Huizinga, Johan. *Homo Ludens. A Study of the Play-Element in Culture*. Boston: Beacon Press, 1955. Print.

Hutcheon, Linda, and Michael Hutcheon. "Smoking in Opera." *Smoke: A Global History of Smoking*, ed. Sanders Gilman and Zhou Xun. London: Reaktion Books, 2004, pp. 230–5. Print.

Ickringill, Steve. "Amateur and Professional: Sport in Britain and America at the Turn of the Twentieth Century." *Sport, Culture, and Politics*, ed. J.C. Binfield and John Stevenson. Sheffield, UK: Sheffield Academic Press, 1993, pp. 30–48. Print.

Illustrazione italiana. Milan: Treves, Garzanti, 1873–1962. Periodical.

L'industria della bicicletta: Giornale tecnico per l'industria e il commercio dei velocipedi. Milan: Stab. Tip. A. Rancati. 1895. Periodical.

Inno alla vita. By Francesco Balilla Pratella. Teatro Costanzi Rome, 21 February 1913. Performance.

International Conference Held at Washington for the Purpose of Fixing a Prime Meridian and a Universal Day, October, 1884: Protocols of the Proceedings. Washington, DC: Gibson Bros., 1884. Print.

Introna, Elena. *Veglia: Un secolo con l'orologio*. Milan: Mazzotta, 1996. Print.

Ioli, Giovanna. *Letteratura e sport: Per una storia delle Olimpiadi*. Novara: Interlinea, 2006. Print.

"Italian Futurism, 1909–1944, EMMA museum." *Beni Culturali.* Web. Accessed 28 February 2017. <http://www.beniculturali.it/mibac/multimedia/MiBAC/documents/1328631674909_ITALIAN_FUTURISM_IMMAGINI.pdf>

Jahier, Piero. *Gino Bianchi: Resultanze in merito alla vita e al carattere.* Florence: Vallecchi, 2007. Print.

James I, King of England. *A Counterblaste to Tobacco.* London: R. Barker, 1604. Print.

Jones, Geoffrey. "The Gramophone Company: An Anglo-American Multinational, 1898–1931." *Business History Review*, vol. 59, no. 1, Spring 1985, pp. 76–100. Periodical.

Judovitz, Dalia. *The Culture of the Body: Genealogies of Modernity.* Ann Arbor: University of Michigan Press, 2001. Print.

Kemp, Tom. *Industrialization in Nineteenth-Century Europe.* 2nd Edition. New York: Routledge, 2013. Print.

Kern, Stephen. *The Culture of Time and Space.* Cambridge, MA: Harvard University Press, 1983. Print.

Kircher, Athanasius. *Ars magna lucis et umbrae.* Rome: Scheus, 1646. Print.

Kittler, Friedrich. *Gramophone, Film, Typewriter.* Stanford, CA: Stanford University Press, 1999. Print.

Kron, Riccardo. "Dal fonografo di Edison al grammofono: Breve storia delle machine parlanti." *Archeofon: Fonografi, grammofoni e radio 1888–1934*, ed. Silvio Fusco and Sandro Mescola. Milan: Electa, 1988, pp. 13–15. Print.

Lacerba: Periodico Quindicinale. Dir. Giovanni Papini. Florence: Tip. Vallecchi, January 1913–May 1915. Periodical.

Laforgue, Jules. *Pierrot fumiste. Oeuvres completes: Mélanges Posthumes.* Paris: 1919. Print.

La Mettrie, Julien Offray de. *L'homme machine.* Leyde, Netherlands: 1748. Print.

La Pietra, Ugo. *Gió Ponti: L'arte s'innamora dell'industria.* Milan: Rizzoli, 2009. Print.

Laxton, Susan. "From Judgment to Process: The Modern Ludic Field." *From Diversion to Subversion: Games, Play and Twentieth-Century Art*, ed. David Getsy. University Park: Pennsylvania State University Press, 2011, pp. 3–24. Print.

Leblanc, Maurice. *Voici des ailes!* Paris: P. Ollendorff, 1898. Print.

Ledeen, Michael. *D'Annunzio: The First Duce.* New Brunswick, NJ: Transaction Publishers, 2009. Print.

Lees-Maffei, Grace. "The Production-Consumption-Mediation Paradigm." *Journal of Design History*, vol. 22, no. 4, 2009, pp. 351–6. Periodical.

Lees-Maffei, Grace, and Kjetil Fallan. *Made in Italy: Rethinking a Century of Italian Design.* New York: Bloomsbury, 2014. Print.

Leo XIII, Pope. *Carmina et inscriptiones cum accessionibus novissimis.* Editio V Utinensis (exemplaribus C). Tip. Del Patronato, 1894. Print.

– *Rerum novarum: Encyclical Letter of Pope Leo XIII on the Condition of the Working Classes; Study Edition.* Ed. Joseph Kirwan. London: Catholic Truth Society, 1983. Print.

– "Testem benevolentiae nostrae: Concerning New Opinions, Virtue, Nature and Grace, with Regard to Americanism." *Papal Encyclicals Online.* Web. Accessed 7 February 2017.

Leopardi, Giacomo. *Canti: Poems.* Trans. Jonathan Galassi. New York: Farrar, Straus & Giroux, 2010. Print.

– *Zibaldone di pensieri.* 3 vols. Ed. Giuseppe Pacella. Milan: Garzanti, 1991. Print.

Lidel. Dir. Lydia De Liguoro. Milan: 1919–35. Periodical.

Lista, Giovanni. *I futuristi e la fotografia: Creazione fotografica e immagine quotidiana.* Modena: Panini, 1986. Print.

Locke, John. *Some Thoughts Concerning Education.* New York: Oxford University Press, 1989. Print.

Loisy, Alfred. *L'évangile et l'église.* Bellevue, France, 1904. Print.

Lolla, Maria Grazia. "Spectres of Photography: Photography, Literature, and the Social Sciences in Fin-de-Siècle Italy." *Enlightening Encounters: Photography in Italian Literature,* ed. Giorgia Alù and Nancy Pedri. Toronto: University of Toronto Press, 2015, pp. 27–50. Print.

Lombardini, A. *La bicicletta.* Livorno: Stab. Tip. Di Gius. Meucci, 1897. Print.

Lombardo, Antonio. *Pierre de Coubertin: Saggio storico sulle Olimpiadi moderne 1880–1914.* Rome: RAI ERI, 2000. Print.

Lombardo, Giovanni Pietro, and Renato Foschi. *La psicologia italiana e il Novecento: Le prospettive emergenti nella prima metà del secolo.* Milan: Franco Angeli, 1997. Print.

Lombroso, Cesare. "Il ciclismo nel delitto." *La nuova antologia,* vol. 86, 1900, pp. 5–14. Periodical.

Lopez, Guido. "Da bici per pochi a bici per tutti." *Uomo a due ruote: Avventura, storia,* passione, ed. Guido Vergani. Milan: Electa, 1987, pp. 58–79. Print.

Macey, Samuel. *Clocks and the Cosmos: The Time in Western Life and Thought.* Hamden, CT: Archon Books, 1980. Print.

Maffioli, Monica. "I Fratelli Alinari: Una famiglia di fotografi 1852–1920." *Fratelli Alinari, fotografi in Firenze: 150 anni che illustrarono il mondo,* ed. Arturo Carlo Quintavalle and Monica Maffioli. Florence: Alinari, 2003, pp. 21–56. Print.

– *I macchiaioli e la fotografia.* Fratelli Alinari: Florence, 2008. Print.

Mallach, Alan. *The Autumn of Italian Opera: From Verismo to Modernism, 1890–1915.* Boston: Northeastern University Press, 2007. Print.

Mandell, Richard. *Sport: A Cultural History.* New York: Columbia University Press, 1984. Print.

"Manifesto degli intellettuali fascisti." *Il popolo d'Italia,* 21 April 1925. Periodical.

Manodori, Alberto. "La fotografia archeologica: Appunti e riflessioni per una categoria delle scienze archeologiche." *Archeologia in posa: Dal Colosseo a Cecilia Metella nell'antica documentazione fotografica,* ed. Barbara Tellini Santoni. Milan: Electa, 1998, pp. 3–12. Print.

Mantegazza, Paolo. *Atlante della espressione del dolore.* Florence: Brogi, 1876. Print.

Manzato, Eugenio. *La voce del padrone: Manifesti e fonografi.* Treviso: Canova, 1996. Print.

Marcenaro, Giuseppe. *Fotografia come letteratura.* Milan: B. Mondadori, 2004. Print.

Marchesini, Daniele, Benito Mazzi, and Romano Spada, editors. *Pàlmer, borraccia e via! Storia e leggende della bicicletta e del ciclismo.* Portogruaro: Ediciclo, 2001. Print.

Margherita: Giornale delle signore italiane, di gran lusso, di mode e letteratura. Milan: Treves, 1878–1921. Periodical.

Mari, Carlo. "Putting the Italians on Bicycles: Marketing at Bianchi, 1885–1955." *Journal of Historical Research in Marketing,* vol. 7, no. 1, 2015, pp. 133–58. Periodical.

Marien, Mary Warner. *Photography: A Cultural History.* Upper Saddle River, NJ: Pearson Prentice Hall, 2011. Print.

Marinetti, Filippo Tommaso. *La battaglia di Tripoli.* Milan: Edizioni Futuriste di *Poesia,* 1912. Print.

–"Contro il lusso femminile." *Roma futurista,* vol. 3, no. 77, 4 April 1920. Periodical.

– *Critical Writings.* Ed. Günter Berghaus. Trans. Doug Thompson. New York: Farrar, Strauss & Giroux, 2006. (Reference edition for "Manifesto of Futurism," "Technical Manifesto of Futurist Literature," "Answers to Objections," "Destruction of Syntax," "Geometrical Splendour," "Variety Theatre," and "New Ethical Religion of Speed."). Print.

– *Elettricità sessuale.* Milan: Facchi, 1920. Print.

– *Mafarka il futurista.* Milan: Edizioni Futuriste di *Poesia,* 1910. Print.

– "Manifesto sul tattilismo." Milan: Direzione del Movimento Futurista, 11 January 1921. Print.

– *Teoria e invenzione futurista.* Ed. Luciano De Maria. Milan: Mondadori, 1968. Print.

– *Zang Tumb Tumb: Teoria e invenzione futurista,* ed. Luciano De Maria. Milan: Mondadori, 1968, pp. 561–710. Print.

Marinetti, F.T., and Pino Masnata. "La radia." *La gazzetta di Torino,* 22 September 1933. Periodical.

Marra, Claudio. *Fotografia e pittura nel Novecento: Una storia 'senza combattimento.'* Milan: B. Mondadori, 2000. Print.

Martineau, Paul. *Still Life in Photography.* Los Angeles: Paul Getty Museum, 2010. Print.

Martinelli, Federico. *Plinio Codognato: L'arte della pubblicitá tra Verona, Milano e Torino.* Verona: Quinta Parete Editore, 2011. Print.

Massoni, Elisa. "Camparisoda: L'aperitivo futurista." *Design Stories: Cinque casi di aziende di design e una piattaforma di comunicazione,* ed. Vittoria Morganti. Milan: Franco Angeli, 2011, pp. 27–50. Print.

Matteoni, Dario. "Gió Ponti, una modernità sospesa." *Gió Ponti, il fascino della ceramica,*, ed. Dario Matteoni. Milan: Silvana Editoriale, 2011, pp. 18–31. Print.

Mazzacurati, Giancarlo. *Effetto Sterne: La narrazione umoristica in Italia da Foscolo a Pirandello.* Pisa: Nistri-Lischi, 1990. Print.

McQuire, Scott. *Visions of Modernity: Representation, Memory, Time and Space in the Age of the Camera.* Thousand Oaks, CA: Sage Publications, 1998. Print.

Meano, Cesare. *Commentario dizionario italiano della moda.* Turin: ENM, 1936. Print.

Melville-Mason, Graham. "The Gramophone as Furniture." *Phonographs & Gramophones.* Edinburgh: Royal Scottish Museum, 1977, pp. 117–38. Print.

Merjian, Ara. "A Future by Design: Giacomo Balla and the Domestication of Transcendence." *Oxford Art Journal,* vol. 35, no. 2, 2012, pp. 121–46. Periodical.

Merli, Stefano. "La classe operaia di fabbrica verso la fine del secolo XIX." *L'industrializzazione in Italia (1861–1900),* ed. Giorgio Mori. Bologna: Il Mulino, 1977, pp. 93–116. Print.

"The Metre Convention." *Bureau International des Poids et Mesures (BIPM).* Web. Accessed 27 January 2017.

Mignemi, Adolfo. *Lo sguardo e l'immagine: La fotografia come documento storico.* Turin: Bollati Boringhieri, 2003. Print.

Minghelli, Giuliana. "Eternal Speed/Omnipresent Immobility." *Stillness in Motion: Italy, Photography and the Meanings of Modernity,* ed. Sally Hill and Giuliana Minghelli. University of Toronto Press: Toronto, 2014, pp. 97–130. Print.

– *In the Shadow of the Mammoth: Italo Svevo and the Emergence of Modernism.* Toronto: University of Toronto Press, 2002. Print.

– "L'occhio di Verga: La pratica fotografica nel Verismo italiano." *L'anello che non tiene,* vol. 20, no. 1, Spring–Fall, 2008–9, pp. 59–86. Periodical.

Ministero delle finanze. *Cenni storico-statistici sul monopolio del tabacco in Italia.* Rome: Tipografia Nazionale di G. Bertero, 1900. Print.

Miraglia, Marina. *Culture fotografiche e società a Torino 1839–1911.* Turin: Umberto Allemandi & Co., 1990. Print.

– "D'Annunzio e la fotografia." *D'Annunzio e la promozione delle arti*, ed. Rossana Bossaglia and Mario Quesada. Milan: Mondadori, 1988, pp. 55–9. Print.

Miscellanea d'arte: Rivista di storia medievale e moderna. Dir. Igino B. Supino. Florence: Alinari, 1903. Periodical.

Misner, Paul. "Catholic Anti-Modernism: The Ecclesiastical Setting." *Catholicism Contending with Modernity: Roman Catholic Modernism and Anti-Modernism in Historical Context*, ed. Jodock Darrell. New York: Cambridge University Press, 2000, pp. 56–87. Print.

Mitchell, Dolores. "Women and Nineteenth-Century Images of Smoking." *Smoke: A Global History of Smoking*, ed. Sanders Gilman and Zhou Xun. London: Reaktion Books, 2004, pp. 294–303. Print.

Modena, Gustavo. "Psicopatologia ed etiologia dei fenomeni psiconevrotici: Contributo alla dottrina di Freud." *Rivista sperimentale di freniatria*, vol. 34, 1908–1909, pp. 657–70; vol. 35, 1909, pp. 204–17. Periodical.

Monaco, Vincenzo. *Il velocipedismo (storia, igiene e pratica).* Bologna: Zanichelli, 1894. Print.

Montale, Eugenio. *Ossi di seppia.* Turin: Gobetti editore [tipografia Carlo Accame], 1925. Print.

Monteleone, Franco. *Storia della radio e della televisione in Italia: Un secolo di costume, società e politica.* Venice: Marsilio, 2001. Print.

Montessori, Maria. *Il metodo della pedagogia scientifica applicato all'educazione infantile nelle Case dei Bambini.* Città di Castello: Casa Editrice S. Lapi, 1909. Print.

Monti, Silvia. *Il tabacco fa male? Medicina, ideologia, letteratura nella polemica sulla diffusione di un prodotto del Nuovo Mondo.* Milan: Franco Angeli, 1987. Print.

Morandi, Rodolfo. *Storia della grande industria in Italia.* Bari: Laterza, 1931. Print.

Morasso, Mario. "La fine del Giro d'Italia." *Illustrazione italiana*, 6 June 1909. Periodical.

– "Il Giro d'Italia." *Illustrazione italiana*, 16 May 1909. Periodical.

– "Mentre si compie il Giro d'Italia." *Illustrazione italiana*, 23 May 1909. Periodical.

– *La nuova arma: La macchina.* Turin: Centro Studi Piemontesi, 1994. Print.

– "Le ultime tappe del Giro d'Italia." *Illustrazione italiana*, 30 May 1909. Periodical.

Mori, Giorgio. *L'industrializzazione in Italia (1861–1900).* Bologna: Il Mulino, 1977. Print.

– "Industrie senza industrializzazione." *Studi storici*, vol. 39, 1989, pp. 608–11. Periodical.

Moroni, Mario, and Luca Somigli. *Italian Modernism: Italian Culture between Decadentism and Avant-Garde.* Toronto: University of Toronto Press, 2004. Print.

Mortara, Giorgio. "La nascita di un gigante: L'industria elettrica dal 1883 al 1900." *L'industrializzazione in Italia (1861–1900)*, ed. Giorgio Mori. Bologna: Il Mulino, 1977, pp. 357–70. Print.

Morton, David. *Sound Recording: The Life Story of a Technology*. Westport, CT: Greenwood Press, 2004. Print.

Motori, cicli e sports. Dir. Mario Morasso. Milan, 1908–11, 1915–19. Periodical.

Muffone, Giovanni. *Come il sole dipinge: Manuale di fotografia per i dilettanti*. Milan: Hoepli, 1887. Print.

Mughini, Giampiero. "The Art Nouveau Period: The Golden Age of the Italian Poster." *Il manifesto pubblicitario italiano: Da Dudovich a Depero 1890–1940*, ed. Giampiero Mughini and Maurizio Scudiero. Milan: Nuove Arti Grafiche Ricordi, 1997, pp. vii–xiv. Print.

Mutti, Roberto. "Lo scrittore che 'catturava' la realtà." *Giovanni Verga scrittore fotografo*, ed. Roberto Mutti. Novara: De Agostini, 2004, pp. 15–22. Print.

Nemiz, Andrea. *Capuana, Verga, De Roberto: Fotografi*. Palermo: Edikronos, 1982. Print.

Newton, Isaac. *Philosophiae naturalis principia mathematica*. London: Streater, 1687. Print.

Nicoli, Marina. *The Rise and Fall of the Italian Film Industry*. New York: Routledge, 2017. Print.

Nietzsche, Friedrich. *The Birth of Tragedy*. Trans. Douglas Smith. New York: Oxford University Press, 2000. Print.

– *Thus Spoke Zarathustra: A Book for Everyone and for Nobody*. Trans. Graham Parkes. 1st ed. Oxford: Oxford University Press, 2005. Print.

– *Twilight of the Idols*. New York: Penguin, 1968. Print.

900: Cahiers d'Italie et d'Europe. Dir. Guido Bontempelli and Curzio Malaparte. Rome: La Voce, 1926–9. Periodical.

Omegna, Roberto, director. *The Last Days of Pompeii*. Società Anonima Ambrosio, 1908. Film.

Oriani, Alfredo. *La bicicletta*. Ravenna: Longo, 2002. Print.

Orlando, Francesco. *Gli oggetti desueti nelle immagini della letteratura: Rovine, reliquie, rarità, robaccia, luoghi inabitati e tesori nascosti*. Turin: Einaudi, 2015. Print.

Ormezzano, Gian Paolo. "Dal celerifero alla bicicletta." *Uomo a due ruote: Avventura, storia, passione*, ed. Guido Vergani. Milan: Electa, 1987, pp. 28–39. Print.

Osborne, Richard. *Vinyl: A History of the Analogue Record*. New York: Ashgate, 2012. Print.

Palazzeschi, Aldo. *Il codice di Perelà. Tutti i romanzi*, vol. 1, by Aldo Palazzeschi, ed. Gino Tellini. Milan: Mondadori, 2004. Print.

– "Controdolore." *Lacerba*, 15 January 1914, p. 19. Periodical.

– *Due imperi mancati*. Milan: Mondadori, 2000. Print.

– *Man of Smoke.* Trans. Nicolas J. Perella and Ruggero Stefanini. New York: Italica Press, 1992. Print.

– *Perelà: Uomo di fumo.* Florence: Vallecchi, 1954. Print.

– *:riflessi. Tutti i romanzi,* vol. 1, by Aldo Palazzeschi, ed. Gino Tellini. Milan: Mondadori, 2004. Print.

– *Tutte le poesie,* ed. Adele Dei. Milan: A. Mondadori, 2002. (Reference text for "Chi sono?," "L'incendiario," "L'orologio," and "La passeggiata"). Print.

Palmesi, Vincenzo. *Del tabacco: Specialmente del tabacco da fumo.* Reggio Emilia: Bondavalli, 1876. Print.

Pampel, Fred. *Tobacco Industry and Smoking.* New York: Facts on File, 2009. Print.

Panzeri, Daria. *Lombardia Beni Culturali.* Web. Accessed 8 February 2017. Reference for "La bicicletta" by Aleardo Villa, and "Raid Parigi-Roma-Torino" by Aldo Mazza).

Panzini, Alfredo. *La lanterna di Diogene.* Milan: Treves, 1907. Print.

Paolini, Federico. "Designing a New Society: A Social History of Italian Car Design." *Made in Italy: Rethinking a Century of Italian Design,* ed. Grace Lees-Maffei and Kjetil Fallan. New York: Bloomsbury, 2014, pp. 211–24. Print.

Papini, Giovanni. "La filosofia del cinematografo." *La stampa,* 18 May 1907, pp. 1–2. Periodical.

– "L'orologio fermo alle sette." *Il Pilota Cieco,* by Giovanni Papini. Florence: Vallecchi, 1907, pp. 189–95. Print.

– *Un uomo finito. Opere: Dal "Leonardo" al Futurismo,* by Giovanni Papini, ed. Luigi Baldacci. Milan: Mondadori, 1977. Print.

Pascoli, Giovanni. *Canti di Castelvecchio.* Bologna: Zanichelli, 1903. Print.

– *Primi poemetti.* Bologna: Zanichelli, 1904. Print.

– *Tutte le poesie.* Milan: Mondadori, 1958. (Reference text for "L'Avemaria," "La bicicletta," and "L'ora di Barga."). Print.

Pastrone, Giovanni, director. *Cabiria.* Itala Films, 1914. Film.

Paulicelli, Eugenia. "Art in Modern Italy: From the Macchiaioli to the Transavanguardia." *The Cambridge Companion to Modern Italian Culture,* ed. Zygmunt Baranski and Rebecca West. Cambridge: Cambridge University Press, 2001, pp. 243–64. Print.

– "Fashion: Narration and Nation." *The Cambridge Companion to Modern Italian Culture,* ed. Zygmunt Baranski and Rebecca West. Cambridge: Cambridge University Press, 2001, pp. 282–92. Print.

– *Fashion under Fascism: Beyond the Black Shirt.* New York: Berg, 2004. Print.

– *La moda è una cosa seria: Su Rosa Genoni.* Milan: Deleyva, 2015. Print.

Pavanello, Giovanna. "Edoardo Weiss." *Il contributo italiano alla storia del pensiero: Scienze. Treccani,* 2013. Web. Accessed 2 February 2017.

Pavia, Alessandro. *L'album dei Mille di Alessandro Pavia.* Ed. Marco Pizzo. Rome: Gangemi, 2004. Print.

Payne, Alina. *From Ornament to Object: Genealogies of Architectural Modernism.* New Haven, CT: Yale University Press, 2012. Print.

Pedroni, Matteo. "Poesia ciclistica delle origini: Betteloni, Cannizzaro, Gozzano, Pascoli, Stecchetti." *Versants,* vol. 40, 2001, pp. 185–205. Periodical.

Pelizzari, Maria Antonella. *Photography and Italy.* London: Reaktion Books, 2011. Print.

Perella, Nicolas. *Midday in Italian Literature: Variations of an Archetypal Theme.* Princeton, NJ: Princeton University Press, 1979. Print.

Perella, Nicolas, and Ruggero Stefanini. "Aldo Palazzeschi's Code of Lightness." *Forum Italicum,* vol. 26, no. 1, 1992, pp. 94–120. Periodical.

Pezza, Gian Luigi. "Storia della radio: Sviluppo storico dell'invenzione della radio." *Istituto di Pubblicismo.* Web. Accessed 9 March 2016.

Pietromarchi, Antonello. *Un occhio di riguardo: Il conte Primoli e l'immagine della Belle-époque.* Florence: Ponte alle Grazie, 1990. Print.

Pirandello, Luigi. *L'escusa.* Turin: Einaudi, 1995. Print.

– *Il fu Mattia Pascal.* Milan: Mondadori, 1988. Print.

– *Novelle per un anno.* Rome: Newton & Compton, 1994. (Reference text for "Ciaula scopre la luna," "Leonora addio!," "Mondo di carta," "Pallottoline!," and "Paura del sonno"). Print.

– *L'umorismo.* Ed. Daniela Marcheschi. Milan: Mondadori, 2001. Print.

Pius X, Pope. *Pascendi dominici gregis: Sugli errori del modernismo.* Siena: Edizioni Cantagalli, 2007. Print.

Pivato, Stefano. *La Bicicletta e il sol dell'avvenire: Sport e tempo libero nel socialismo della Belle Epoque.* Florence: Ponte alle Grazie, 1992. Print.

– "The Bicycle as a Political Symbol: Italy 1885–1955." *International Journal of the History of Sport,* vol. 7, no. 2, September 1990, pp. 173–87. Periodical.

– *Il Touring Club Italiano.* Bologna: Il Mulino, 2006. Print.

Pizzo, Marco. *Lo stivale di Garibaldi: Il Risorgimento in fotografia.* Milan: Mondadori, 2011. Print.

Plato. *The Laws of Plato.* Ed. Thomas Pangle. Chicago: University of Chicago Press, 1988. Print.

Poesia: Rassegna internazionale. Dir. Filippo T. Marinetti. Milan: Tip. Parma, 1905–20. Periodical.

Poggioli, Renato. *Teoria dell'arte d'avanguardia.* Bologna: Il Mulino, 1962. Print.

Il politecnico: Repertorio mensile di studj applicati alla prosperità e coltura sociale. Dir. Carlo Cattaneo. Milan. 1839–44, 1859–69. Periodical.

Pollard, Tanya. "The Pleasures and Perils of Smoking in Early Modern England." *Smoke: A Global History of Smoking,* ed. Sanders Gilman and Zhou Xun. London: Reaktion Books, 2004, pp. 38–45. Print.

Pollina, Elvira. "Greenwich festeggia i 125 anni." *Corriere della sera,* 22 October 2009. Web. Accessed 26 January 2017.

Ponti, Gió. "Arte e industria." *Domus,* vol. 4, no. 54, June 1932. Periodical.

– "La casa all'italiana." *Domus,* vol. 1, no. 1, January 1928. Periodical.

– "La casa di moda." *Domus,* vol. 1, no. 8, August 1928. Periodical.

– "Morte e vita della tradizione." *Domus,* vol. 4, no. 51, March 1932. Periodical.

Il popolo d'Italia. Milan, 1914–43. Periodical.

Portaluppi, Piero. *Aedilitia.* Milan: Bestetti & Tumminelli, 1924 (vol. 1); 1930 (vol. 2). Print.

Pratella, Francesco Balillla. "Contro il 'grazioso' in musica." *Lacerba,* 15 May 1913, pp. 103–6. Periodical.

– "Manifesto of Futurist Musicians." *Futurist Manifestos,* ed. Umbro Apollonio. Boston: Art Works, 2001, pp. 31–8. Print.

Prato, Paolo. *Suoni in scatola: Sociologia della musica registrata; Dal fonografo a internet.* Ancona and Milan: Editori Associati, 1999. Print.

Prezzolini, Giuseppe. *L'arte di persuadere.* Florence: Lumachi, 1907. Print.

Il progresso ciclistico: Rivista mensile illustrata tecnica e sportiva di ciclismo e automobilismo. Milan: Stab. Tip. Lit. Colombo e Tarra, 1896–7. Periodical.

Il progresso fotografico. Milan: Progresso, 1894–2002. Periodical.

Puorto, Elvira. *Fotografia tra arte e storia: Il "Bullettino della Società Fotografica Italiana," 1889–1914.* Naples: A. Guida, 1996. Print.

Quintavalle, Arturo. *Gli Alinari.* Florence: Alinari, 2003. Print.

Raboy, Marc. *Marconi: The Man Who Networked the World.* New York: Oxford University Press, 2016. Print.

Raimondi, Gloria. "Maria Monaci." *Dizionario biografico. Treccani.* Web. Accessed 2 February 2017.

Rainey, Lawrence, et al. *Futurism: An Anthology.* New Haven, CT: Yale University Press, 2009. Print.

Rebora, Clemente. *Le poesie (1913–1957).* Ed. Gianni Mussini and Vanni Scheiwiller. Milan: Garzanti, 1998. Print.

Renzi, Renzo. *Il cinematografo al campo: L'arma nuova nel primo conflitto mondiale.* Ancona: Transeuropa, 1993. Print.

Il resto del carlino. Bologna, 1885– . Periodical.

Richter, Gregory. *Albert Giraud's Pierrot Lunaire.* Kirksville, MO: Truman State University Press, 2011. Print.

Riello, Giorgio. *La moda: Una storia dal Medioevo a oggi.* Rome: Laterza, 2012. Print.

Rigoletto. Composed by Giuseppe Verdi. Perf. Enrico Caruso. Metropolitan Opera House, New York, 23 November 1903. Performance.

Rinaldi, Marco. "Italian Still Life, Italian Lifestyle: Lo sguardo degli oggetti." *Natura morta: Rappresentazione dell'oggetto, oggetto come rappresentazione,* ed. Dalma Frascarelli and Costanza Barbieri. Naples, 2010, pp. 145–50. Print.

Risveglio di una città, Si pranza sulla terrazza del Kursaal, Convegno di automobili e aeroplani. By Luigi Russolo. Teatro Dal Verme Milano, 21 April 1914; Politeama Genova, 20 May 1914. Performance.

Rivista fono-cinematografica e degli automatici, istrumenti pneumatici e affini. Dir. Gualtiero Fabbri. Turin: 1907–8. Periodical.

Rivista fotografica universale. Dir. Antonio Montagna. Brindisi: Tipografia Adriatico-Orientale, 1870–6. Periodical.

Rocchetti, Lucio. *Brevetti di invenzione sul doppio fondo fotografico, sul colorito istantaneo e sulla vernice preservativa Sistema Crozat: Tecnica e diffusione tra i fotografi italiani nel 1870.* Rome: Gangemi Editore, 2016. Print.

Rocchini, Claudia. "Il vate, il museo e la fotografia." *Fotografia Reflex,* July 2010, pp. 78–81. Periodical.

Rocco, Vanessa. "Exhibiting Exhibitions: Designing and Displaying Fascism." *Made in Italy: Rethinking a Century of Italian Design,* ed. Grace Lees-Maffei and Kjetil Fallan. New York: Bloomsbury, 2014, pp. 179–92. Print.

Roche, Maurice. *Mega-Events and Modernity: Olympics and Expos in the Growth of Global Culture.* New York: Routledge, 2000. Print.

Rossi, Luigi. *L'anarchico delle due ruote: Luigi Masetti; Il primo cicloviaggiatore italiano.* Portogruaro: Ediciclo, 2008. Print.

Roster, Alessandro, and Alfredo Orlandini. *La pratica del velocipede e la tecnica dell'allenamento.* Florence: Stab. Tip. Fiorentino, 1895. Print.

Rousseau, Jean-Jacques. *Emile or on Education.* New York: Basic Books, 1979. Print.

Rubino, Antonio. *Tic e Tac, ovverossia l'orologio di Pampalona: Romanzo per bambini di tutte le età.* Florence: La Nuova Italia, 1931. Print.

Russolo, Luigi. *The Art of Noises.* New York: Pendragon, 1986. Print.

Ruttmann, Walter, director. *Acciaio.* Screenplay by Luigi Pirandello. Cines, 1933. Film.

Saccone, Antonio. *L'occhio narrante: Tre studi sul primo Palazzeschi.* Naples: Liguori, 1987. Print.

Sacre du printemps. By Igor Stravinsky. Théâtre des Champs-Élysées, Paris, 29 May 1913. Performance.

Salaris, Claudia. *Il Futurismo e la pubblicità: Dalla pubblicità dell'arte all'arte della pubblicità.* Milan: Lupetti, 1986. Print.

Salerno, Luigi. *Tabacco e fumo nella pittura.* Turin: [s.l.], 1954. Print.

Salgari, Emilio. *Al polo australe in velocipede.* Ed. Vittorio Sarti. Milan: Mondadori, 2002. Print.

Salsa, Carlo. *Trincee: Confidenze di un fante.* Milan: Mursia, 1982. Print.

Samaritani, Fausta. "Bicicletta o bicicletto?" *Repubblica letteraria.* Web. Accessed 9 March 2016. <http://archive.is/20130413084507/www.repubblicaletteraria.it/Bicicletta_rivista.htm>

Sangiovanni, Andrea. *Le parole e le figure: Storia dei media in Italia dall'età liberale alla seconda guerra mondiale.* Rome: Donzelli, 2012. Print.

Sansone, Luigi. *F.T. Marinetti = Futurismo.* Milan: F. Motta, 2009. Print.

Sant'Elia. Antonio. "Manifesto of Futurist Architecture." *Futurist Manifestos,* ed. Umbro Apollonio. Boston: Art Works, 2001, pp. 160–71. Print.

Sarfatti, Margherita. *Dux.* Milan: Mondadori, 1926. Print.

Savelli, Giulio. "Ultima sigaretta: Eventi e storia nella Coscienza di Zeno." *MLN,* vol. 105, no. 1, January 1990, pp. 87–104. Periodical.

Sbarbaro, Camillo. *Trucioli.* Florence: Vallecchi, 1920. Print.

Sbolenfi, Argia (alias Olindo Guerrini). *Rime di Argia Sbolenfi: Con prefazione di Lorenzo Stecchetti.* Bologna: Monti, 1897. Print.

Scalzi, Francesco. *Il consumo del tabacco da fumo in Roma in attinenza alla salute pubblica.* Rome: 1868. Print.

Schiller, Friedrich. *On the Aesthetic Education of Man, in a Series of Letters.* New York: F. Ungar, 1965. Print.

Schirinzi, Claudio. "Non piaceva a Bava Beccaris." *Uomo a due ruote: Avventura, storia, passione,* ed. Guido Vergani. Milan: Electa, 1987, pp. 40–5. Print.

Schmitz, Dawn. "Only Flossy, High-Society Dudes Would Smoke 'Em: Gender and Cigarette Advertising in the Nineteenth Century." *Turning the Century: Essays in Media and Cultural Studies,* ed. Carol Stabile. Boulder, CO: Westview Press, 2000, pp. 100–21. Print.

Schnapp, Jeffrey. *Modernitalia.* Ed. Francesca Santovetti. Oxford: Peter Lang, 2012. Print.

– "Propeller Talk." *Modernism/Modernity,* vol. 1, no. 3, 1994, pp. 153–78. Periodical.

– "The Romance of Caffeine and Aluminum." *Critical Inquiry,* vol. 28, no. 1, Autumn 2001, pp. 244–69. Periodical.

Scott, Michael. *The Great Caruso.* New York: Alfred Knopf, 1988. Print.

Scudiero, Maurizio. *Depero.* Florence: Giunti, 2009. Print.

– "From Belle Époque to Balilla: The Italian Manifesto between the Two Wars." *Il manifesto pubblicitario italiano: Da Dudovich a Depero 1890–1940,* ed. Giampiero Mughini and Maurizio Scudiero. Milan: Nuove Arti Grafiche Ricordi, 1997, pp. xv–xxv. Print.

Secrest, Meryle. *Elsa Schiaparelli: A Biography.* New York: Alfred Knopf, 2014. Print.

Segel, Harold. *Pinocchio's Progeny: Puppets, Marionettes, Automatons and Robots in Modernist and Avant-Garde Drama.* Baltimore, MD: Johns Hopkins University Press, 1995. Print.

Segreto, Luciano. "Storia d'Italia e storia dell'industria." *Storia d'Italia: Annali; L'industria,* vol. 15, ed Franco Amatori, et al. Turin: Einaudi, 1999, pp. 7–86. Print.

Selvatico, Pietro. *Storia estetico-critica delle arti del disegno, ovvero l'architettura, la pittura e la statuaria considerate nelle correlazioni tra loro e negli svolgimenti storici, estetici e tecnici.* Venice: Naratovich, 1852. Print.

Semeonoff, Boris. "The Gramophone and Opera." *Phonographs & Gramophones.* Edinburgh: Royal Scottish Museum, 1977, pp. 101–06. Print.

Septième conference géodésique internationale. Rome: Imprimerie Royale D. Ripamonti, 1883. Print.

Serao, Matilde. *La conquista di Roma.* Ed. Wanda De Nunzio Schilardi. Rome: Bulzoni, 1997. Print.

– *Saper vivere: Norme di buona creanza.* Florence: Passigli, 1989. Print.

– *La virtù di Checchina.* Lecce: Piero Manni, 2000. Print.

Sessa, Ornella. *L'automobile italiana: Tutti i modelli dalle origini a oggi.* Florence: Giunti, 2006. Print.

Sewall, Charles. *Wireless Telegraphy: Its Origins, Development, Inventions, and Apparatus.* New York: Van Nostrand, 1904. Print.

Shechter, Relli. *Smoking, Culture, and Economy in the Middle East: The Egyptian Tobacco Market, 1850–2000.* London: Tauris, 2006. Print.

Sighele, Scipio. *L'intelligenza della folla.* Turin: F.lli Bocca, 1922. Print.

Simmel, Georg. "The Metropolis and Mental life." *On Individuality and Social Forms: Selected Writings,* by Georg Simmel, ed. Donald Levine. Chicago: University of Chicago Press, 1971, pp. 324–39. Print.

Simoni, Antonio. *Orologi italiani dal cinquecento all'ottocento.* Milan: Vallardi, 1965. Print.

Smil, Vaclav. *Creating the Twentieth Century: Technical Innovations of 1867–1914 and Their Lasting Impact.* Oxford: Oxford University Press, 1995. Print.

Smith, Michael Llewellyn. *Olympics in Athens 1896: The Invention of the Modern Olympic Games.* London: Profile Books, 2004. Print.

Soffici, Ardengo. *BÏF§ ZF+18 simultaneità e chimismi lirici.* Florence: Vallecchi, 1915. Print.

– *Kobilek: Giornale di battaglia.* London: Forgotten Books, 2013. Print.

Sontag, Susan. *On Photography.* New York: Doubleday, 1990. Print.

Sorge, Paola. *D'Annunzio e la magia della moda.* Rome: Edizioni Elliot, 2015. Print.

Spackman, Barbara. "Fascist Puerility," *Qui parle,* vol. 13, no. 1, Fall/Winter 2001, pp. 13–28. Periodical.

– *Fascist Virilities: Rhetoric, Ideology, and Social Fantasy in Italy.* Minneapolis: University of Minnesota Press, 1996. Print.

Spagnoletti, Giacinto. *Svevo: Ironia e nevrosi.* Massa: Memoranda, 1986. Print.

Sparke, Penny. *Italian Design: 1870 to the Present.* London: Thames & Hudson, 1988. Print.

– "A Modern Identity for a New Nation: Design in Italy since 1860." *The Cambridge Companion to Modern Italian Culture*, ed. Zygmunt Baranski and Rebecca West. Cambridge: Cambridge University Press, 2001, pp. 265–81. Print.

Starobinski, Jean. *Portrait de l'artiste en saltimbanque.* Geneva: Skira, 1970. Print.

Stecchetti, Lorenzo (alias Olindo Guerrini). *Le rime di Lorenzo Stecchetti.* 2nd ed. Bologna: Zanichelli, 1902. Print.

Steele, Valerie. *Fashion, Italian Style.* New Haven, CT: Yale University Press, 2003. Print.

Stefanoni, Luigi. "Note Geodetiche." *Corriere della sera*, 2–3 November 1883, pp. 1–2. Periodical.

Steffen, David. *From Edison to Marconi: The First Thirty Years of Recorded Music.* Jefferson, NC: McFarland, 2005. Print.

Stella, Sabino. *La bicicletta.* Turin: Roux Frassati & C., 1896. Print.

Sterne, Lawrence. *A Sentimental Journey.* New York: Penguin Classics, 2001. Print.

Stewart-Steinberg, Suzanne. *The Pinocchio Effect: On Making Italians, 1860–1920.* Chicago: University of Chicago Press, 2007. Print.

Stoppani, Antonio. *Il bel paese: Conversazioni sulle bellezze naturali, la geologia e la geografia fisica d'Italia.* Milan: Cogliati, 1910. Print.

Strazzi, Marco. *Lancette & C.: Un secolo di orologi da polso; La storia, la tecnica, il design.* La Chaux-de-Fonds: Pressision, 2005. Print.

Sutton-Smith, Brian. *The Ambiguity of Play.* Cambridge, MA: Harvard University Press, 1997. Print.

Svevo, Italo. *La coscienza di Zeno.* Ed. Giovanna Joli. Turin: UTET, 1990. Print.

– *Senilità.* Brescia: La Scuola, 1986. Print.

– *Una vita.* Milan: Rizzoli, 2001. Print.

Il tabacco: Organo dell'industria e del commercio del tabacco. Rome, 1897–1973. Periodical.

Tacchi, Francesca. *Storia illustrata del Fascismo.* Florence: Giunti, 2000. Print.

Tagg, John. *The Burden of Representation: Essays on Photographies and Histories.* Amherst: University of Massachusetts Press, 1988. Print.

Taroni, Paolo. *Bergson, Einstein e il tempo: La filosofia della durata bergsoniana nel dibattito sulla teoria della relatività.* Urbino: QuattroVenti, 1998. Print.

Tayłor, Timothy. "The Role of Opera in the Rise of Radio in the United States." *Music and the Broadcast Experience: Performance, Production, and Audiences*, ed. Christina Baade and James Deaville. New York: Oxford University Press, 2016, pp. 69–89. Print.

Tempel, Benno. "Symbol and Image: Smoking in Art since the Seventeenth Century." *Smoke: A Global History of Smoking*, ed. Sanders Gilman and Zhou Xun. London: Reaktion Books, 2004, pp. 206–17. Print.

Thomas, Ronald. "The Legacy of Victorian Spectacle: The Map of Time and the Architecture of Empty Space." *Functions of Victorian Culture at the Present Time*, ed. Christine Krueger. Athens: Ohio University Press, 2002, pp. 18–35. Print.

Thompson, Christopher. *The Tour de France: A Cultural History.* Berkeley: University of California Press, 2006. Print.

Tiffany, Daniel. *Toy Medium: Materialism and Modern Lyric.* Los Angeles: University of California Press, 2000. Print.

Tosca. Composed by Giacomo Puccini. Dir. Leopoldo Mugnone. Teatro Costanzi, Rome, 14 January 1900. Performance.

– Composed by Giacomo Puccini. Dir. Arturo Toscanini. Teatro La Scala, Milan, 17 March 1900. Performance.

Tozzi, Federigo. *Con gli occhi chiusi.* Milan: Mondadori, 1994. Print.

– *Le Novelle.* 2 vols. Ed. Glauco Tozzi. Milan: Rizzoli, 2003. Print.

Triossi, Amanda. *Bulgari: From 1884 to 2009; 125 Years of Italian Jewels.* Milan: Skira, 2009. Print.

La Tripletta: Giornale bisettimanale di ciclismo ed altri sport. Dir. Eugenio Camillo Costamagna. Turin: Tip. Il Nuovo Giornale, 1895–6. Periodical.

Tucker, Jean. *The Modernist Still Life – Photographed.* St Louis: University of Missouri-St Louis, 1989. Print.

Turandot. Composed by Giacomo Puccini. Teatro La Scala, Milan, 25 April 1926. Performance.

Turani, Giuseppe. "La saga dei Borletti." *Repubblica*, 11 September 1987. Web. Accessed 25 January 2017.

Uglietti, Alberto. *Orologi da polso.* Milan: Federico Motta Editore, 2009. Print.

Ulmann, Jacques. *Corps et civilization: Education physique, médecine, sport.* Paris: Libraire Philosophique J. Vrin, 1993. Print.

Ungaretti, Giuseppe. *Allegria di naufragi.* Florence: Vallecchi, 1919. Print.

Usioli, Arnoldo. "Novità della scienza." *Illustrazione italiana*, 28 October 1883. Periodical.

Valori plastici: Rivista d'arte. Rome: Stabilimento Poligrafico Editoriale Romano, 1918–22. Periodical.

Vannucci, Marcello. *Mario Nunes Vais: Gentiluomo fotografo.* Florence: Bonechi, 1976. Print.

Veglia, Marco. "Sigaro, sigarette e oggetti da fumo." *Oggetti della letteratura italiana*, ed. Gian Mario Anselmi and Gino Ruozzi. Rome: Carocci, 2008, pp. 166–75. Print.

Vercelloni, Matteo. *Breve storia del design italiano.* Rome: Carocci, 2008. Print.

Verga, Giovanni. *I Malavoglia.* Turin: Einaudi, 1995. Print.

– *Novelle.* Ed. Francesco Spera. Milan: Feltrinelli, 1992. (Reference text for "L'amante di Gramigna" and "Rosso Malpelo."). Print.

Vergani, Guido. *Il piacere del corpo: D'Annunzio e lo sport.* Milan: Electa, 1999. Print.

– *Uomo a due ruote: Avventura, storia, passione.* Milan: Electa, 1987. Print.

Verne, Jules. *Le tour du monde en quatre-vingts jours.* Paris: Pierre-Jules Hetzel, 1873. Print.

Volt (alias Vincenzo Fani). "Futurist Manifesto of Women's Fashion." *Roma futurista,* 29 February 1920. Periodical.

Vota, Giuseppe. *I sessant'anni del Touring Club Italiano, 1894–1954.* Milan: Touring Club Italiano, 1955. Print.

Walker, Deborah. *Giorgio De Chirico and the Real: Art, Enigma, and Nietzschean Innocence.* Saarbrücken, Germany: VDM Verlag Dr. Müller, 2008. Print.

Weber, Max, et al. *The Rational and Social Foundations of Music.* Carbondale: Southern Illinois University Press, 1958. Print.

Weber, Robert. *Forks, Phonographs and Hot Air Balloons: A Field Guide to Inventive Thinking.* New York: Oxford University Press, 1992. Print.

Weidlich, Carlo. *Ciclismo e letteratura.* Palermo: Libreria Editrice Domino, 1932. Print.

Weiermair, Peter. *The Nature of Still Life: From Fox Talbot to the Present Day.* Milan: Electa, 2001. Print.

Weill, Alain. *The Poster: A Worldwide Survey and History.* Boston: G.K. Hall, 1984. Print.

Weiss, Edoardo. *Elementi di psicanalisi.* Pordenone: Edizioni Studio Tesi, 1985. Print.

Welch, Walter, et al. *From Tinfoil to Stereo: The Acoustic Years of the Recording Industry, 1877–1929.* Gainesville: University Press of Florida, 1994. Print.

Wheeler, George. *Report upon the Third International Geographic Congress and Exhibition at Venice, Italy, 1881.* Washington, DC: Government Printing Office, 1885. Print.

Wittman, Laura. *The Tomb of the Unknown Soldier, Modern Mourning, and the Reinvention of the Mystical Body.* Toronto: University of Toronto Press, 2011. Print.

Woolf, Virginia. *Mr. Bennett and Mrs. Brown.* London: Hogarth Press, 1924. Print.

Yorick (alias Pietro Coccoluto Ferrigni). *La storia dei burattini.* Florence: Tip. Editrice del Fieramosca, 1884. Print.

– *Su e giù per Firenze: Monografia fiorentina.* Florence: Bàrbera, 1877. Print.

Young, David. *The Modern Olympics: A Struggle for Revival.* Baltimore, MD: Johns Hopkins University Press, 2002. Print.

Zamagni, Vera. *The Economic History of Italy, 1860–1990.* New York: Oxford University Press, 1993. Print.

Zanghieri, Tancredi. *La fotografia turistica.* Milan: Hoepli, 1908. Print.

Zannier, Italo. *Cultura fotografica in Italia: Antologia di testi sulla fotografia 1839–1949.* Milan: Franco Angeli, 1985. Print.

– *Leggere la fotografia: Le riviste specializzate in Italia (1863–1990).* Rome: NIS, 1993. Print.

– "Pittorialismo fatale." *Pittorialismo e cento anni di fotografia pittorica in Italia,* ed. Italo Zannier. Florence: Alinari, 2004, pp. 7–18. Print.

– *Venice, the Naya Collection.* Venice: O. Bohm publishers, 1981. Print.

Zapperi, Roberto. *Freud e Mussolini: La psicanalisi in Italia durante il regime fascista.* Milan: Franco Angeli, 2013. Print.

Zevi, Bruno. "Gruppo 7: The Rise and Fall of Italian Rationalism." *Architectural Design,* vol. 51, no. 1/2, 1981, pp. 41–3. Periodical.

Zola, Émile. *Au bonheur des dames.* Ed. Colette Becker. Paris: Garnier-Flammarion, 1971. Print.

Index